W9-APY-875

STRANGE GLOW

STRANGE GLOW

The Story of Radiation

TIMOTHY J. JORGENSEN

PRINCETON UNIVERSITY PRESS
Princeton and Oxford

Published by Princeton University Press, 41 William Street,
Princeton, New Jersey 08540
In the United Kingdom: Princeton University Press, 6 Oxford
Street, Woodstock, Oxfordshire OX20 1TR

press.princeton.edu
Jacket art by Jessica Massabrook

ISBN 978-0-691-16503-5
Library of Congress Control Number 2015959168

British Library Cataloging-in-Publication Data is available

This book has been composed in Montserrat & Sabon LT Std.

Printed on acid-free paper ∞

Printed in the United States of America

1 3 5 7 9 10 8 6 4 2

DEDICATED TO MY PARENTS,

Charles and Marion Jorgensen

A tribute of my respect,
admiration, and love

CONTENTS

PREFACE

Classic—a book which people praise and don't read.

—Mark Twain

Things that are complex are not useful. Things that are useful are simple.

—Michail Kalashnikov, inventor of the simple and reliable Automatic Kalashnikov-1947 (AK-47) assault rifle, which has only eight moving parts.

This book was not meant to be a classic; it was meant to be useful. I have, therefore, written it using straightforward language largely devoid of scientific jargon. In so doing, it is my wish that the book will be accessible to the widest possible audience of readers, regardless of whether they have any technical background. If I have done my job well, readers of this book will learn a tremendous amount about radiation and will find this information useful in many practical ways.

People like to get their learning in the form of stories.[1] As actress Audrey Hepburn once said, "Everything I've learned, I've learned from the movies." Although this is a book, and not a movie, the point is well taken. If you tell an engaging and compelling story, be it through movie or book, people will learn something from it. So that's what I attempt to do here. This book is the story of people's encounters with radiation, and of how mankind has been transformed by the experience. The story is, therefore, told with an emphasis on the human aspects, and it is told from a health-centric perspective. The goal is to integrate the technological aspects of radiation with the human experience and thereby remove some of the mystery and misunderstanding that surround radiation. Nevertheless, this is not a book about lessening your fear of radiation. Fear is a very subjective emotion, driven by many factors. The only thing that can be achieved here is to present the facts about radiation as objectively and evenhandedly as possible, leaving you to decide which aspects to fear.

Another purpose of this book is to dispel the myth that the subject of radiation risks is so complicated that it is beyond the capability of ordinary people to grasp, leaving reliance on radiation "experts" as their only recourse. This is simply not true. Intelligent people, even those lacking any technical background, should be able to understand the fundamental principles that drive radiation risk and then make their own decisions about how large a threat radiation poses to them personally and collectively. This book seeks both to convince people that they can be masters of their own radiation fate, and to empower them to make their own well-informed decisions about their personal radiation exposures.

Lastly, this book is an experiment in risk communication. The open question is whether radiation risks can be characterized accurately and effectively without reliance on a lot of mathematics, tables, and graphs. These highly quantitative approaches have proved to be largely ineffective in communicating the essence of risk to the public.[2] This book is devoid of graphs and tables and keeps the mathematics to a minimum. Instead, it tries to instill a sense of the magnitude of the threat through a historical scientific narrative about the people who encountered radiation of various types and dose levels, and the health consequences of those exposures. In this way, we can get an accurate sense of the level of the radiation hazard even if we don't have a detailed understanding of the underlying technology.

Can all this be achieved? I don't see why not. It's been done before—in the case of electricity. Electricity was a technological innovation introduced to society shortly before radiation. Initially, it was greatly feared as a deadly and invisible threat to health. With time, however, people began to understand that a flashlight battery didn't pose the same danger as a downed power line. Even people who couldn't explain the difference between an amp and a volt began to understand that, although there were risks of death, those risks could be managed so that the benefits of electricity could be maximized and the risks minimized. Now no one speaks of being pro- or anti-electricity. They understand that electricity is here to stay. There is no going back. All we can do is manage electricity to minimize its dangers relative to its benefits.

But we haven't yet reached a similar place with radiation. People today often react to radiation in the same way that people reacted to electricity over a hundred years ago. Our understanding of radiation needs to advance to the point that we develop the same good sense about radiation risks that we have for electricity. Radiation, like electricity, is a technology that is here to stay. So the more that people learn about radiation, the better off we'll all be. This book seeks to increase public understanding of radiation, in much the same way that people gradually came to understand electricity.

You will get the most from this book if you appreciate how it is organized. After a brief opening chapter that sets the stage for the topic (chapter 1), the book is divided into three parts.

Part One (chapters 2–4) tells the story of how radiation was discovered, and how society immediately put that discovery to practical use. You'll learn how a chance observation of a glowing fluorescent screen in a laboratory in Edinburgh, Scotland, saved a man's leg from amputation in Montreal, Canada, just a few weeks later. You'll learn about Thomas Edison's initial enthusiasm for x-ray tubes, why he soon became afraid of them, and the heavy price his assistant paid for acting carelessly. You'll also learn how a cloudy day in Paris resulted in the discovery of radioactivity. Along the way, you will be introduced to a few physics concepts that are important to understanding how radiation affects health. Ideas about radiation and its relevant physics are introduced progressively and systematically, while the health aspects of radiation make cameo appearances in the form of anecdotes about problems suffered by the early scientific pioneers of radiation research. A comparison is made to electricity, as an example of an earlier technology that was originally regarded by the public with even greater suspicion.

Part Two (chapters 5–11) introduces the effects of radiation on human health. It begins with the story of miners in Germany who unknowingly suffered from a radiation-related illness before radiation had even been discovered. You'll find out what their illness had to do with the mysterious deaths of women who painted watch dials in the United States. And you'll learn how it was discovered that radiation can cure cancer. You'll also learn about radiation sickness and why most medical doctors have never seen a case of it. And you'll find out why you shouldn't be drinking milk after a nuclear power plant accident. Evolving notions about how human cells and tissues react to radiation are introduced, with a focus on how radiation's health effects are measured, culminating in what we now know about their underlying causes. The radiation biology related to health issues, rather than the radiation physics, is dealt with systematically, and the concept of equating "safety" with "low risk" is introduced anecdotally. This is done as a prelude to Part Three, where the role of risk/benefit analysis in making decisions regarding radiation use is explored.

Part Three (chapters 12–17) is a collection of chapters narrowly focused on radiation topics of popular interest. To mention just a few, you'll learn how dangerous the radon in your basement is, how hazardous it is to eat food contaminated with radioactivity, and how risky it is to live next to a nuclear power plant. Although you may be tempted to cherry-pick these chapters, and read only those of particular interest, you should resist that temptation. Embedded within each chapter is an illustration of a specific risk assessment concept, and the chapters are ordered so that they progressively reveal the value of considering both risks and benefits when making health decisions, as well as the importance of weighing alternative risks. Also included is systematic discussion of how uncertainty affects the validity of our radiation decisions. It is possible to read the chapters in this part out of numerical order without any loss of narrative continuity, but the developing story of risk assessment and its relationship to safety will be garbled, so a nonsequential approach is not recommended.

The epilogue contains some final thoughts regarding all that we've learned about radiation and how best to apply that information to everyday life. It also includes one final story about radiation that has a very important take-home message.

With this overview, you are now ready to begin your exploration of the world of radiation. You will likely find it both interesting and enjoyable, but also a little scary. Nevertheless, in the end, you'll be much better equipped to deal with any radiation issues that you encounter during your travels through life in a modern technological society, where radiation presents itself at every turn. Good luck on your journey.

TIMOTHY J. JORGENSEN
WASHINGTON, DC

STRANGE GLOW

CHAPTER 1

NUCLEAR JAGUARS

In the dark, all cats are jaguars.

—*Anonymous Proverb*

As a rule, men worry more about what they can't see than about what they can.

—*Julius Caesar*

The common denominator of most radiation exposure scenarios is fear. Just mention the word radiation, and you instill fear—a perfectly understandable response given the images of mushroom clouds and cancerous tumors that immediately come to mind. Those images would justifiably cause anyone to be anxious. Nevertheless, some people have also become highly afraid of diagnostic x-rays, luggage scanners, cell phones, and microwave ovens. This extreme level of anxiety is unwarranted, and potentially dangerous.

When people are fearful, they tend to exaggerate risk. Research has shown that people's perception of risk is tightly linked to their fear level.[1] They tend to overestimate the risk of hazards that they fear, while underestimating the risk of hazards they identify as being less scary. Often their risk perception has little to do with the facts, and the facts might not even be of interest to them. For example, many Americans are terrified of black widow spiders, which are found throughout the United States. They are uninterested in the reality that fewer than two people die from black widow bites each year, while over 1,000 people suffer serious illness and death annually from mosquito bites. Mosquitoes are just too commonplace to worry about. Likewise, the risk of commercial airplane crashes is tiny compared to motorcycle crashes, but many a biker is afraid to fly.

The point is that risk perception drives our decision making, and these perceptions often do not correspond to the real risk levels, because irrational fear is taking our brains hostage. When irrational fear enters the picture, it is difficult to objectively weigh risks. Ironically, health decisions driven by fear may actually cause us to make choices that increase, rather than decrease, our risks.

Fear of radiation is particularly problematic considering the trend in radiation exposures. Since 1980, the background radiation exposure level for Americans has doubled, and is likely to continue to climb.[2] Similar patterns are occurring in all of the developed and developing countries. This increase in background radiation is almost entirely due to the expanding use of radiation procedures in medicine. The benefits of diagnostic radiology in identifying disease and monitoring treatment progress have been significant; however, radiation has also been overused in many circumstances, conveying little or no benefits to patients while still subjecting them to increased risks. Furthermore, medical radiation is not distributed evenly across the population. While some people are getting no medical radiation exposure at all, others are receiving substantial doses. Under such circumstances, the "average" background radiation level means little to the individual. People need to be aware of their personal radiation exposures and weigh the risks and benefits before agreeing to subject themselves to medical radiation procedures.

In addition to the medical exposures, people receive radiation doses from a variety of consumer products, commercial radiation activities, and natural radioactivity sources in our environment. Some of these exposures are low level and low risk, while others can be at a high level and potentially hazardous. People need to be aware of these different radiation exposure hazards, and protect themselves when necessary.

In the pages that follow, we will explore the story of radiation with a specific focus on health. We will investigate what we know about radiation, and how we know it. We will weigh the risks and the benefits and characterize the uncertainties. We will identify the information needed to make rational health decisions about radiation, and we will uncover the limits of that information.

Since we typically cannot see radiation, we tend to be both intrigued by and afraid of it. We have endowed radiation with magical transformative powers that produce the superheroes of our comic books, such as the Incredible Hulk (exposed to gamma rays), Spider-Man (bitten by a radioactive spider), and the Teenage Mutant Ninja Turtles (overexposure to radiation). Yet, even superheroes are ambivalent about radiation. Superman is thankful for his x-ray vision, but scared to death of kryptonite (a fictional radioactive element).

In the end, we don't know what to think. Nevertheless, we have practical decisions facing us at the individual, community, state, national, and international levels. Questions range from whether one should agree to have a dental x-ray today, to whether more nuclear power plants should be built for energy needs 20 years from now. Such questions must be answered both individually and collectively, and answered now. We cannot postpone our decisions until more data are accumulated or more research is done, and we cannot relegate the responsibility to scientists or politicians. These decisions must be made by the electorate and the stakeholders; that is, by every person in our society.

The problem is that there are different types of radiation, and they are not all equally dangerous. Regrettably, most types are invisible, and we tend to fear things that we cannot see. Furthermore, we've lumped the invisible ones all together as equally hazardous. They are not, and we need to be able to tell the difference. All cats are not jaguars.

Which types of radiation should be feared? People must decide that for themselves. But that decision needs to be based on the facts. Although some things remain unknown, obscure, or uncertain, we cannot pretend that we know nothing about radiation and its health effects. The scientific and medical communities know a great deal about the effects of exposure after more than a century of experience with radiation. In fact, we know more about radiation than any other environmental hazard.

Although seeing is believing, not seeing shouldn't mean not believing. This is particularly true for radiation, since most forms cannot be seen directly. There is, however, one type of radiation that can be seen. This is light. Fortunately, light possesses many of the same properties as the invisible types of radiation. So we can overcome at least one barrier to understanding radiation by starting with the kind that we all can see, and then moving on from there to the invisible types that lurk ominously in the dark. We'll begin our radiation journey by looking under the lamppost, where reality is revealed . . . and where jaguars dare not roam.

PART ONE

RADIATION 101: THE BASICS

CHAPTER 2
NOW YOU SEE IT: RADIATION REVEALED

A man should learn to detect and watch that gleam of light which flashes across his mind from within.

—Ralph Waldo Emerson

CHASING RAINBOWS

Radiation is simply energy on the move, be it through solid matter or free space. For the most part, it is invisible to us and can only be detected with instruments. But there is one type of radiation that can be seen with the naked eye. This visible radiation is called light.

Who isn't fascinated by light? Whether as sunrises, fireworks, or lasers, light mesmerizes us *because we can see it*. Most of us have at least a rudimentary understanding of the physical principles of light because we have practical experience with it. We know that light casts shadows, bounces off mirrors, and comes in different colors. Other types of radiation have similar properties, but we don't see them. They exist apart from our everyday experience, and thus seem mysterious—ever present but never seen. Yet, the physical properties of light have much in common with these other types of radiation. Hence, by examining what's going on with light, we can actually get a glimpse of what is happening with its unseen cousins.

Light is just a tiny part of the world of radiation, but a very important part. We would be remiss if we neglected light; the story of radiation is strongly tied to its story. So as not to be remiss, let's briefly review what we know about light.

Throughout prehistory, light came almost exclusively from the sun and fire. From these observations of light, primitive humans deduced that the sun was a massive ball of fire. This deduction, which was likely one of mankind's first scientific conclusions, turned out to be absolutely correct.

Fire produces both heat and light, but warm things aren't always bright (e.g., body heat), and bright things aren't always warm (e.g., fireflies). These facts suggest that heat and light, though often found together and in some ways related, are fundamentally different phenomena that can be uncoupled and studied separately.

One of the things people noticed about light was that it could bend when it moved through water. Consider the appearance of an object protruding through the surface of clear water. The object seems to be displaced. That is, the position where the object, such as a stick, appears to enter the air seems to be different from where it exits the water; and thus it seems to be broken. (We all see this when we look at a drinking straw in a glass of water.) This occurs because light passing through liquid bends. The part of the object in the water is actually in a different position than where it appears to be to the human eye.

This property of light (called *refraction*) had very practical implications even for primitive peoples. For example, archers hunting for fish with a bow and arrow knew that if they aimed directly at the fish in the water they would always miss. Rather, if they aimed slightly below the fish, that fish would be dinner. The same thing did not work for prey in the air. Shooting arrows under birds did nothing but scare them.

Likewise, refraction is responsible for the magnification produced by droplets of water on leaves or other surfaces. Following the discovery of clear glass,[1] the magnification of water droplets could be simulated (and made permanent) by using droplets of glass. Through forming the clear glass droplets into different shapes, their magnification properties could be altered at will, and a variety of these droplets (i.e., lenses) could be made and used individually or in combination to produce telescopes and other visual instruments. Thus was born the field of optics—the branch of physics that studies the properties of light and its interactions with matter—and the path was made clear for research into the properties of light, the only type of radiation known at that time.

Before that falling apple and gravity got his attention, the first scientific passion of Isaac Newton (1643–1727) was optics.[2] In his laboratory, Newton made a number of novel observations about the properties of light. For example, he discovered that white light is composed of colored light blended together.[3] As he demonstrated, white light can be separated into its colored components with the use of a glass prism. This occurs by virtue of a light-bending effect known as *dispersive refraction*, whereby various wavelengths of light (i.e., different colors) bend slightly differently when passing through the prism; and thus the wavelengths separate from one another.

Newton used manmade prisms as light-bending tools for many of his optical experiments and demonstrations on the properties of light. However, the most dramatic evidence of a prism effect is the natural separation of sunlight into the colors of the rainbow by a large volume of water droplets, as can be seen in the sky following a storm. Unfortunately, Newton's experiments with light rays were limited by his senses. He couldn't measure what his eyes couldn't see. He didn't know that beyond those colored rays there was a universe of invisible rays that he could not sense or detect.

Later, scientists were fascinated by another phenomenon often associated with storms. This was electricity. Ever since the time of the apocryphal story of Benjamin Franklin flying a kite in the lightning storm, the public had been generally aware of the existence of electricity and what it could do. Besides killing people and burning down barns, in weaker forms it could produce sparks and even make the hair on one's head stand up. Electricity could also be generated simply by rubbing two different materials together (i.e., static electricity). So one didn't need to wait for a storm in order to play with electricity. In fact, small static-electricity machines were highly popular parlor toys for the aristocrats of Franklin's day.

Investigations of light and electricity ran in parallel paths. Yet, it wasn't until the late 1800s that the connection between light and electricity would begin to be understood and then developed into practical uses for the general public.

A BRIGHT IDEA

Contrary to common wisdom, Thomas Alva Edison (1846–1931) did not invent the first electric light bulb. That had been invented by Humphry Davy (1778–1829) in approximately 1805, and was called an arc lamp. Arc lamps produce light in a glass bulb in the form of intense brilliant white sparks that are produced in rapid succession. Arc lamps were suitable for outside illumination from tall lampposts, and such lampposts were in common use in public areas of major cities in Edison's day. But the arc lamp was simply too bright for use in the home.[4] A different approach would be needed to bring electric lighting into homes.

Edison knew, as did others, that running electricity through a variety of materials could make those materials glow—a process called *incandescence*—thereby producing a light source that could be used as an alternative to candles and natural gas lamps. The problem was that the glowing material (the *filament*) would degrade after a short while, making its use as a household lighting device impractical. Not knowing any of the physical principles by which electricity destroyed the filament, Edison simply tried every material he could to see if one would glow brightly, yet resist burning out. After trying 300 different materials, including cotton and turtle shell, he happened upon carbonized bamboo, which turned out to be the filament of choice (to the joy of turtles everywhere).[5] When used in an air-evacuated bulb (i.e., a vacuum tube), the carbonized bamboo outshone and lasted much longer than any of the other tested filaments. Edison had his light bulb. Although tungsten soon replaced carbonized bamboo in home light bulbs, illumination by incandescence became the predominant mode of interior lighting for many decades to follow.

Despite its potential for lighting homes, the electricity that powered light bulbs was initially looked upon with great fear and suspicion by the public. And the public had good reason to be wary. Newspaper reports of people being electrocuted were common. Pedestrians were, understandably, frightened by the spider webs of telegraph and electrical wires that loomed ominously over city streets, just waiting to fall on an unsuspecting passerby and end his life in a literal flash. In fact, many witnessed an accidental electrocution in midtown Manhattan in April 1888. A 17-year-old boy was struck dead when a dangling wire from overhead brushed against his body as he walked along Broadway. The *New York World* disgustedly reported, "Right on Broadway, where thousands and thousands of people are passing during all hours of the day, scores of wires are swinging in mid-air, any one of which is likely to become dislodged!"[6]

One of the paradoxes about electricity was that, despite its extremely swift lethality, it usually left its victims unmarked. It seemed to be a mysterious and spooky kind of death, where the life forces were simply sucked out of its victim without visible damage to organs or body parts. It was remarked that one victim seemed to have died simply from a hole in his thumb, where he had touched the wire. There was little understanding of the mechanisms of death by electrocution, and this heightened the fear of it.[7] This lack of understanding likely explains the enormous public outrage regarding accidental electrocutions, while other technology-related deaths went without notice. For example, in 1888, in addition to that 17-year-old boy, there were only four other electrocution deaths in New York City, while there were 23 deaths due to natural gas, and 64 deaths caused by streetcars. The gas and streetcar deaths were simply too ordinary to raise public concern.[8]

Edison, however, saw the public's fear as an opportunity. He had already tried to use fear of gas fires to promote electric lights over gaslights. Now he would use the fear of electrocution against business rival George Westinghouse (1846–1914).

Westinghouse was Edison's chief competitor in the delivery of electricity to homes, and was the king of overhead street wires. He found that wiring overhead, rather than underground, was vastly cheaper to install and maintain. Edison had chosen to use underground wiring for his electrical transmission lines, and the cost of digging trenches for wires was driving him out of business. Westinghouse was also exploiting another technological innovation to Edison's detriment by using alternating current (AC) rather than direct current (DC) for power transmission.[9] The AC could travel much greater distances than DC, thus reducing the number of generating stations that were required for any geographic area and also allowing transmission of electricity to areas remote from any generator. Westinghouse had bought a large number patents for AC-based inventions from another of Edison's rivals, Nikola Tesla (1856–1943), all of which would become worthless if DC became the standard.[10] Since the public's fear of electricity was closely associated with overhead wires, and those wires were largely conducting Westinghouse's AC, Edison began a campaign to redirect that fear to the AC current rather than the overhead wiring. According to Edison, AC was deadly, and DC current (his own product) was a much safer alternative.

To demonstrate AC's lethality, Edison performed a number of public electrocutions of dogs and other animals. In 1903, he even went so far as to use AC to euthanize a temperamental female circus elephant named Topsy. The elephant had been brought to America as a juvenile in 1885 to perform in the Adam Forepaugh Circus, which was a competitor of Ringling Brothers. Unfortunately, as she grew, she became temperamental and killed three of her handlers. The circus planned to put her down, but not before finding up-front buyers for her various body parts. Once the financial deals were complete, the circus needed a means to kill her without damaging the body parts they had already sold. They decided on a public hanging, and even built an enormous scaffold at Coney Island, New York, to perform the deed. But the Society for the Prevention of Cruelty to Animals (SPCA) soon got wind of the plan and intervened.[11] The SPCA contended that it was impossible to kill a six-ton elephant quickly by hanging, and the job was sure to be botched, resulting in pain and suffering to the elephant. At this point, the story reached the newspapers and the attention of Edison. His solution for the circus's predicament was AC electrocution.

Edison sent a team of workers to Coney Island with copper-clad sandals for the elephant's feet. After being affixed with the sandals, the elephant was jolted with 6,000 volts of AC. Before a crowd of 1,500 onlookers, Topsy staggered and fell to the ground dead, amidst smoke rising from her feet.[12]

The crowd was duly impressed with AC current's power to kill. They might have been less impressed, however, had they known that the elephant was also fed carrots laced with cyanide minutes before the electrocution. Whether by cyanide or AC, the elephant died quickly and Edison was pleased, as were the body-part merchants. Despite Topsy's feet being burnt, they were in good enough shape to make nice umbrella stands. To further give the event an aura of scientific authenticity, the unsellable internal organs were donated for anatomical studies to Professor Charles F. W. McClure (1852–1944) of the biology department at Princeton University.

TOPSY-TURVY

Westinghouse responded to Edison's accusation about the lethality of AC with his own unfounded accusations about the dangers of DC (e.g., higher household voltages required for DC would result in more deaths). Ironically, both men also claimed health benefits for their own currents. Westinghouse claimed the AC's momentary reversal of current "prevents decomposition of tissue."[13] Edison once marketed a DC-powered product, called an *inductorium*, that produced mild electric shocks to the body; it was claimed to be a "specific cure for rheumatism."[14]

As the AC/DC health hazard debate began to wind down, Edison made a last ditch effort to sway public opinion on a national level. The execution of Topsy had been preserved as a very short movie by Edison's motion picture crew. Edison released it to the public as a silent film entitled "Electrocution of an Elephant," and it was widely distributed to audiences throughout the United States.[15] (It still can be found as a video on the Internet.[16])

Unfortunately for Edison, his plan to scare the public away from AC backfired. All of his public electrocutions of animals proved too much for many to stomach. In addition, Edison had even gone so far as to promote AC as a means of executing humans and had designed an electric chair for the purpose. The chair was first used to execute a condemned prisoner at Auburn Prison in New York State, in 1890. Although Edison had claimed the chair would produce a rapid and painless death, it did neither. The Auburn electrocution did not go as well as Topsy's at Coney Island. The prisoner had to be jolted multiple times before he died, and his body was badly burned in a slow process that took over eight minutes to complete. Edison was heavily criticized for his prediction that the prisoner would die painlessly in 1/10,000th of a second. Westinghouse said, "They could have done better with an axe." The bungled execution received wide coverage in the press, and the public was outraged.[17]

Rather than turn against AC, people turned against Edison. There was a tremendous backlash to these horror spectacles. The great Edison—father of the incandescent light bulb—was subsequently vilified. In the end, Westinghouse's AC current ultimately won the day as the favored mode of electrical transmission.

Ironically, Topsy fared much better than Edison in the court of public opinion. Her serial killing spree was forgiven, and she came to be seen as a poor victim of technology gone amuck. She became a posthumous celebrity, a poster child for the anti-animal abuse movement, and the subject of the song, "Coney Island Funeral."[18] Most recently, she had a monument dedicated to her memory at the Coney Island Museum in 2003.

The ghost of Topsy may even have exacted its toll on Luna Park, the amusement park where she met her demise. On August 12, 1944, Luna Park burned completely to the ground in a mysterious fire that the locals called "Topsy's Revenge." The cause of the fire remains unknown, but it may have been electrical.

With time, Edison regained his stature as a great inventor, and electric wiring in the home—whether AC or DC—gained wide acceptance. It wasn't that people necessarily became less fearful of electricity, but rather, as they became more familiar with it, they began to believe that the risks could be managed with some safety precautions. People began to accept the trade-off of the risk of accidental electrocution for better and cheaper illumination and work-saving electrical appliances. They simultaneously experienced a lower risk of candle and gaslight fires. They even ignored a dying gaslight industry's warning to its few remaining customers that incandescent light projected a toxic ray that would turn their skin green and increase their death rate.[19] But this fabricated claim was also seen by the public as the scare tactic that it was—no better than Edison's rants about AC—and they were unmoved. Gas lighting in homes soon disappeared, and the death rate from house fires decreased accordingly.

As for radiation in the form of electrically produced light, the public feared the invisible electricity, rather than the electric light itself. Once electricity was tamed for human use, however, it was available to generate invisible forms of radiation—some benign and some not. Still, none of these types of radiation would engender as much fear as the electricity itself had . . . at least not in the beginning.

DOT DASH DOT DOT DASH DASH DOT DOT DOT DOT DASH DASH DASH

Many think that Guglielmo Marconi (1874–1937) discovered radio waves. As is the case with Edison's electric light bulb, this story is also only half true. Radio waves were predicted and searched for by a number of scientists working over a number of years. Marconi, however, had the good fortune of showing up just as everything was coming together and, more importantly, he immediately grasped the practical implications of the work. Thus, he took a lot of the fame and glory, as well as the 1909 Nobel Prize in Physics. The other important players, including Oliver Joseph Lodge (1851–1940), a physics professor at University College of Liverpool, England, remain more obscure. Regrettably, such is the life of a scientist. Still, it is important to reflect on why the scientists suspected the existence of radio waves and why they began to hunt them down.

It all began with mathematical calculations performed in 1867 by physicist James Clerk Maxwell (1831–1879). Maxwell noticed some similarity between the wavelike characteristics of light and the properties of electrical and magnetic fields. From this observation, he formulated a set of equations that described waves of what is now called *electromagnetism* traveling through empty space. For the shorter wavelengths,[20] his equations accurately described the known properties of light, so it was concluded that light was, in fact, a form of electromagnetism. Likewise, if you plugged values corresponding to longer wavelengths into Maxwell's equations, they also described some physical properties. But of what? That wasn't at all clear, because the existence of electromagnetic waves with wavelengths longer than light's was unknown at the time. Did such electromagnetic waves really exist, or was it all just mathematics carried too far? The hunt was on, and it lasted for over 20 years. Finally, in 1888, Heinrich Hertz (1857–1894) was able to experimentally produce and detect long wavelength electromagnetic waves (radio waves) in his laboratory. Unfortunately, Hertz died young and was never able to carry his discovery to its next logical phase.

When Hertz died, his scientific achievements were reviewed in the popular press, and caught the eye of Marconi, a young electricity enthusiast. Only 20 years old at the time, Marconi, like no other, fully appreciated the potential of Hertz's work. Though he had minimal formal training in electricity, Marconi started with that little knowledge and jumped wholeheartedly into the field of radio waves.

It was the chance discovery, in 1891, of an apparently unrelated phenomenon by French scientist Édouard Eugène Désiré Branly

(1844–1940), that opened the door to radio communications. Branly found that an electric spark could increase the electrical conductivity of nearby metal filings enclosed within a glass tube. If the tube were then tapped with a finger, the filings would become disoriented again and conductivity would stop. Furthermore, the phenomenon could be demonstrated to occur even when the spark and the tube were on opposite sides of the room. It wasn't long before an electric bell was hooked up to the tube, and public demonstrations were being made of sparks ringing bells from across the room through apparently empty space. Thus, a new scientific parlor trick was born.

Lodge and his colleagues first made the speculation, which turned out to be correct, that the phenomenon was due to Hertz's waves being emitted by the spark, but they failed to see the practical significance of this phenomenon for communications. (At least they failed to appreciate it initially.) Only Marconi saw it for what it was—a new means to transmit telegraphic signals without wires.[21]

Working initially in his home attic in Italy with funding from his father—a financially successful businessman and farmer—Marconi was able to transmit wireless telegraphic signals short distances. He methodically improved the technology to lengthen the distance signals could travel, and subsequently solicited commercial backing. Eventually, on December 12, 1901, Marconi made the first transatlantic transmission (from Poldhu, England, to St. John's, Newfoundland, Canada). With the success of that transmission, wireless telegraphy instantly became a contender to the undersea transatlantic cable as a communication medium between the United States and Europe.[22] Marconi also made the technology portable and marketed it to shipowners through his business, the Marconi International Marine Communication Company. The *Titanic* was equipped with the latest of Marconi's marine telegraphs when it hit an iceberg during its maiden voyage on April 14, 1912. Wireless distress signals in Morse code, transmitted from the sinking ship, brought rescue vessels. Unfortunately, they arrived too late for the many who froze to death in the frigid sea due to insufficient lifeboats; but they were a godsend to the rest.

Although Marconi's only interest in radio waves was for wireless telecommunications for the good of humanity (and profit for himself), others saw more sinister uses. He was once asked by a journalist: "Could you not from this room explode a box of gunpowder placed across the street in that house yonder?" Marconi answered that, as long as the gunpowder had been prewired for electrical ignition, it could easily be achieved.[23] Still others saw the potential for radio communication use in warfare, and Marconi was soon visited by German spies seeking to obtain a telecommunication advantage in the looming hostilities that would become the First World War. No one, however, foresaw *radar*, one of the most powerful military uses for radio waves. Radar would become the radio wave technology of the Second World War.

Marconi's telegraphic radio technology was further developed by engineers to allow full audio transmission rather than simply the intermittent clicking of Morse code.[24] This innovation allowed radio to become a major vehicle for the news and entertainment industries. By the 1930s, a console radio could be found in virtually every home in America.

How is it that Marconi was able to transmit radio waves across the entire ocean? Marconi's basic strategy for long-distance radio transmission was to build very high antennae and increase electrical power as much as possible. His transatlantic transmission of December 12, 1901, utilized up to 300,000 watts of power[25]—enough to power 5,000 sixty-watt light bulbs[26]—to propel his wireless telegraph message across the ocean. His approach worked, but all that power wasn't necessary. He subsequently learned that if he had shortened the wavelength of the radio waves just a small amount, he could have greatly increased transmission distances without all that excess wattage.[27] He later said that his fixation on wattage, rather than also attending to wavelength, was the biggest mistake of his career.[28] He had underappreciated the great importance of wavelength in transmission distance. We will soon see that wavelength is the primary factor determining radiation's health effects as well.

Marconi and his coworkers realized that the electrical power levels they were dealing with were very dangerous, and they took extensive precautions to protect themselves from electrocution during their experiments.[29] In contrast, they were never wary of radio waves and took no precautions at all to protect themselves from the possible health effects. Likewise, the public seemed equally unconcerned about the potential health consequences of radio waves . . . with one notable exception.

Before his transatlantic attempt, Marconi first sought to demonstrate transmission across a humbler body of water; namely, the English Channel. In preparing for a transmission on the French side of the channel in the little town of Wimereux, Marconi's work crew was interrupted one night by a man who barged into the transmission room wielding a handgun. The man claimed to be suffering from some sort of internal pain, which he attributed to the radio wave transmissions, because the symptoms apparently started when Marconi's crew began their experiments. He demanded that all further transmission stop, and his gun proved to be very persuasive.

Fortunately, one of Marconi's engineers, W. W. Bradfield, thought quickly and told the man that all would be made right if he would just put down his gun and allow himself to be immunized against radio waves. Although skeptical, the gun-wielding man acquiesced to the treatment, and Bradfield gave him a slight electrical shock as an "immunization." Bradfield told him the immunization would convey lifelong resistance to radio-wave illnesses. Satisfied, the man left. Apparently his symptoms abated, because the workers never heard from him again.[30]

Marconi's crew did suffer one transmission-related death, though. On Sunday, August 21, 1898, on an island off of Ireland, a young technician was preparing a transmission station in a house on the edge of a seaside cliff by stringing transmission wires outside one of the bedroom windows. No one knew the details of the mishap, as he was unwisely working alone. But the next day his body was found at the base of the 300-foot cliff. Presumably, he had somehow slipped and fallen to his death.[31]

Marconi himself died in Rome on July 20, 1937, at the age of 63, after a brief illness. The exact cause of his death is unknown. He was given a state funeral by the Italian government, and, the next day, radio stations everywhere in the world participated in two minutes of silence in tribute to the man who had made it all possible.[32]

A BULLET DODGED

Unfortunately for hospital patient Toulson Cunning in Montreal, Canada, the standard of care was amputation. Cunning had the ill luck to have been shot in the leg on Christmas Day, 1895. Despite painful surgical explorations by his doctors, the bullet could not be located. Amputation appeared imminent. But, as fate would have it, Cunning would become the first beneficiary of the medical use of x-rays. His leg would be spared the knife.

While Cunning writhed in his bed in pain, Professor Wilhelm Conrad Roentgen (1845–1923),[33] in Würzburg, Germany, had his own worries that Christmas night. Just days earlier, he had made what would prove to be one of the most momentous discoveries of the century, but at the time he was concerned that he might have overlooked something and been badly mistaken. He confided to one of his colleagues at the University of Würzburg, "I have discovered something interesting, but I do not know whether or not my observations are correct."[34] They were.

Roentgen had discovered invisible rays that could pass through solid objects. This was an outrageous claim at the time, and Roentgen likely wouldn't have believed it himself if he hadn't seen it with his own eyes. It seems that he and most everyone else were unaware that another scientist, Hermann von Helmholtz (1821–1894), had predicted in 1893 that rays with wavelengths shorter than visible light, if they existed, would be able to pass through matter.[35] Had Roentgen known of von Helmholtz's theoretical work, he might have been less worried about his own experimental findings, but not much less.

Roentgen was not the type of scientist who put much stock in theorizing and grand hypotheses. He was an experimentalist. He

FIGURE 2.1. WILHELM CONRAD ROENTGEN. Roentgen discovered x-rays in 1895, and won the first Nobel Prize in Physics for his efforts (1901). X-rays became an immediate sensation among scientists and lay people alike because of their mysterious ability to penetrate solid objects, and for their potential utility in diagnosing disease.

believed in the existence of only those things that he could directly measure and test in his laboratory. Roentgen's philosophy was that all knowledge was acquired empirically. To him, science advanced through many hours of hard trial-and-error work, coupled with a little luck. By putting in the laboratory time, the perceptive scientist afforded himself the chance to make an important discovery. A casual observer might have discounted the strange glow coming from the fluorescent screen hanging in his laboratory. But Roentgen realized that the glow on the screen coincided precisely with the electricity experiments he was performing on the other side of the room. But exactly what was it that caused this strange glow?

Roentgen's experiments involved running an electric current through a vacuum tube, commonly known as a Crookes tube after its inventor, William Crookes (1832–1919). Crookes was a self-styled chemist who never worked for a university. Rather, he supported himself by publishing scientific periodicals. He was primarily interested in chemical elements, and his seminal contributions to the chemistries of selenium and thallium won him admittance into the Royal Society—the United Kingdom's most prestigious body of scientists—at the young age of 31.

Crookes was also fond of making gadgets from glass. He spent a lot of time playing with a novelty item invented by the famous German glassblower Heinrich Geissler (1814–1879). Geissler found that neon and some other gasses confined within a tube glowed in different colors when electricity passed through them. By blowing the glass tubes into various shapes, brightly colored artwork could be produced. These tubes eventually developed into what we now know as neon lighting. Crookes decided to study the science underlying Geissler's tubes, and ended up creating his own namesake tube that would provide him with worldwide fame even greater than Geissler's.

A *Crookes tube* can be thought of as a light bulb lacking a filament. It's typically in the form of a clear glass tube in the shape of a pear with electrodes at each end, and is about the size of an American football. William Crookes found that by applying high voltages across the tube he could produce mysterious cathode rays, which could be seen if the Crookes tube were coated with fluorescent material, or if the cathode ray beam were pointed at a fluorescent screen (a sheet of glass or cardboard coated with some fluorescent chemical).

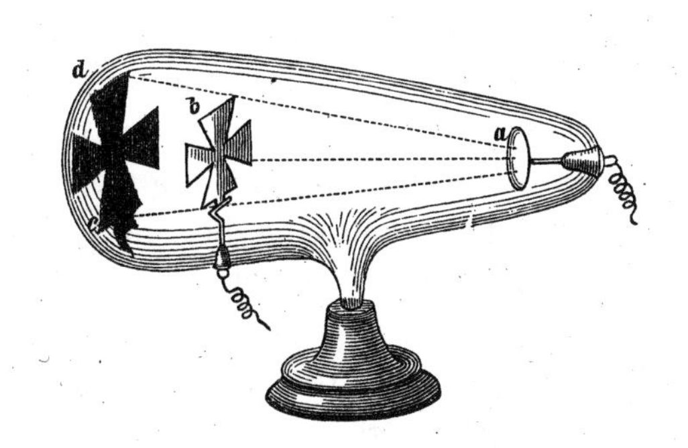

FIGURE 2.2. CROOKES TUBE. Crookes tubes were widely used in physics laboratories in the late 1800s to study the rays emitted from the cathode (negative electrode) of the tube. Roentgen was studying the properties of cathode rays when he accidentally discovered that the anode (positive electrode) of the tube, typically shaped like a Maltese cross, emitted its own rays, but of an undetermined type. Since their identity was a mystery, Roentgen called them "x" rays. (*Source*: Fig. 244 in *Lehrbook der Experimentalphysik* 2nd ed., by E. von Lommel, Leipzig: Johann Ambrosius Barth, 1895)

It turned out that cathode rays are really just electrons jumping through empty space. Crookes tubes are vacuum tubes. That is, they contain no air or any other gas, just empty space. Since there are no gas molecules or filament to conduct them, electron flow is not possible in a Crooke's tube until very high voltages are applied across the electrodes, at which point electrons begin to jump from one electrode (i.e., the cathode) to the other (i.e., the anode) across the empty space. It was just these jumping electrons that were the explanation for what Crookes was calling cathode rays. The study of cathode rays was a hot topic among physicists in Roentgen's day, and virtually all physicists had Crookes tubes in their laboratories. Roentgen was no different.

Remarkably, whenever Roentgen experimented with his Crookes tube, he noticed a faint glow even on fluorescent screens that were not in the vicinity of the cathode ray beam. The glow persisted even when the Crookes tube was completely blocked from the fluorescent screen with objects handy in his laboratory, including cardboard, books, and rubber. It was as though some type of invisible rays were coming from the tube, penetrating these materials, and hitting the fluorescent screen. Thinking these mysterious rays were some new form of light, he tried to bend them with a prism just as Newton would have undoubtedly attempted, but to no avail. They could not be bent with a prism, so they weren't rays of light. Thinking alternatively that perhaps they might just be stray electrons escaping from the tube, he tried to bend their path with a magnet.[36] That didn't work either, thus ruling out electrons as the source of the glow. All rays known at the time (i.e., light rays and cathode rays) were deflectable by either prisms or magnets, but these mysterious "rays" were affected by neither . . . and they penetrated solid objects! It was simply amazing. Whatever was going on was something new that had never been described before. What were these invisible rays, how were they being formed, and how did they penetrate the objects? He didn't know. So Roentgen called them "x" rays.

Roentgen continued to investigate the penetrating ability of x-rays and found that metal could block them, while they readily passed through wood. This allowed him to play a little game whereby he could "see" coins concealed in a wooden box by detecting their shadow on the fluorescent screen. These stunts raised the question: If wood was transparent to x-rays, what about human flesh? When he placed his hand in front of the screen, he could see, to his astonishment, the shadow of his own bones! His bones blocked the x-rays while his flesh did not. Thus, the medical potential of this new discovery was immediately apparent to Roentgen.

An avid nature photographer and outdoor enthusiast, Roentgen lugged his cameras on annual hiking vacations throughout Switzerland. He stored a lot of photographic equipment in his laboratory. This was a convenient location, since he lived with his family in an apartment above his lab. Roentgen decided to see what x-rays would do to his photographic film. When he replaced the fluorescent screen with his film, he found after developing it that he could

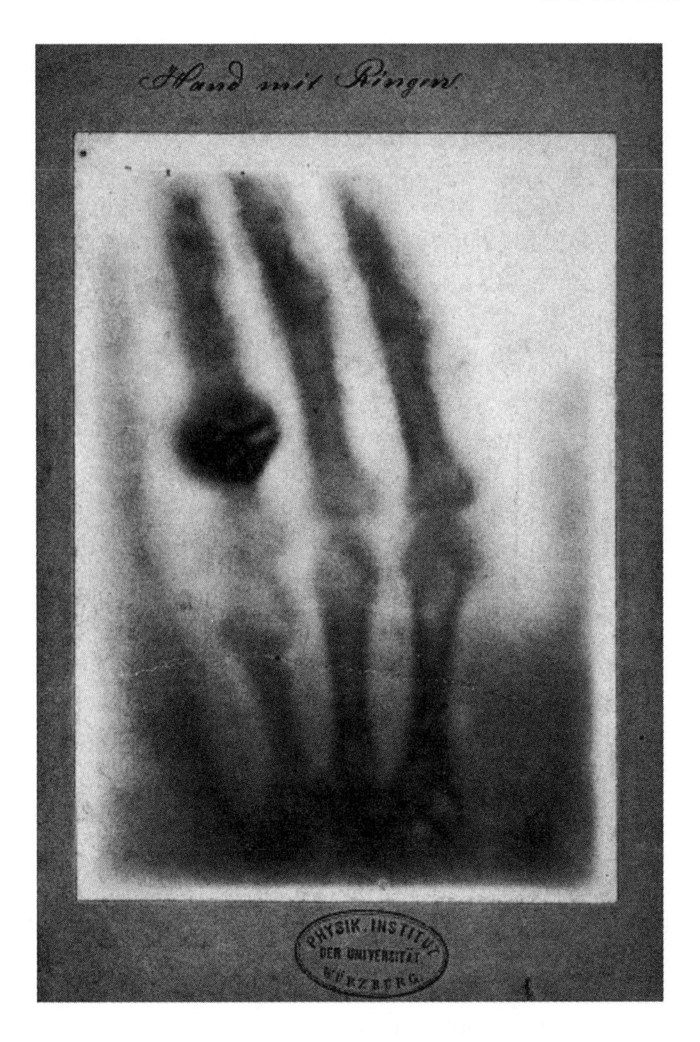

FIGURE 2.3. X-RAY OF ROENTGEN'S WIFE'S RINGED HAND. This x-ray photograph was one of the very first ever taken. Such photographs, showing shadows of human bones, both amazed and terrified people.

produce a permanent image similar to the shadowy ones he saw on the fluorescent screen.[37]

On December 22, 1895, Roentgen took a trusted friend—his wife—into his confidence. He called her down from their apartment to his laboratory. He demonstrated his discovery with the glowing screen, and then he took a "photograph" of her hand. In a short while, he developed the film and showed it to her. She was both astounded and frightened by what she saw.

It should be kept in mind that this was a time when skeletal bones were only seen after a person died and their flesh had decayed away; an image of a skeleton was then the universal depiction of death. In fact, when Roentgen showed his wife the image of the bones in her hand, she is said to have exclaimed, "I have seen my own death!"

After producing the image of his wife's hand, things moved very quickly. Roentgen knew he had to publish as soon as possible to establish the primacy of his discovery, and he thought the fastest route would be through a local scientific journal. He approached a friend who was the president of the Würzburg Physical Medical Society with a handwritten manuscript. Although the Society typically only published the proceedings of oral presentations delivered at their meetings, in this instance the editor decided to publish first and schedule the oral presentation for later. The manuscript was quickly sent for typesetting and just made the December 1895 issue of the society's journal. The article was entitled "On a New Kind of Rays," and it appeared on December 28, 1895.[38]

Although the paper contained no illustrations, a reprint with the x-ray of Roentgen's wife's hand was sent to a colleague in Berlin, who presented it as a poster at the Berlin Physical Society meeting on January 4, 1896. It was the first public viewing of an x-ray image.[39]

Another set of reprints and x-ray images was sent to colleagues in Vienna, one of whom happened to be the son of the editor of Vienna's newspaper *Neue Freie Presse*. The son showed the material to his father, who rushed the story into the January 5 issue, along with the x-ray image and a discussion of the potential medical utility of the discovery.[40] The next day, the *Chronicle* in London picked up the story. Although the more prestigious *Times of London* initially passed on publishing—assuming the finding to be some sort of minor photographic advancement—even they had to eventually publish Roentgen's discovery (albeit after being scooped by many other news organizations). From Europe, the story spread like wildfire throughout the world. It was published in the *New York Sun* on January 6, and the *New York Times* on January 12.[41] So both the scientific community and the general public became enamored with x-rays and their potential at virtually the same time, and Roentgen became an overnight celebrity. Not until 50 years later, when President Harry S. Truman (1884–1972) announced the creation of the atomic bomb, would a scientific event of any kind receive as much media attention as the discovery of x-rays.[42]

The coupling of the discovery with a practical medical use likely contributed to the enormous fame that Roentgen immediately enjoyed, but that had eluded Hertz and his radio wave discovery just a few years before. The journal article on x-rays also had the distinction of being one of the few physics papers that contained no mathematical calculations, so the work could be easily understood by nonscientists, including the news media and much of the public.

Seeing is believing, and Roentgen's x-ray photographs were evidence that was hard to refute. Most people regarded them as proof of the discovery. Besides, Crookes tubes were widely available to scientists and engineers throughout the world, so any scientist who doubted Roentgen's claims could get his own Crookes tube and replicate Roentgen's experiments to his satisfaction. And replicate they did. Within a few short weeks, Roentgen's results were replicated at Harvard (January 31), Dartmouth (February 3), and Princeton (February 6) universities.[43] By mid-February 1896, no credible scientist in the world doubted the soundness of Roentgen's work. His worries had been for nothing.

At McGill University, in Montreal, Professor John Cox, the Director of the Macdonald Physics Laboratory, was also fascinated by Roentgen's discovery and had replicated the x-ray photographic imaging on February 3. Hearing of this, Robert C. Kirkpatrick (1863–1897), the doctor treating the unfortunate Mr. Cunning for his gunshot wound, prevailed upon Cox to produce an x-ray image of Cunning's injured leg. The obliging professor took a 45-minute exposure of the leg with his Crookes tube. Even with such a long exposure time, the image of the leg was still somewhat underexposed. Nevertheless, on February 7, 1896, the bullet was found lodged against the tibia bone, and the surgeons quickly removed it. A report of the successful operation appeared in the *Montreal Daily Witness* the next day, which was barely six weeks after Roentgen had made his momentous discovery.[44] Never before or since has any scientific discovery moved from bench to patient bedside so quickly.[45]

Within months of his important discovery, Roentgen became internationally acclaimed, and was awarded the Nobel Prize in Physics in 1901, which was the first Nobel in physics ever awarded. Later, he would be further honored by having a unit of radiation exposure named after him, as well as a newly discovered element (roentgenium; atomic number 111). Some even sought to rename x-rays roentgen rays, and his x-ray photographs roentgenographs, but the modest Roentgen resisted these changes. In fact, ever the self-effacing academic, he never patented his discovery, and he bequeathed all of his Nobel Prize money to the University of Würzburg.[46] He considered himself to be a pure scientist, and the practical applications of his work to be the purview of others. Inventions and business were of no interest to him.

Even so, Roentgen did not just discover x-rays, collect his honors, and walk away. He subsequently dedicated his career to studying the properties of x-rays. In fact, famed British physicist Silvanus Phillips Thompson (1851–1916),[47] a staunch admirer of Roentgen, once remarked, "Roentgen has so thoroughly explored x-rays that there remains little for others to do beyond elaborating on his work."[48] This would prove to be a gross overstatement. Many novel discoveries about x-rays were forthcoming. But for years, it was not far from the truth.

Unlike some other radiation pioneers who spent considerable time working with the rays, Roentgen always took precautions to shield himself from the beam. It is unclear why he did. Perhaps he was just a cautious man by nature,[49] or maybe he took his wife's remark about seeing her own death as some kind of premonition. In any event, he was wise to do so. X-ray exposures in those early years were typically quite high. For example, it took 15 minutes of exposure to produce the x-ray photograph of Roentgen's wife's hand—something that can be accomplished in 1/50th of a second today. Roentgen's self-protection paid off. He lived many years and apparently suffered no ill effects from his work with x-rays.

Roentgen obviously had not set out to produce electromagnetic radiation from his Crookes tube. He was interested in the properties of cathode rays flying free through space. So how was it that Roentgen was able to produce x-rays from what was little more than a souped-up light bulb, and what underlying physical principle did his experiments reveal?

Cathode rays (i.e., flying electrons) could be deflected by magnetic fields, and this is the phenomenon that Roentgen had hoped to study. What he failed to appreciate was that the very high voltage would cause the flying electrons to bring along with them a tremendous amount of energy that they would necessarily impart upon the anode when they hit it. All that energy would need to go somewhere, and quickly. That surplus energy left the tube in the form of x-rays. Essentially, that's all that Roentgen was witnessing, but let's add a few more details.

It turns out that some of the excess energy carried by the flying electrons is simply dissipated as heat, which explains why the tubes get hot. In addition, however, another energy dispersing process occurs. As a high-speed electron passes near the electrically charged orbitals of a metal atom in the anode, it gets deflected from its straight path and then abruptly slows to a stop, like a speeding car swerving to miss a pedestrian and hitting a brick wall. When this happens, the energy of the fast-moving electron is immediately dissipated in the form of an electromagnetic wave, which Roentgen was calling an x-ray.[50] Instead of the flying bricks that the crashing car produces, an electron crashing into an anode generates flying x-rays. Since Roentgen's time, volumes have been written on the physics of x-rays, and you could spend a lifetime studying it. But we don't have to go into all that here. This is as technical as we need to get to consider the effects of x-rays on health.

☢

Ironically, the great William Crookes himself realized in retrospect that he, too, had witnessed an effect of x-rays years before but never had recognized it. He was constantly returning photographic plates that he stored in has laboratory back to the vendor, complaining that they were "fogged." After he saw Roentgen's paper, he knew why they were fogged; they had been exposed by x-rays as a consequence of his Crookes tube experiments. He also understood that he had missed the greatest scientific opportunity of his life.

Interestingly, Edison's enterprises with electromagnetic radiation didn't end with his light bulb. Three weeks after Roentgen's discovery of x-rays and its effects on fluorescent screens, Edison started working on a medical invention he called the *fluoroscope*—a device that coupled a fluorescent screen with a Crookes tube to produce a live image of a patient's internal structures.[51] Edison demonstrated his fluoroscope at the National Electric Exposition in New York City's Grand Central Palace Hotel, in May 1896.[52] It proved to be an extremely useful medical device, and is still in widespread use today. When President William McKinley was shot twice in the abdomen in Buffalo on September 5, 1901, his doctors called on Edison to send a fluoroscope, to guide them in the surgical removal of the bullets. Edison sent a fluoroscope at once, along with two technicians to operate it.[53] But it was never used because the surgeons decided it was safer to leave the bullets where they lay, rather than extract them. McKinley then took a turn for the worse and died on September 14, and the technicians brought the unused fluoroscope back to Edison's laboratory in New Jersey. Although Edison's fluoroscope did not save McKinley's life, another one of his devices helped dispense justice; on October 19, McKinley's assassin, Leon Frank Czolgosz (1873–1901), was executed in the electric chair.

Despite Edison's expressed concern about the dangers of AC electricity, he apparently had no such reservations about x-rays. Unlike Roentgen, he took no precautions to protect himself from what amounted to extremely high doses of x-rays produced by his

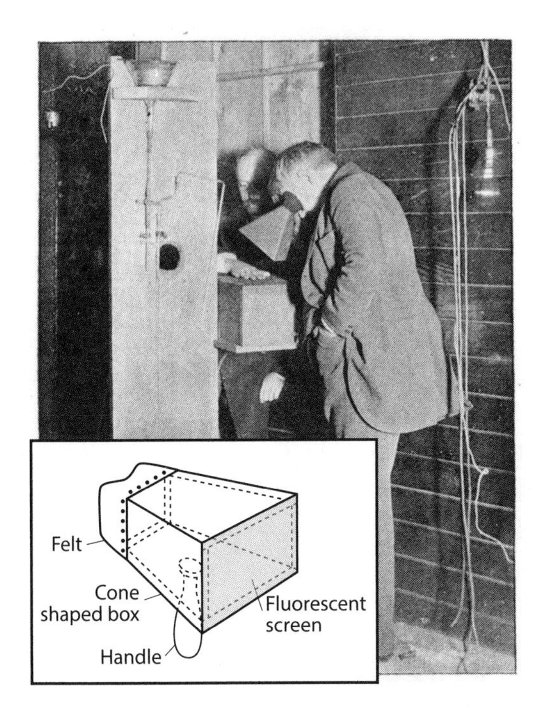

FIGURE 2.4. THOMAS EDISON VIEWING THE BONES IN HIS ASSISTANT CLARENCE DALLY'S HAND WITH THEIR LATEST INVENTION, THE FLUOROSCOPE. Edison's fluoroscope was simply a Crookes tube mounted within a wooden box. The box provided a stage on which the hand or other body parts could be viewed. This was done by means of a fluorescent screen mounted within a hand-held, light-shielding viewer. The fluoroscope provided a real-time moving image of the bones to the viewer, with no need to wait for development of photographic film.

experiments. Furthermore, his assistant, Clarence Madison Dally (1865–1904), eagerly and frequently volunteered to have his hands imaged with the fluoroscope. When one hand became badly burned by the beam, he would just switch to the other, allowing the burned one time to recover. Unfortunately, this unwise practice eventually caught up with him. He ultimately suffered severe hand ulcerations resulting in the loss of four fingers from his right hand and amputation of his left hand. It was soon found that cancer from his hands was spreading up his arms, so he had both his arms amputated as well. All these surgeries, however, were to no avail. The cancer had spread to his chest and he died in October 1904.[54]

Edison himself didn't escape unscathed. He nearly lost his eyesight from the high x-ray exposures his eyes sustained through the viewfinder of his fluoroscope. Traumatized by the whole experience, Edison abandoned all work with x-rays. When later asked about that work, he replied, "Don't talk to me about x-rays, I'm afraid of them!"

Not all investigators were as cavalier as Edison and Dally when working with x-rays. Soon after Roentgen's discovery became public, two brothers-in-law and close friends, Francis H. Williams (1852–1936) and William H. Rollins (1852–1929)—a physician and a dentist, respectively—began collaborating in Boston with the goal of developing x-rays for medical diagnostics. In fact, many consider Williams and Rollins the fathers of diagnostic radiology.[55] Yet, they always protected themselves from the beam, even before reports of injuries began to appear. Asked later why they did so, Williams said, "I thought that rays having such power to penetrate matter, as x-rays had, must have some effect upon the system, and therefore I protected myself."[56] Simple as that.

As early as 1898, Rollins had designed a metal box with a diaphragm aperture, like a camera shutter, to contain their Crookes tubes and thus limit their exposure to stray x-rays. Williams and Rollins promoted the use of the box, as well as other protective devices, among medical x-ray workers. Nevertheless, five years later, Rollins lamented, "Most of these precautions are neglected at the present time . . . partly due to attempts to ignore . . . the [known] effects of x-[rays]."[57] In fact, some investigators were in complete denial that x-rays could do them harm, claiming that the skin burns were not associated with the Crookes tubes themselves, but rather from electricity leaking from the generators that powered them (so-called *brush discharges*). Their remedy was to stop worrying about Crookes tube x-rays and just replace the generator. It seems that long after the big AC/DC controversy, some people still remained focused on the hazards of electricity, to the exclusion of all else.

Still, Edison and Dally could plainly see that the x-ray beam itself was burning their exposed tissues, and they likely knew about the protective devices that Rollins and Williams had been heavily promoting since 1898. In 1901, Williams had even published a massive volume, entitled *The Roentgen Rays in Medicine and Surgery: As an Aid in Diagnosis and as a Therapeutic Agent*, in which he noted the dangers of x-rays and recommended specific protective measures for both patient and physician. It immediately sold out, and was republished again in 1902, and then again in 1903, with editions in other languages. Edison and Dally could not have been unaware of the potential health hazards, yet they seemed to lack Williams's common sense. They ignored the safety precautions and paid the price with their health. Rollins and Williams, in contrast, never suffered any radiation injuries, and lived to 77 and 84, respectively.

It truly can be said that most of the radiation injuries suffered by early radiation workers were more the result of negligence than ignorance. There was ample evidence that x-rays could damage tissues, and there were ready means available to protect one's body from x-ray exposure while working with Crookes tubes. Unfortunately, few chose to take the threat seriously.[58]

SIZE MATTERS: THE IMPORTANCE OF WAVELENGTH

Fluorescent screens and photographic films are sensitive to more types of electromagnetic radiation than is the human eye. It was the fluorescent screen and photographic film that provided Roentgen with the tools that Newton lacked. We now have many tools besides fluorescent screens and photographic film to detect "invisible" rays, and these tools have allowed us to learn a tremendous amount about x-rays and other types of the electromagnetic waves that we now simply call radiation. Today, scientists understand that all electromagnetic radiation travels in the form of waves at the same unchanging speed as waves of light (i.e., Einstein's universal constant, c = the speed of light, from his famous $E = mc^2$ equation).[59] In fact, the difference between radio waves, x-rays, and visible light is simply the distances between the crests of their waves (i.e., their wavelengths). At the extremes, x-rays have very short distances between their crests—much less than the width of a human hair—while the distance between the crests of radio waves can be as long as an American football field.

These differences in wavelength drive all of the various properties of the multiple types of electromagnetic radiation and, most importantly, determine their energies. The shorter the wavelength, the greater the energy. This can be compared to ocean waves that strike the beach. When the distance between the crests of the ocean waves is short, many more waves hit the beach during any period of time than when the wave crests are farther apart. If more waves are hitting the beach, more energy is deposited on the beach during that time. Thus, shorter wavelengths deliver more energy and longer wavelengths less. This analogy of electromagnetic waves to beach waves is very crude and can even be misleading if carried to extreme. But it does provide a good mental picture of an electromagnetic wave that cannot otherwise be experienced or observed by our senses and is, therefore, innately hard to comprehend. Even though electromagnetic waves cannot be experienced directly by human senses, they can be detected with the use of technology; they are just as real as ocean waves, and just as powerful a force of nature.

This concept is so important to understanding radiation's effects on health that it warrants repeating. *Radiation often takes the form of an electromagnetic wave, and the wavelength of the radiation determines exactly what properties that radiation will have.* These properties include everything from how penetrating the radiation is, to whether or not it causes cancer. So, as we move ahead, remember that even though the different types of radiation discussed above are all composed of electromagnetic waves, the physical, chemical, and biological effects of different wavelengths can be as different as apples and oranges—or rather, as different as apples and baseballs. The wavelength is the key.

To illustrate the importance of wavelength, let's return to visible light and explore exactly why it is we can see it. The light we see falls in the middle of the spectrum of possible electromagnetic wavelengths. It has an extremely narrow range of wavelengths (just

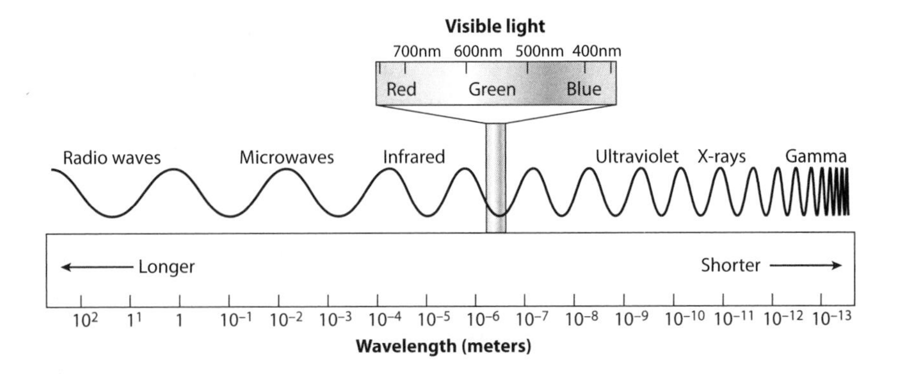

FIGURE 2.5. WAVELENGTH ALONE DISTINGUISHES THE DIFFERENT TYPES OF ELECTROMAGNETIC RADIATION. Most types of radiation in commercial use are electromagnetic; that is, they come in the form of waves that differ only by their wavelengths. The radiation types with the shorter wavelengths (e.g., x-rays) carry more energy than the types with longer wavelengths (e.g., radio waves). Only a very narrow band of wavelengths near the middle of the electromagnetic spectrum is visible to the human eye.

380 to 740 nanometers), which corresponds to approximately 1/100th the width of a human hair.[60] Electromagnetic radiation within these wavelengths can be seen by the human eye because there is a biological chemical in the retina of the eye, called *retinal*, that bends (i.e., *photo-isomerizes*) into a different molecular shape when exposed to radiation within this specific range of wavelengths. The nerve cells in the eye detect retinal's bending and then send a signal to the brain through the optic nerve. The brain processes the spatial signals from the retina into a visual image. This process is what we call sight. Radiations with wavelengths that are too large or too small are not capable of bending the retinal molecule, and, consequently, we do not see them. Still, that doesn't mean that such radiation is not there; it's just that it's invisible to us. The way we distinguish colors is a little more complicated, and not all animal species are able to see color. Color perception is beyond what we need to know about here, but let's just say that when the wavelengths start to get too long to bend the retinal, we call them *infrared* because they are just below the visible color red.

When the wavelengths are a little too short to be seen we call them *ultraviolet* because they are just beyond the visible color violet.[61] Visible light is squeezed in between the invisible infrared and ultraviolet wavelengths.

If we consider visible light to be the dividing line within the universe of invisible electromagnetic waves, what can we say about wavelengths on either side of the visible? Those wavelengths shorter than visible light carry more energy. The shorter wavelength radiations (think x-rays) carry enough energy that they go beyond simply bending molecules; they can actually break them. And it is the breaking of biological molecules that results in radiation's adverse biological effects. In contrast, those radiations with longer wavelengths carry much less energy than light (think radio waves), not even enough to bend retinal or other biological molecules. If these low energy radiations have biological effects, their mechanism is more obscure and beyond what we currently understand about biology.

So the dividing line of visible light turns out to be the dividing line for adverse biological effects. Scientists call the radiations with shorter wavelengths (i.e., higher energies) the *ionizing radiations* because they damage molecules by ripping off electrons and producing ions.[62] The longer wavelength (i.e., lower energy) radiations, in contrast, are called nonionizing because they don't have sufficient energy to ionize anything. That's not to say that nonionizing radiation can't affect molecules, but their effect is primarily to shake them up so that they start bumping into each other. Scientist call this bumping of molecules *kinetic energy*—energy that is commonly experienced as an increase in temperature and the production of heat. So it is possible to heat something (and even burn it) with nonionizing radiation (think microwave ovens). Nevertheless, these nonionizing radiations usually do not permanently damage the molecules in the way that the ionizing radiations do. Consequently, the nonionizing types of radiation have much less, or no, potential to produce the kinds of adverse biological effects that we are most concerned about, such as genetic damage and cancer.

Before we go any further with this line of thought, it needs to be mentioned that ultraviolet (UV) radiation is a special case. The wavelengths of UV radiation lie just at the cusp between the ionizing radiations and visible light and, therefore, share properties with each. Although UV radiation does not have energy sufficient to ionize atoms, it does have sufficient energy to move the atom's electrons to higher energy states. Such atoms are said to be in an *excited state*. As with all excitement, it subsides with time. But in the meantime, such atoms can undergo excited-state chemical reactions that don't normally occur when the atoms are not excited (i.e., when they are in their so-called *ground state*). Excited-state reactions can produce specific types of biochemical damage that are harmful to cells. Although UV radiation is not able to penetrate tissues to any extent, it is able to cause damage at the surface. So high exposures to UV radiation can produce skin burns (i.e., sunburns) and skin cancer. Because the hazards of UV radiation are in many ways unique and largely limited to skin and eye damage, the health risk associated with UV exposure is a topic unto itself.

We need not dwell on these technical aspects of wavelength now. We will revisit wavelength issues in our discussions of health effects. Just know for now that when the question is in regard to radiation risks, wavelength matters.

WAVES AREN'T THE END OF IT

Thus far, we have discussed radiation solely in terms of electromagnetic waves and, in fact, much of the radiation that humans are exposed to is of the electromagnetic kind. There is, however, another subclass of ionizing radiation that comes in the form of high-speed atomic particles, called *particulate* radiation. The particulate radiations are ionizing, just like x-rays. As such, their effects on health are quite similar, as we shall see. These particles are typically produced through the decay of unstable atoms—a phenomenon known as radioactivity—and are best described in terms of their connection with radioactivity. We will now turn our sights to radioactivity.

CHAPTER 3

SEEK AND YOU SHALL FIND: RADIOACTIVITY EVERYWHERE

Name the greatest of all inventors. Accident.

—*Mark Twain*

A SILVER LINING

Antoine Henri Becquerel (1852–1908) was not interested in Roentgen's x-rays or even in seeing images of his bones. But he was fascinated with fluorescence.[1] Becquerel was specifically interested in determining exactly how chemicals could store visible light and release it later. In fact, his family had worked on this exact problem for decades.

Becquerel held the chair of the Department of Physics at the Musée d'Histoire Naturelle in Paris, like both his grandfather and his father had before him. All of them had studied fluorescence.[2] Over many years of pursuing this multigenerational scientific passion, the Becquerel family had accumulated a large collection of minerals that had the ability to fluoresce. It was this collection that enabled Antoine Henri Becquerel to make his most important discovery, and win the Nobel Prize. To Becquerel's dismay, however, his momentous discovery had nothing to do with fluorescence.

Becquerel had fluorescent screens and photographic films, but no Crookes tube; nor did he want one. Nevertheless, Roentgen's report of invisible x-rays accompanying the faint visible glow from his Crookes tube made Becquerel wonder whether x-rays were also mixed in with the visible glow from fluorescence. That is, were fluorescent materials emitting those penetrating x-rays along with their nonpenetrating visible light? He designed a series of simple experiments to test his idea.

He took a piece of photographic film and sealed it tightly within thick black paper so that no light could enter to expose the film. Then he sprinkled on the black paper granules of the fluorescent material he wanted to test. He then put the whole set-up out in bright sunlight, to stimulate fluorescent emissions. At the end of the day, he developed the film hoping to find an image of the overlying fluorescent material on the developed film. He reasoned that, if the penetrating x-rays were being emitted along with the fluorescent rays, the x-rays alone and not the visible fluorescent light should penetrate the dark paper and develop the film, just as they had done for Roentgen. Consequently, if images of the granules appeared on the developed film, he would have evidence of x-ray emissions from fluorescent chemicals. Methodically he tested his entire collection of fluorescent chemicals for evidence of x-rays.

To Becquerel's chagrin, no images were forthcoming from any of his fluorescent chemicals until he tested uranium sulfate. This compound gave him a dark image of the granules. Elated that his hypothesis was proving correct in at least one case, he proceeded to do more experiments with the uranium sulfate. One day he prepared it and the film packets as usual. But when he went to place them out in the sunlight, the skies had turned cloudy, so he decided to postpone the experiment until the next day. Unfortunately, the poor weather persisted. Impatient to get an answer, he decided to develop the film in the hope that even the weak room lighting might induce a faint image. Instead, he found a very intense image—much more intense than could be explained by room light. At this point he realized that he had neglected an essential experimental control: he had never shown that the image from uranium sulfate granules actually required prior exposure to light. Consequently, he prepared a control where he completely covered the uranium sulfate so it would experience no light exposure at all. To his astonishment, the resulting photographic film still had the same image of the granules.

Adding to the mystery, Becquerel soon found that even nonfluorescent uranium compounds exposed the films. The only common denominator in the experiments seemed to be the presence of uranium atoms. Neither prior sun exposure nor fluorescent properties of the compound were required. The presence of uranium atoms alone was necessary and sufficient to expose the film.[3]

By this time Becquerel knew his hypothesis about x-rays and fluorescence was wrong. But exactly how was it wrong? He was confused. Ultimately, Becquerel had no other option but to conclude that uranium atoms spontaneously emit some type of invisible radiation with properties similar to x-rays. Further experiments confirmed this supposition and supported the notion that the invisible penetrating radiation and the visible nonpenetrating fluorescence were completely unrelated phenomena; the invisible penetrating radiation resulted from a nuclear process, and the visible nonpenetrating radiation from a chemical process. The co-occurrence of these two properties in uranium sulfate was a fortuitous coincidence.[4] That coincidence allowed Becquerel to discover radioactivity.

Luck had once again favored a prepared mind, and Becquerel's place in scientific history was secure. In 1903, he was awarded the Nobel Prize in Physics for his discovery, just two years after Roentgen received his for discovering x-rays. To further honor him, the standard international unit for measuring radioactivity would eventually be named a *becquerel*.[5]

Still, Becquerel never became the celebrity that Roentgen was. This was probably for three reasons. First, he shared his 1903 Nobel Prize with Marie and Pierre Curie, who were the colorful husband and wife scientific team that discovered radium—a highly radioactive element that they found in uranium ore—and would soon capture the public's imagination.[6] Second, unlike the easily obtainable Crookes tubes, uranium was very scarce, so few scientists could easily perform the same kinds of experiments that Becquerel had. In fact, uranium was thought to be scarce even up to the time of the invention of the atomic bomb.[7] We now know it to be one of the more common and widely distributed elements on Earth; more common than silver or gold. Third, the photographic images produced by uranium radioactivity were very diffuse, and the medical utility of radioactivity not immediately apparent. It wasn't until radium was purified and separated from the uranium ore that the use of radioactive substances in medicine started to grow.[8]

Although Becquerel appreciated the parallels between his discovery of radioactivity and Roentgen's discovery of x-rays, he either was not aware of the danger of x-rays or he did not realize that he had enough radioactivity on hand to produce similar health effects. He continued to work with radioactivity without any protection. One day, he put a vial of radioactive material in his vest pocket where it remained for several hours. Later, he found the skin on his abdomen had been burned.[9] He suddenly had a new respect for the stuff.

Becquerel's biggest stroke of personal luck may have been that he stopped working with radioactivity very soon after he discovered it. Radioactivity had been only a side interest of his, and he thought the big discoveries with radiation had all been made. Having discovered radioactivity in 1896, he was largely out of the radioactivity business within a couple of years. He published his last paper on the subject in 1897 and moved on to new things.[10] He left it to others to hammer out the scientific details. Other than that one-time radiation burn on his belly, Becquerel was not known to have suffered any ill health due to his brief work with radioactivity. His short tenure in that research field, coupled with the fact that his uranium samples had relative weak radioactivity levels, likely meant that his body's lifetime radiation dose from his exposures to radioactivity was quite modest, and was probably restricted mostly to his hands.

HOT ROCKS: URANIUM

By definition, radioactivity is the ability of an atom to spontaneously release radiation without any stimulation from light, electricity, or any other form of energy. It is a property intrinsic to the nucleus of an atom and resistant to modification by any outside forces. Uranium was the first radioactive substance to be discovered, and still is one of the best known, but dozens of elements exist in hundreds of different radioactive forms known as radioisotopes. Many of these radioisotopes are mixed with and cannot be easily separated from their nonradioactive forms. And some of the elements with natural radioisotopes are essential to life, including carbon, the element that plays a critical role in all biochemical synthesis pathways, and potassium, an essential element in cells and blood. Thus all living things are radioactive, to some extent. Radioactivity inside and outside of human bodies exposes humans to a continuous bath of low-level radiation.

Becquerel possessed uranium sulfate within his collection of minerals simply because it fluoresced. It was the only mineral in his fluorescent mineral collection that also happened to be significantly radioactive, so Becquerel would not have discovered radioactivity if he didn't have uranium sulfate. But there is another coincidence about uranium that Becquerel did not discover, and it would turn out to be even more important. Uranium is one of the very few radioactive elements that can also undergo spontaneous nuclear fission—a natural process where an atom's nucleus suddenly splits into two or more smaller parts.[11] Spontaneous nuclear fission is the property that made uranium the key to the development of a nuclear bomb. Thus, this convergence of fluorescence, radioactivity, and fission into a single element could be viewed as one of the luckiest coincidences of nature . . . or one of the unluckiest. Had uranium's radioactive properties not been discovered in 1896, it is highly unlikely that the atomic bomb could have been developed in time for deployment during World War II by any nation, and humanity's loss of innocence regarding nuclear warfare would have either been averted entirely or at least greatly postponed.

RADIOACTIVE DECAY

Why are some elements radioactive and others not? To answer this question we must first consider what makes an atom stable (i.e., nonradioactive). As most high-school science students know, the nucleus of the atom contains a mixture of two subatomic particles. These are the positively charged particles called protons and the uncharged particles termed neutrons; the latter are electrically neutral (i.e., they have no charge). Positive charges repel each other, so the nucleus would not remain together were it not for the presence of the neutrons, which dilute the positive charge to the point that the nucleus can remain intact and stable.

As it turns out, for most atoms, the nucleus is stable when the number of protons nearly equals the number of neutrons.[12] If an atom's nucleus diverges too much from this one-to-one ratio in either direction, the nucleus is likely to be unstable. The farther from a one-to-one ratio, the more unstable it will be, and the more likely the situation will be corrected. If an atom has too many protons, a proton is converted into a neutron. If it has too many neutrons, a neutron is converted into a proton. This process is called *radioactive decay*. By decaying, the atom moves closer to a one-to-one ratio, and closer to stability. If after one decay, the atom still has an excess of either protons or neutrons, the process can repeat itself until ultimately the nucleus is stable. (Such a sequence of decays of radioactive elements on the way to a stable atomic nucleus is called a *decay chain*.)

To illustrate this concept, let's look at some different nuclear forms of carbon. Stable carbon has six protons and six neutrons (known as carbon-12, because it has a total of 12 nuclear particles). If a carbon atom has six protons and seven neutrons (carbon-13), the extra neutron can be tolerated and the atom is still stable. However, if the carbon atom has six protons and eight neutrons (carbon-14), the neutron burden is too high, the atom is radioactive, and it decays by converting a neutron into a proton.[13] This results in seven protons and seven neutrons—a one-to-one ratio—and a stable nucleus. Ironically, the new stable atom is no longer carbon! This is because an element's chemical identity is determined specifically by its number of protons. All carbon atoms have six protons, with a varying number of neutrons. Because our new stable atom now has seven protons it has become a nitrogen atom. So, the consequence of this radioactive decay is that an unstable carbon-14 atom has been converted into a stable nitrogen-14 atom. This, fundamentally, is how radioactive decay works, and atoms that can potentially decay because of their unstable nuclei are said to be *radioactive*.

We will soon learn more about neutrons in the atomic nucleus. But while we are discussing carbon-12, carbon-13, and carbon-14, we might as well clarify some terminology that is commonly used, but frequently confused. These three different versions of the same element (carbon) are called isotopes, from the Greek word meaning "in the same place." They are in the same place in that they all have six protons and are all, therefore, carbon, as explained previously. Nonetheless, they differ from one another in their neutron numbers—six, seven, and eight, for carbon-12, carbon-13, and carbon-14, respectively. Elements that differ only in their neutron numbers are said to be *isotopes* of one another. These isotopes can be either stable (e.g., carbon-12 and carbon-13) or unstable and, therefore, radioactive (e.g., carbon-14). The radioactive isotopes are often called *radioisotopes*. So as we move through our story, know that radioisotopes are just the versions of elements that happen to be radioactive because they have an unstable ratio of neutrons to protons in their nucleus. Remember also that all elements have proton-neutron combinations that are radioisotopes. Some are very common and easily found in our environment, while others are so rare and fleeting that they need to be artificially produced in order to study them.

PARTICULATE RADIATION

You might well have noticed that as carbon-14 decays to nitrogen-14, the charge of the nucleus has gone from 6^+ to 7^+ (i.e., six protons become seven). You may also remember from your high-school physics that charge can neither be created nor destroyed (law of conservation of charge). So how can the creation of an additional positive charge in this atom be accounted for?

What has not yet been mentioned is that when the carbon-14 decays, its surplus neutron (charge = 0) changes into both a proton (charge = 1^+) and an electron (charge = 1^-). So the net charge is zero ($1^+ + 1^- = 0$), and there actually is a conservation of charge. But the negatively charged electron is ejected from the atom in the form a high-speed particle known as a *beta particle.* This beta particle is but one representative of the family of radiation types called particulate radiation.

We will deal more with beta particles and other particulate radiations in later chapters, but for now it is sufficient to know that there are two subclasses of ionizing radiation—the electromagnetic short-wavelength radiations, like x-rays, and the particulate radiations, such as beta particles. Both have energies sufficient to generate ions by dislodging the orbital electrons of neighboring atoms, and thereby producing comparable biological effects. Consequently, they are simply grouped together as ionizing radiation for the purposes of understanding radiation biology. All radioisotopes emit ionizing radiation.

GAMMA RAYS

As we have just seen for carbon-14, radioactive decay often involves release of a high-energy beta particle that is ejected from the atom. In the case of carbon-14, particulate radiation is the only form of ionizing radiation that is released. For some radioisotopes, however, the energy that must be dissipated by their decay is too great to be carried by the energy of the particle alone. In those cases, an electromagnetic wave is released concurrent with the particle. The wave typically leaves the nucleus simultaneously with the particle but moves in a direction independent of the particle. We call these electromagnetic waves *gamma rays.* It was the gamma rays from uranium that were exposing Becquerel's film.[14]

Such gamma rays are typically indistinguishable from x-rays, although their wavelengths tend to be shorter and thus have higher energies.[15] (Remember, as we said in chapter 2, the shorter the wavelength, the higher the energy.) The only difference between a gamma ray and an x-ray is that gamma rays emanate from the atom's nucleus while x-rays emanate from the atom's electron orbitals. So a gamma ray can simply be thought of as an x-ray coming from an atomic nucleus. And because, under normal circumstances, nuclear decay is required for it to be produced, a gamma ray is exclusively associated with radioactivity.

HALF-LIFE

The concept of half-life is a useful way to comprehend exactly how unstable a radioisotope is. As mentioned earlier, the stability of a radioisotope is an intrinsic property of the atomic nucleus that cannot be altered, so all radioisotopes have their own unique half-lives. A radioisotope with a long half-life is relatively stable, and a short half-life indicates instability.

We'll use carbon-14 again to illustrate. The half-life of carbon-14 is 5,730 years. That means if we had one gram (about a quarter of a teaspoon) of carbon-14, in 5,730 years we would have 0.5 gram of carbon-14 and 0.5 gram of nitrogen-14, which had been produced from the cumulative decay of carbon-14. In another 5,730 years, we would then have 0.25 gram of carbon-14 (half again of the remaining 0.5 gram), and a cumulative 0.75 gram of nitrogen-14. After 10 half-lives (57,300 years) there would only be trace quantities of carbon-14 left and nearly a full gram of the nitrogen-14. In contrast, during this entire time, one gram of stable carbon-12 would remain as one gram of carbon-12.

We can use knowledge of radioisotope half-lives in a number of practical ways. For example, knowledge of carbon-14's half-life allows scientists to determine the age of ancient biological artifacts (e.g., wooden tools). Since the ratio of carbon-14 to carbon-12 in our environment is constant, all living things have the same ratio of carbon-14 to carbon-12. When a plant or animal dies, however, the exchange of carbon between the environment and its tissues stops, and the carbon in the tissue at the time of death remains trapped there forever. With time, the carbon-14 in the tissue decays away while the carbon-12 does not. So the ratio of carbon-14 to carbon-12 drops with time, and drops at a predictable rate due to the constancy of carbon-14's half-life. By measuring the ratio of carbon-14 to carbon-12 in a biological artifact a scientist is able to calculate how long ago that plant or animal died. This method of determining the age of artifacts is called radiocarbon dating (or simply carbon dating) and has contributed greatly to advancements in archeology, anthropology, and paleontology.[16]

The only thing that we need to remember about half-lives for health-related purposes is that all radioisotopes have their own unique half-lives, and the shorter they are, the more radioactive they are. Some highly radioactive elements have half-lives on the order of minutes or seconds and, therefore, do not survive long enough to have any significant impact on our environmental radiation exposure levels. In contrast, others have half-lives so long (e.g., tens of thousands of years) that they, too, contribute little radiation to the environment. But those with intermediate lengths of half-lives persist long enough to contribute to our environmental radiation burden. We will talk more about these environmentally significant radioisotopes later. But first, we should consider the stories of the greatest radioactivity hunters of all time, and what they discovered about radioactivity.

THE FRENCH TRIFECTA: BECQUEREL AND THE CURIES

As mentioned, Becquerel had to share his Nobel Prize in 1903 with two other French scientists who ended up being even more famous—Marie Curie (1867–1934) and Pierre Curie (1859–1906). This husband and wife scientific team contributed mightily to the characterization of radioactivity. In fact, they were the ones who introduced the term "radioactive," and they ended up going far beyond Becquerel with their studies.

The Curies also realized something that Becquerel had overlooked. They realized that uranium ore—the crude material that contained elemental uranium—had more radioactivity in it than could be accounted for by its uranium content alone. And they thought they knew the reason. They correctly surmised that uranium ore contained other radioactive elements even more radioactive than uranium.

Starting with a couple of tons of a tarry substance called pitchblende, the major mineral component of uranium ore, the Curies ultimately purified just 0.1 gram (about one-third of an aspirin tablet) of radium. The whole process involved the use of a radioactivity meter that Pierre designed, and segregating the nonradioactive components from the radioactive ones through various chemical processes.

The Curies ultimately showed that uranium ore actually contains at least three radioactive elements. In addition to the known uranium, there were also two previously unknown elements. One they called *polonium*, in honor of Marie's native Poland,[17] which was being subjugated by the Russian Empire at the time; they called the other *radium*, a name derived from the Latin word for "ray."

What the Curies accomplished was the result of a Herculean effort on their part. Unlike Roentgen and Becquerel, they didn't expose a few photographic films and wait for their Nobel Prizes to arrive. The Curies earned their awards through hard physical labor. They purified new radioactive elements from a mountain of rock.[18]

The most distinguishing thing about the scientific contribution made by the Curies, as opposed to Becquerel's, was that the former had actually discovered two previously unknown elements, polonium and radium, that both had radioactive properties. Becquerel simply discovered that an already known element, uranium, was radioactive. Since all the elements Becquerel tested were from his stash of known fluorescent elements and compounds, it was impossible for him to discover a completely new element. The Curies, however, traced the radioactivity in raw ore to its source by purifying it away from the nonradioactive minerals, and ended up adding two new elements to the periodic table of elements.[19] So theirs was both a physical and chemical achievement. While the physicists continued to work on the mechanisms of radioactive decay, the chemists now had a couple of new elements to study, with their own novel chemistries.

What the chemists soon learned about radium was that it fit into the second column of the periodic table; the so-called alkaline earth metals. This meant that radium shared chemical properties with another element in that column, calcium, which happens to be the major constituent of bone. The implications of this to human health would prove to be immense, but at that time little attention was paid to it, not even by the Curies, who worked with high levels of radium on a daily basis,[20] and thus had the most to lose. A premature death in a horse cart accident, in 1906, would spare Pierre the worst health consequences of his work. Marie, however, would keep working for nearly three more decades until the radiation got the best of her, as we shall see.

CUTTING THE PIONEERS SOME SLACK

The straightforward interpretation of the discoveries surrounding radiation and radioactivity, as explained here, is enriched by the benefit of modern hindsight. Although our current understanding of the nature of radioactive decay is enlightened by our knowledge of the structure of the nucleus, these early radiation pioneers had no such information through which to interpret their own findings and discoveries. This tormented them and forced them toward explanations that even they themselves knew were seriously lacking. For example, Becquerel clung for some time to the idea that radioactivity represented some long-lived fluorescence that released energy from much earlier exposure of the radioactive material to light. Marie Curie proposed that heavy elements (e.g., uranium, polonium, and radium) could absorb background levels of x-rays in our environment and release them later as radioactivity, akin to an "invisible fluorescence" process produced by x-rays rather than visible light. Even Crookes, the father of the Crookes tube, promoted his own theory in which radioactive elements extracted kinetic energy from air molecules and then released it all at once in a radioactive decay event. (This idea was particularly easy to kill since it was quickly shown that radioactive elements displayed the same level of radioactivity both in and out of a vacuum.) The issue that haunted all these scientists, and caused them to doubt what their own eyes had seen, was the embarrassing problem of explaining where the energy released by radioactive elements came from. They all well knew, or at least thought they knew, that energy could neither be created nor destroyed. It could only be moved around.[21]

But one can't be expected to interpret new discoveries in the context of later discoveries. So the pioneers should be forgiven if they really didn't understand their own discoveries. Besides, one of the pioneers even publicly owned up to it. Marconi, in his acceptance speech for the 1909 Nobel Prize received for his work with radio waves, freely admitted, with some embarrassment, that he had no idea how he was able to transmit radio waves across the entire Atlantic Ocean. The fact that he had even tried was a testimony to his ignorance. Classical physics had predicted it should not have been possible because electromagnetic waves traveled in straight lines, so their transmission distance should have been limited to less than 100 miles for a 1,000-foot-tall radio tower,[22] due to the curvature of Earth.[23] He told his audience in humble understatement, "Many facts connected with the transmission of electric waves over great distances await explanation."[24] It seems that, despite his apparent scientific ignorance, Marconi's greatest genius was that he did not take the scientific dogma of the moment too seriously.[25] He understood better than most that all dogma is ephemeral.

In Marconi's case, it turned out that radio waves can actually skip around the globe by reflecting off an inner layer of the upper atmosphere.[26] This reflective layer is a stratum of ionized gas, unknown in Marconi's day, that was discovered later by Oliver Heaviside (1850–1925), an electrical engineer and physicist.[27] Heaviside had come to the rescue of Marconi. Similarly, the radioactivity pioneers would soon have their own knight in shining armor who would help them make sense of all that they had found. In fact, he would be a real knight—Sir Ernest Rutherford—and he would use his sword to cut open the atomic nucleus and reveal its contents to all who wished to see. And many did.

CHAPTER 4
SPLITTING HAIRS: ATOMIC PARTICLES AND NUCLEAR FISSION

Nothing exists except atoms and empty space; everything else is opinion.

—*Democritus, fifth century BC*

PLUM PUDDING AND THE CONSERVATION OF CHARGE

In 1904, Joseph John (J. J.) Thomson (1856–1940) unveiled a model of the atom in which electrons were described as being negatively charged plums floating around in a pudding of positive charge. The British love their plum pudding, so the image of all physical matter being an assembly of little plum puddings, as proposed by this British scientist, appealed to both their senses and their national pride. But it was an image not easily swallowed by everyone, even within Britain, and it would soon be shown to be wrong.[1]

Nevertheless, Thomson was no fool. He had actually discovered the electron, and in 1906 he would be awarded the Nobel Prize for his work in the electrical conductivity of gases.[2] So, he had a mind to be reckoned with, and that was the way Thomson's mind envisioned the structure of the atom. He visualized the atom as simply a little ball of goop with electrons floating inside. It was a model that had served him well and allowed him to make his discoveries. But, by the end of the decade, the pudding was getting stale. Other scientists soon appreciated that they had gotten all they could out of the plum pudding model. They began to search for a new and better model of the atom. By 1910, physicists were beginning to understand that, rather than being pudding, an atom was mostly just empty space. At its center was a very small, positively charged nucleus, and flying around that nucleus were even smaller negatively charged electrons.

We have since learned that the nucleus is incredibly small, even when compared to the dimensions of the atom itself. Consider this: If an atom were the size of a major league baseball stadium, with its center at the pitcher's mound, the nucleus would be the size of the baseball in the pitcher's hand. And the atom's outermost electrons, each the size of a grain of sand, would be randomly moving around from seat to seat somewhere in the upper decks. All the rest of the stadium would just be empty space.[3]

THE IMPORTANCE OF BEING ERNEST

Eventually scientists learned that the nucleus of an atom is made up of a mixture of protons and neutrons in varying numbers, but this understanding of the nucleus's architecture was hard won. And it was won mostly through the efforts of Ernest Rutherford (1871–1937), one of Thomson's former students.[4]

Rutherford was born and raised on a farm in New Zealand. He was more comfortable hunting pigeons and digging potatoes on his family's farm than hobnobbing with intellectuals.[5] Nevertheless, he was brilliant, and his family struggled to provide him with a first-rate scientific education. But there were few opportunities for the

FIGURE 4.1. ERNEST RUTHERFORD. The brilliant young Rutherford would become the father of nuclear physics. Fascinated by Antoine Henri Becquerel's discovery of radioactivity, he picked up where Becquerel left off, pioneering the use of particulate radiations to probe the structure of the atomic nucleus. He was even able to show that when an atom radioactively decays it changes from one element into another, something that scientists had previously thought impossible. (*Source*: Photograph courtesy of Professor John Campbell)

expression of his brilliance in New Zealand, and eventually he found himself at Cambridge University in England, in the laboratory of J. J. Thomson. At Cambridge he encountered some prejudice and belittlement because of his provincial roots. But messing with this muscular farmer was not without risk. In a letter home complaining of disparaging treatment from graduate teaching assistants, he wrote, "I'd like to do a Maori war-dance on the chest of one, and will do that in the future, if things don't mend."[6] Things mended.

Initially, Rutherford was fascinated by radio waves, just as Marconi was, and delighted in demonstrating the bell-ringing tricks of Édouard Branly to his friends and roommates. From half a mile away, he was able to ring a bell in his living room, to everyone's astonishment.[7] But when Becquerel discovered radioactivity in 1896, Rutherford's interests turned to radioactivity.

Rutherford decided to move to McGill University in Canada to start his professional academic career and to focus his research specifically on radioactivity. McGill was a good choice. John Cox, the physics professor and x-ray researcher who had performed the first diagnostic x-ray on gunshot victim Toulson Cunning, would be working in the same research group as Rutherford. McGill also had Frederick Soddy (1877–1956), a brilliant chemistry professor who was as interested in radioactivity as Rutherford was.[8] He and Rutherford would soon become close research collaborators.

It was Rutherford who discovered that all radioisotopes have distinct half-lives.[9] He also determined that radioactive decay could involve the change of an atom from one element into another (e.g., C-14 to N-14; see chapter 3) in a process he called nuclear *transmutation*. Rutherford used the word with some trepidation because he was well aware that the term was previously associated with the discredited alchemists—the medieval practitioners who sought to transmute lead into gold.[10] Yet, that is exactly what was happening with radioactive decay. One element was transmuting into another!

All of this work with radioactivity earned Rutherford the Nobel Prize in Chemistry in 1908. Nevertheless, his best work was yet to come. He would go on to describe the structure of the atom's nucleus and propose a new model of the atom, now known as the Rutherford model, which would replace plum pudding and survive in substance, if not detail, to the present day. And although he never was awarded another Nobel Prize himself, he would mentor other scientists who would earn their own Nobel Prizes, in no small part due to his guidance.[11]

In 1919, Rutherford succeeded his old professor, J. J. Thomson, as head of the Cavendish Laboratory, which was essentially the physics department of Cambridge University. First opened as a physics teaching laboratory in 1874, the Cavendish became the intellectual home of some of the greatest physics minds of all time. Even James Clerk Maxwell, author of the equations that had predicted the existence of radio waves, had also once headed the Cavendish.[12] The Cavendish scientists took a special interest in anything to do with radiation and radioactivity, and much of what we know about radiation today can trace its roots to research first done at that laboratory.[13]

The first known radioisotopes (uranium, polonium, and radium) all emitted a relatively large type of particle of unknown nature. Rutherford discovered that these large particles were essentially the same as a helium atom's nucleus (i.e., a helium atom devoid of its electrons) traveling at very high speed. When they ultimately slowed to a stop, they picked up some electrons from the environment and formed helium gas, the lighter-than-air gas that makes party balloons float.[14] He named these large particles alpha particles, to distinguish them from much smaller beta particles, which he also discovered and named. (A beta particle, as we've seen, is simply a high-speed electron that is ejected from the nucleus when an atom decays.)

Alpha particles have short ranges, high energies, and a double positive charge (2^{+}). They can be attracted toward negative electrodes, and they can be deflected from their paths by magnetic fields. Thus, they are highly amenable to experimental manipulation. Further, they are so large that you can actually follow their paths under the microscope with the use of a tiny fluorescent screen. In short, they make an excellent atomic tool with which to explore the nature of the nucleus. That's exactly how Rutherford intended to use them.

Rutherford hoped to prove the existence of the atomic nucleus and measure its size using the alpha particles as an atomic probe. Of course, neither the atomic nuclei nor the alpha particles can be seen directly, so making direct measurements was not possible. Nevertheless, since an individual alpha particle's path can be seen with the help of a fluorescent screen, he had the ability to count them. When he pointed a beam of alpha particles from a radium source at a piece of thin gold foil, most of the particles passed

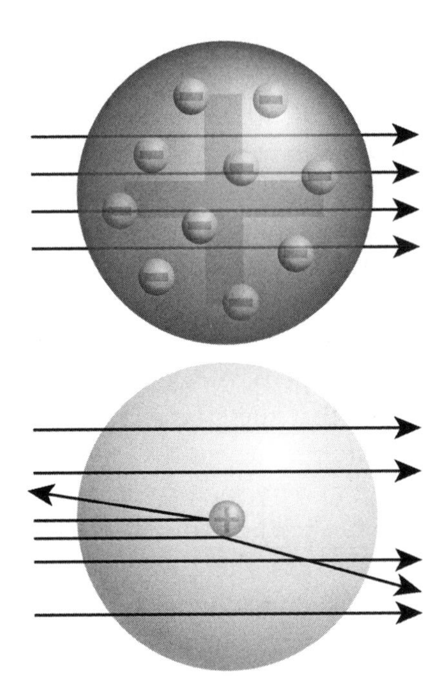

FIGURE 4.2. TWO COMPETING MODELS OF ATOMIC STRUCTURE. J. J. Thomson's "plum pudding" model of the atom (*top*), where electrons are immersed in a "pudding" of positive charge could not be reconciled with Rutherford's bounce-back experiments using gold foil. Rutherford found that when high-energy alpha particles (arrows) were shot at a very thin foil made of gold, most alpha particles just passed through, but a very small number bounced back. Rutherford's explanation was that the mass of an atom is concentrated in a central, positively charged nucleus, surrounded by a nearly massless cloud of electrons (*bottom*). An alpha particle bounces back only when it makes a direct hit on a gold nucleus. The implication of the experiment was that most of the volume of an atom is just empty space.

through the empty space of the atoms, pulling the gold's electrons as they went but otherwise moving unobstructed through the foil.[15] But some of them—and just some of them—bounced back toward their source. He used a miniature fluorescent screen and a microscope to detect the alpha particles that bounced back, visible as flashes on the screen. Rutherford and his assistants took turns sitting in a dark room looking through a microscope for many hours, meticulously recording the number of flashes they saw per hour of observation time (i.e., the bounce-back rate). They ended up recording a very small, but significant, number. Rutherford was elated. He was later to say that this bounce-back finding was "quite the most incredible event in my life . . . as incredible as if you fired a 15-inch shell at a piece of tissue paper and it came back to hit you."[16] But exactly why did those alpha particles bounce back?

Rutherford correctly reasoned that an alpha particle bounced back whenever it made a direct hit with a nucleus of the gold atoms. He further correctly reasoned that the probability of hitting a nucleus was dependent upon how large that nucleus was. That is, the bigger the nucleus, the more likely an alpha particle would hit it and bounce back. Since their counts of flashes had shown that the probability of hitting the nucleus was extremely low (i.e., most of the particles did not bounce back), the nucleus of the gold atom must be extremely tiny. Assuming that the nucleus's size was directly proportional to the probability of a bounce back, he estimated the size of the gold nucleus to be 14 femtometers (0.000000000000014 meters) across. Yes, it was small, very small indeed.

After he had determined its size, Rutherford next focused on determining what the nucleus was made of. He knew that there must be protons in it because of its overall positive charge. But what else was in there? As it turned out, the chemists were able to provide some insight.

Chemists had been able to measure the relative masses of atomic nuclei of different elements by cleverly employing what they already well knew about a fundamental property of gases. Under standard temperature and pressure conditions, equal volumes of gases have equal numbers of atoms.[17] So the difference in masses per unit volume must be due to the differences in relative masses of their nuclei.[18]

Employing the above approach, the chemists were able to show that the masses of the nuclei for all elements they measured seemed to differ by a multiple of the mass of a single proton. For example, hydrogen weighed the same as one proton, helium's mass equaled four protons, and carbon's mass equaled twelve protons. This discovery led to the conclusion that atomic nuclei are composed exclusively of protons.

All of this seemed to fit, but it didn't sit well with Rutherford. He recognized one huge problem that he would struggle with for years. It was that pesky "conservation of charge" issue. Most atoms were known to be electrically neutral; that is, the negative charges of their orbital electrons were balanced by the same number of positive charges in the nucleus. If atoms were composed of all those protons, they would have more protons than electrons and, therefore, have an excess number of positive charges, which they obviously did not have. So how could nuclei be composed of only protons? That simply could not be true. Rutherford's answer was to describe a new particle that he called the *neutron*. It was a particle that weighed the same as a proton but had no charge (i.e., it was neutral). It solved the problem by providing the required nuclear mass without adding surplus charge.[19]

To many, Rutherford's neutron sounded more like a creative accounting gimmick than a real physical particle, so the burden was on Rutherford to prove the neutron's existence. This would be no small feat since all atomic particles known up to this time (alpha particles, beta particles, electrons, and protons) were charged and could, therefore, be detected and measured exclusively based on their electrical properties. Detecting a particle that had no charge was a major problem. Unless they could be detected and shown to be real physical objects, neutrons would be considered just a figment of Rutherford's overactive imagination.

GERMANY AND FRANCE WEIGH IN

While Rutherford and his Cavendish colleague James Chadwick (1891–1974) were pondering how they might detect neutrons, an intriguing new report got their attention. German physicist Walther Bothe (1891–1957) and his student Herbert Becker had been able to show that when high-energy alpha particles from polonium hit a target made of beryllium, some of the alpha particles were absorbed by the beryllium atoms, which in turn emitted extremely powerful rays of some kind. The rays had no charge, so Bothe and Becker assumed that they must have been of the electromagnetic rather than the particulate type. They suggested that excess energy from the absorption of the alpha particle into the beryllium was being emitted as a powerful gamma ray.[20]

French scientists Irène Curie—the daughter of Marie and Pierre Curie—and her husband, Frédéric Joliot, were able to replicate the experiment and, thereby, validate the findings. They also showed that if these same gamma rays then went through a layer of paraffin wax, the rays were further changed into very high-energy protons. Since paraffin has a high concentration of hydrogen atoms, the interpretation was that the gamma rays were knocking out the hydrogen nuclei in the form of positively charged proton particles.[21] What's more, they calculated the energy needed for a gamma ray to expel the hydrogen nuclei at the observed proton energies as being 55 million electron volts![22]

These experimental findings from two reputable laboratories were not in question, but their interpretations of what was happening in the nucleus were too fantastic for Rutherford and Chadwick to accept. Extremely high-energy gamma rays like this had never been described before and their existence was not thought possible.

Still, Chadwick's imagination was set into motion by these tantalizing clues. He soon developed another interpretation of what was going on, which he was ultimately able to demonstrate experimentally. But it would not come easily. Joliot and Curie had a major advantage over him in conducting their experiments: they were using polonium that they had purified themselves. Polonium was a rare and precious radioisotope that the Cavendish scientists could not purify or purchase. And polonium was vastly superior to radium as a source of alpha particles because it emitted high-energy alpha particles, while radium emitted lower-energy alpha particles. It also released about 5,000 times more alpha particles than the same quantity of radium.[23] The Cavendish scientists had no polonium at their disposal. Help was on the way, however, and it was coming from the United States.

AMERICA SENDS REINFORCEMENTS

Norman Feather (1904–1978), a former Cambridge student who had done a research stint at the Cavendish, had developed a strong interest in nuclear architecture and radioactivity. In fact, he had worked briefly with Chadwick on a project to measure the penetration ability of alpha particles. They developed a mutual appreciation for each other's scientific talents, and remained in close contact. Capitalizing on a growing interest in the field of radioactivity among American academic physicists,[24] Feather came to the United States in 1929 with a one-year appointment to conduct radioactivity research at the Johns Hopkins University in Baltimore.[25]

Searching for radioactive sources to work with, he was happy to find that nearby Kelly Hospital had an abundance of the radioactive gas radon, encapsulated in small sealed glass ampules. The ampules were no longer of value to the hospital because the radon they contained had decayed to the point that it could no longer be used for radiation therapy (a subject that we will explore in chapter 6).[26] So the hospital gave him a couple of spent radon ampules for his radioactivity research.[27] Even partially decayed radon was very rare and expensive because the purified radium needed to generate the radon was in very limited supply. At the time, the Kelly Hospital had the largest supply of pure radium in the world—5 grams (about one teaspoon full)—which they used to produce radon gas for encapsulation in therapeutic ampules. The gift of the old radon ampules was a research windfall for Feather.

After working at Johns Hopkins for the year, Feather was preparing to return to the Cavendish when he stopped by the Kelly Hospital to see if they could spare a few more spent ampules. They gave him 300! He packed them in his luggage and returned to the Cavendish, riches in hand.

At some point Feather realized that he had more than just a stash of partially decayed radon in his possession. Radon-222, he knew, ultimately decayed into polonium-210, the very radioisotope that Marie Curie had discovered and that Chadwick so badly needed. He knew that the old radon ampules should have accumulated within them appreciable amounts of polonium. It wouldn't be long before Chadwick would have his precious polonium.

Over a period of a few months Chadwick purified the polonium from the radon ampules and encased it at the end of a short metal barrel, effectively producing an alpha particle shotgun. He would use this gun to reveal the elusive neutron and prove that it was more than the just a nuclear accounting gimmick.

RUTHERFORD'S NEUTRON PREDICTION IS VINDICATED

Using the polonium alpha particles and various target elements, Chadwick was able to show a pattern of emissions that were not at all consistent with a high-energy gamma ray but could easily be explained as the result of a proton lacking charge, which was precisely what a neutron was defined to be! The mysterious neutron predicted by Rutherford had been found, its existence inferred despite its lack of charge, simply by its interactions with other particles that were charged, and, therefore, measurable. In other words, a ghost particle had been revealed by its shadow.

Now, in 1932—12 years after the neutron had first been predicted by his mentor, Rutherford—Chadwick had finally confirmed the existence of this mysterious particle. With his discovery, all the component parts of the atom were in place. The equally massed protons and neutrons together constituted the nucleus's mass, and much smaller electrons traveled in orbits around it. There were no electrons in the nucleus, and there was a lot of empty space. For his achievement, Chadwick was awarded the Nobel Prize in Physics in 1935. Plum pudding was dead.

With the elusive neutron now in hand, all the major particulate radiations had also been identified. We now know that the smaller particulate radiations represent free-moving protons, neutrons, and electrons (beta particles). In contrast, the larger alpha particles, which Rutherford first identified as simply fast-moving helium nuclei, consist of exactly two protons and two neutrons (just like the nuclei of helium gas). All of these different types of particles are flying through space with great energy and dissipating that energy by interacting with various molecules they encounter along their paths. The future would reveal fission products and cosmic radiations as other forms of particulate radiation, but by 1932, the major particulate radiations were all defined in terms of their subatomic counterparts. These particles are very different from each other in their sizes, energies, and charges, and we will see that these differences can have varying consequences on health.

By this time, too, the higher-energy electromagnetic radiations that we're familiar with today were all known. X-rays emanate from the electron shells of the atom, while relatively higher-energy gamma rays emanate from the nucleus. X-rays can be produced artificially by using a high-voltage electric current as Roentgen did, while gamma rays are a result of natural radioactive decay and derive specifically from the decaying atom's nucleus. Apart from their origins in different locations within the atom, x-rays and gamma rays are fundamentally the same thing—electromagnetic waves with very short wavelengths. As might be expected, since they have wavelengths of similar size, they produce similar effects on health. We shall soon learn more about this.

SPLITTING ATOMS

While Chadwick was busy discovering the neutron with his little alpha particle shotgun, two other Cavendish scientists, John Cockcroft (1897–1967) and Ernest Walton (1903–1995), were busy supersizing things. Not content with just ejecting protons from nuclei, they had a larger goal. They wanted to split the atom.[28]

Cockcroft and Walton had decided that the best way to look inside the nucleus would be to split the nucleus apart and see what the fragments were made of. During Rutherford's earlier bounce-back experiments, they had witnessed, in addition to heavy fluorescent tracks from the alpha particles, fainter but distinct fluorescent tracks from some other type of particle. They correctly postulated that the other particle tracks represented individual protons that had been chipped away from the nucleus by the alpha particle bombardment. They calculated the yields of these small particles and found them to be very low. Then they calculated the amount of polonium that they would need to significantly increase alpha particle bombardment in order to chip more of the atom away; it was depressingly high. In addition, they didn't simply want to chip away at the nucleus; they wanted to blow it apart. Alpha particles, even from polonium, were just not going to do the job.

What the Rutherford team really needed was some sort of particle machine gun; it would help to have an instrument that could rapidly fire hundreds of millions of high-energy particles at nuclei and blast them apart. The radium and polonium shotguns that they had been using were just not up to the task. Cockcroft and Walton began to work on making this machine gun.

The first question was what to use for bullets. That seemed to be straightforward. The element hydrogen is composed of a single proton and a single orbiting electron, making it electrically neutral. If you strip off the electron, a positively charged proton would be left, which would make an excellent bullet. But how could that electron be removed?

As it turns out, the hydrogen nucleus with its single proton has a weak grip on its single orbiting electron. By moving an electrical current through a chamber of hydrogen gas, the hydrogen atoms could be easily scrambled into their constituents—positively charged protons and negatively charged electrons—and the protons could be drawn off by attracting them toward a negatively charged cathode. So a continuous source of protons turned out not to be a problem; all that was needed was a good supply of hydrogen gas and a little electricity to run through it.

The next issue was how to shoot the protons at a target, and that was a trickier problem. Again, since the protons were positively charged, it was theoretically possible to propel (accelerate) them toward a target by placing the target close to the negative electrode (i.e., cathode) within a strong electrical field.[29] All that would be needed would be to feed a stream of protons into a particle acceleration tube. The more particles fed in, the more bullets would shoot out; and the higher the electrical voltage across the tube, the faster (and more energetic) the bullets would be when they hit the target. But how much particle energy would be required to split an atom? Or, more importantly, how high would the voltage across the tube have to be in order to achieve the required particle energies? One thing was certain, the voltage requirements would be substantial. And that was the rub. Where to get the voltage?

So the issue really wasn't whether or not the atom could be split. Most scientists accepted the notion that the atom's nucleus could be split. What they questioned was how much energy it would take to split it. Nearly everyone thought that prohibitively high voltages were needed to break the atomic nucleus apart. In fact, several physicists had calculated the voltages required, using different sets of assumptions. All came up with electrical requirements in excess of one million volts! In Rutherford's day, this was a staggering amount of voltage. So Rutherford's goal of splitting the atom was not viewed as a crazy idea, but rather just impractical, given the technology of the day.

High voltages had been commercially achieved, however, on an industrial scale using massive transformers. These voltages, nevertheless, were attained only momentarily and were highly unstable. Furthermore, the cost and size of these commercial transformers put them well out of reach for experimentation in physics laboratories. So the Cavendish scientists were either going to have to find a whole lot of money, or they were going to have to come up with another idea. They chose the latter.

WUNDERKIND

It was the laws of classical physics that had enabled calculation of the average particle energies needed to split an atom, and those calculations had produced discouragingly high estimates of the voltage requirements. Yet, classical physics was no longer the only game in town. A new probability-based approach to interpreting nuclear events, called *quantum mechanics*, was capturing the imagination of some physicists. Largely the brainchild of Danish physicist Niels Bohr (1885–1962), quantum mechanics was attracting many disciples because it allowed physicists to understand some complicated atomic processes that couldn't be adequately probed with classical physics.

One of the disciples of Bohr and his quantum mechanical approaches was George Gamow (1904–1968), a young man of tremendous intellectual ability who was soon to become a friend and colleague of Rutherford and the other Cavendish scientists.[30] By all accounts Gamow was a child prodigy.[31] Born in Odessa, Ukraine, in 1904 (the same year as the plum pudding model's birth), Gamow's revolutionary ideas would allow the Cavendish scientists to split the atom before he reached his 29th birthday.

Gamow's familiarity with quantum mechanics gave him some insight into splitting the atom that had escaped the classical physicists. He realized that it was not impossible for the lower energy particles to split the atom but, rather, it was just extremely unlikely that they would do so. In other words, it had a very low probability of happening. He pointed out, however, that if exceedingly large numbers of lower energy particles were shot at a lot of nuclei, it was likely that at least some of them would split some atoms, which the experimenters should be able to detect as visible particle tracks on their fluorescent screens. Basically, the problem of finding a needle in a haystack could be overcome just by using a lot of needles.

Gamow's quantum mechanical ideas, known as *tunneling theory*, and their implications for splitting the atom were not a secret. He had published a series of papers and most classical physicists were aware of them. Nevertheless, the classical physicists had trouble fully accepting their implications. Even Albert Einstein (1879–1955) struggled with the probability theories of quantum mechanics and famously expressed his skepticism of its probabilistic nature in his oft quoted assertion, "God does not play dice!" Nevertheless, Gamow was a respected scientist despite his tender age, and his calculations were reason to be optimistic about the possibility of splitting the atom at lower voltages. According to Gamow's calculations, the job could be done with just 300,000 volts,[32] still a respectable amount of voltage but within grasp. And so, despite the potential challenges posed by voltage restrictions, Rutherford and his team decided to keep trying.

Gamow's insight had lowered the bar for voltage. But even attaining 300,000 volts was a formidable goal. Also, a series of fitful starts, technical setbacks, and engineering obstacles delayed progress. Many accelerator designs were tried and found to be unworkable, requiring construction of multiple prototypes, which cost the group more time and money and strained the limits of their laboratory facilities. Yet, they would not be deterred. By 1932, after many failures, Rutherford's group finally constructed a proton accelerator that they hoped would be up to the task. For good measure they had designed it to accommodate 700,000 volts, which was more than twice what Gamow's calculations said were needed to do the job.

Apart from the energy insight provided by Gamow, the scientific team also knew that choosing the right atom to split would be critical to achieving success. They correctly anticipated that using lithium-7 as the target element would improve their prospects another twofold. Lithium has a molecular mass of seven (three protons and four neutrons). If another proton is forced into its nucleus at high energy, it should produce a larger, highly energized nucleus (four protons and four neutrons), which would then likely split to form two alpha particles (each with two protons and two neutrons). Thus, for every one proton absorbed by lithium-7, two alpha particles should be emitted (a nuclear reaction termed "p, 2α"); this was a doubling of alpha particle yield!

This insight about how lithium-7's nucleus would absorb a single high-speed proton and double the alpha particle yield was absolutely spot-on, and this awareness was central to achieving success. As we shall see later, lithium-7 can also, surprisingly, absorb a single high-speed neutron and then emit two neutrons (a nuclear reaction termed "n, 2n"). This other nuclear reaction of lithium-7, which can double the yield of neutrons, was unanticipated by anyone. Unfortunately, it is this unexpected reaction of lithium-7 that would one day take everyone by surprise and contribute to one of the worst nuclear weapon test accidents of the twentieth century. But for now, nuclear weapons were far from anyone's mind.

The scientists had finally reached the moment of truth. It was time to attempt to split the atom. They had their proton machine gun and their lithium-7 target. Next to the target they placed a fluorescent screen, their trusted tool for visually detecting alpha particles. The thought was that, if the proton beam were successful in splitting the lithium atoms, there should be alpha particles released as nuclear fragments that would be visible on the neighboring screen.

The big day was Thursday, April 14, 1932. The accelerator was fired up to the required voltage and the protons were injected. All eyes were on the screen. It glowed! Microscopic inspection of the screen confirmed their supposition: the glow was being produced by plainly visible alpha-particle tracks. When they interrupted the proton flow into the accelerator, the alpha particles stopped, showing that protons were responsible for the particle production. Other controls further validated their greatest hope; the protons were splitting the lithium atoms! History had been made, and Cockcroft and Walton had earned the Cavendish yet another Nobel Prize.[33]

FLYING BY THE SEAT OF THEIR PANTS: RADIATION SAFETY AT THE CAVENDISH

Despite all his work with radioactivity, it is not apparent that Rutherford ever suffered any adverse health effects from radiation. The same cannot be said of his laboratory manager at the Cavendish, George Crowe. The only radiation safety rule at the Cavendish was simply that those handling radioactivity should wear rubber gloves to protect their skin from radiation burns, since by this time the dermal effects of radioactivity were well established. Nevertheless, Crowe typically disregarded the rule because he found the gloves a nuisance when conducting delicate work with radioactive sources. Consequently, he lost some feeling in his fingertips and developed skin sores that refused to heal. He ended up having multiple skin grafts and, ultimately, one finger amputated.[34]

Besides the gloves, little else was done to protect the scientists at the Cavendish from radiation. It wasn't so much that the scientists were cavalier about their own safety. It was, rather, that radiation's protracted risks appeared small to them compared to what they considered their major health threat—immediate death by electrocution. This, as we learned earlier, was the same concern that surrounded Edison's electric light bulb. It was also the concern of Marconi and his team while transmitting radio waves. Marconi's workers routinely wore very large, electrically insulating gloves whenever operating a live telegraph key.

Of particular concern were corona effects, in which electrons tended to accumulate on the sharp edges or points of an electrified apparatus and then jump like a small lightning bolt to something nearby. Unlike radiation, which couldn't be seen during an experiment, coronas could sometimes be observed forming on the surface of glass tubes, as twitching blue lights, or heard as a hissing sound, like a snake about to strike.

Perhaps inspired by Edison's animal electrocutions, a demonstration experiment was ordered by Rutherford to impress on his charges the dangers inherent in their work. Under the supervision of a medical professor, a corona bolt blew a half-inch hole through the skull of a laboratory rat, thereby establishing a healthy respect for electricity among all the scientists who witnessed it.[35] Compared to sudden death by electrocution, the delayed threats posed by radiation seemed quite trivial.

Because of the electrocution hazard, the Cavendish scientists took precautions to remain far away from live electrical apparatuses whenever possible. Still, sometimes they needed to be nearby in order to observe the experiment. Cockcroft and Walton therefore made a wooden hut that they could stay in during their experimental runs. They took pains to electrically insulate the hut so that they were completely safe from electrocution while inside. For good measure, they decided to make the hut radiation safe as well. To deal with the radiation hazard, they lined the hut with lead foil and then hung a fluorescent screen inside. If the screen seemed to be glowing a little too much during an experiment, they simply added another layer of lead foil before the next run.[36] This was radiation protection at its crudest, but it was better than nothing. In any event, no one seemed to be harmed from the radiation and, more importantly to the scientists, nobody in the Cavendish laboratory was ever electrocuted.

THE ATOMIC MAGICIAN: THE CONSERVATION OF MASS TRICK

Besides the dilemma of conservation of charge, there was another annoying problem with the nucleus that needed to be dealt with—the conservation of mass. As the chemists well knew, matter could neither be created nor destroyed. That is, the products of a chemical reaction had to weigh the same as the reactants. Mass must be conserved.[37] But for nuclear reactions, this fundamental scientific law did not seem to hold true. Had God made an exception for nuclear reactions? Not likely.

Let's imagine the old magician's trick of pulling a rabbit out of a hat but with a slight twist. As usual, the magician has secreted a rabbit in a hidden compartment in the hat, and shows the audience an apparently empty hat. But before going further, this magician weighs the hat in front of the audience and shows everyone that the hat weighs 2 kilograms. Then he pulls a rabbit out of the hat and weighs the rabbit: 1 kilogram. Then he weighs the now empty hat. The audience expects the hat to weigh 1 kilogram since everyone knows that 2 minus 1 equals 1. But the scale inexplicably reads: 0.5 kilograms! What's this? How can 2 minus 1 equal 0.5? This is true magic! But where did that extra mass go?

Now, imagine that the rabbit coming out of a hat is equivalent to an alpha particle coming out of an atomic nucleus. That is, when the nucleus releases its alpha particle during radioactive decay, the mass of the alpha particle plus the remaining nucleus weighs less than the original nucleus containing the alpha particle. This is exactly the situation that exists when a radioactive atom decays.

There seems to be a loss of mass, and the result is what is called a *mass deficit*. Where did that mass go? Nuclear magic!

It was Albert Einstein who revealed the secret behind this magic. Einstein understood that the missing mass of the hat had changed into the hop of the rabbit. That is, he recognized that mass could be changed into energy. His theory of relativity suggested it was possible for mass to become energy and, conversely, for energy to become mass. Not only that, he proposed a simple equation that showed the relationship between energy and mass: $E = mc^2$. In words, energy (E) is equal to mass (m) times the speed of light (c) multiplied by itself. This relationship suggests that mass and energy are actually different forms of the same thing. They are, therefore, interchangeable! What this means is that, when an atom decays, it releases energy equivalent to the mass that was lost.

Don't try to wrap your mind around this; even Einstein struggled with it. The bottom line is that it works. If you convert the mass deficit to energy, the resulting quantity is exactly equal to the energy of the radiation emitted during the radioactive decay. In effect, a portion of the mass of the nucleus was converted into energy and escaped from the atom as radiation. Interesting you may say, but what's the big deal?

You'll see how big a deal it is if you take the next logical step and ask the following: What would be the energy released if just one gram of matter (e.g., about one sugar cube) were entirely converted into energy? You can answer this for yourself using Einstein's equation.

Given that the speed of light, c, is a constant (i.e., it doesn't change) and equals 299,792,458 meters per second, you can calculate for yourself the total amount of nuclear energy in one gram (i.e., 0.001 kilograms) of sugar:

$E = mc^2$

$E = \text{mass in kilograms} \times (\text{speed of light in meters per second})^2$

$E = 0.001 \text{ kilograms} \times (299{,}792{,}458 \text{ meters per second})^2$

$E = 90{,}000{,}000{,}000{,}000 \text{ joules (90 trillion joules)}$

Exactly how much energy is 90 trillion joules?[38] Consider this: it's the energy of 10,000 lightning bolts, or the energy needed to heat 1,000 homes for one year, or the energy needed to put 10 space shuttles in orbit, or the energy released by one atomic bomb. As you can plainly see, it's a lot of energy. And it's also one hell of a magic trick!

So the energy released from atoms as radiation is simply the mass gone missing.

Ironically, the very scientists who revealed that matter and energy were interchangeable were slow to realize the practical implications of their work. After Einstein published his famous equation, $E = mc^2$, he received a letter from a layman who had made some calculations similar to the one we just did that shows the huge amount of energy in a gram of matter. The man asked Einstein whether he realized that he had provided the world with the strategy for making a bomb of enormous potential. Einstein had not. In fact, he thought the concept was foolish.[39]

Rutherford was also once queried about the potential for producing electrical power by splitting atoms, but he didn't think it possible. Yes, he said, he fully realized that splitting a single atom with a proton accelerated by 125,000 volts produces atomic fragments with energies equivalent to accelerations by 16,000,000 volts. Nevertheless, he pointed out, only one proton out of ten million was actually entering a target nucleus (à la Gamow's calculations). All of the rest of the protons along with their energies were being wasted just to split that one atom. So in the end, he said, splitting atoms, rather than being a power source, was an energy-losing proposition![40]

It is amazing that these scientists, who had so readily discarded the scientific dogma of their day and overcome tremendous technical obstacles in achieving their own scientific visions, would so easily dismiss as misguided fantasy the readily apparent implications of their own work. But their eyes were very soon opened. On August 12, 1933, just one year after those few atoms of lithium were split for the first time at the cost of tremendous high voltage energy input, a young Hungarian scientist, Leó Szilárd (1898–1964), proposed a way that large uranium atoms might be split with virtually no energy input at all. He suggested that splitting large nuclei (i.e., fission) might be readily achieved by means of a nuclear chain reaction.

THE DOMINO EFFECT: FISSION

When it comes to nuclei, being big is not good. We mentioned earlier that atomic nuclei tend toward having a similar number of protons and neutrons, and atoms that have an excess of one or the other are likely to be unstable and, therefore, radioactive. But there is another condition essential to the stability of nuclei—small size. Remember the idea of the nucleus as a baseball? While baseballs are happy and stable, it turns out that larger balls are not. When atoms start having protons in excess of 100, moving from baseball to basketball size, the density of positive charge becomes so great that even the neutrons can't dilute the nuclear charge enough to keep the nucleus stable. In this situation, the large nuclei tend to decay in a special way. They spontaneously split apart into smaller nuclei. This process is called nuclear fission, and it is a major mode of radioactive decay for very heavy nuclei, like uranium-235. When uranium-235 splits, it produces various fission products, including multiple atoms with smaller nuclei, and it also typically ejects two or three neutrons as particulate radiation.

Just as with other forms of radioactive decay, the sum of the masses of the fission fragments is again less than the whole nucleus before fission, and the amount of energy released is determined by the mass deficit, as we've already discussed. This process is similar to other types of radioactive decay with one huge exception: the emitted neutrons can go on to be absorbed by other large nuclei nearby and induce them to split by fission, thereby releasing more neutrons. Those neutrons then go on to interact with other nuclei, and on and on, ad infinitum. In this way, a chain reaction can be born, and the cumulative mass deficit can translate into a huge release of energy.

Under normal situations, however, a chain reaction doesn't usually occur. That is because neutrons are highly penetrating and more likely to escape from the volume of radioactive material than they are to react with neighboring heavy nuclei. But as the volume (and thus the mass) of fissionable material increases, and the number of fissionable nuclei becomes more concentrated (i.e., enriched), the probability of inducing a chain reaction increases. The point at which a self-sustaining chain fission reaction occurs spontaneously is called *criticality*, and the mass of material required to produce criticality is called the *critical mass*.

THE CHICAGO PILE

The first man-made criticality was achieved by Italian physicist Enrico Fermi (1901–1954). Fermi was a gifted theoretical and experimental physicist, who was awarded the Nobel Prize in 1938 for his discovery of the transuranic elements (i.e., elements that have more protons in their nuclei than uranium does). Unfortunately, in that same year, he was forced to leave Italy with his Jewish wife, because new Italian racial laws were making her a target for persecution. He immigrated to the United States, and was ultimately recruited to work on the Manhattan Project, a secret program to make an atomic bomb.[41]

Fermi was tasked with the responsibility of experimentally demonstrating that a nuclear chain reaction was more than just a theoretical possibility. He started work immediately and built himself a small nuclear reactor under the abandoned west stands of the University of Chicago's Stagg Field Stadium. The reactor was called a pile because it amounted to literally a pile of uranium and graphite blocks. The graphite (i.e., carbon) was used to slow the fast neutrons down so they would be more readily absorbed by the uranium atoms. The reactor had no radiation shielding, nor any type of cooling system, since Fermi claimed to have 100% confidence in the accuracy of his experimental calculations, which predicted that none would be necessary.[42]

The definitive experiment came on December 2, 1942, when criticality was reached by amassing enough uranium to attain a critical mass. As soon as the resulting nuclear chain reaction was achieved, it was then quickly terminated. It was a proof-of-principle experiment, and the principle had now been proved.[43] Szilárd's nuclear chain reaction hypothesis of just 10 years earlier was absolutely correct, and production of an atomic bomb was possible.

Criticality is the driving force behind both nuclear power plants and nuclear bombs. And uncontrolled criticality is the biggest hazard for all nuclear activities. Still, as we will see, nuclear power plants cannot become nuclear bombs even if humans lose all control over the chain reaction. If Fermi had lost control of the Chicago Pile it would have caused a nuclear meltdown, not a nuclear explosion. There are major engineering obstacles to making a nuclear bomb. They cannot be produced by accident; you have to intentionally build one. So power plants are subject to the risk of a meltdown, but they are not latent nuclear bombs waiting to explode.

SUN STROKE: FUSION

We began Part One with the sun and its visible electromagnetic radiation. Fittingly, we have come full circle and will now close out Part One back at the sun, with a brief discussion of the sun's visible particulate radiation. Yes, the sun also produces particulate radiation, and it's all made possible by nuclear fusion.

Fusion is the opposite of fission. Strange as it may seem, while very large atoms can split apart through fission and thus release energy, very small atoms can fuse together and also release energy. The mechanism of a fusion reaction is too complicated for us to deal with here.[44] But an important distinction is that while fission can occur spontaneously, fusion must be forced to occur. That is, it requires an enormous input of energy to initiate a fusion reaction, which ultimately results in even more output energy. While there is a net gain in energy production during the fusion reaction, the input energy needs of fusion are extremely high and require the attainment of temperature levels found only on stars. That is why the only place that fusion naturally occurs in our solar system is on the sun, our closest star, where temperatures are extremely hot. These fusion events release energy that pushes through the sun's surface in the form of solar flares, emitting large quantities of high-energy particles into space. These high-energy particles coming from the sun constitute *cosmic radiation* and are a major source of the radiation that reaches Earth.[45]

Cosmic radiation is part of our background radiation. It is responsible for the nighttime color display that we see at the northern and southern latitudes on Earth, known as the aurora borealis (northern lights) or aurora australis (southern lights), respectively. The auroras occur when cosmic particles are deflected by Earth's magnetic field and shower down through converging entry points in the atmosphere near the poles.

Cosmic particles contribute to our natural background radiation exposure. But even at the poles, cosmic particles contribute only slightly to our annual radiation dose, because Earth's atmosphere shields us from most of the impinging cosmic particles. In contrast, astronauts in outer space lack atmospheric protection, and receive significant cosmic particle exposure during space missions. Some of these cosmic particles are so large that when one crosses the retina of an astronaut's eye, they can see a small flash. This is reminiscent of Rutherford's microscope experiments in which he could actually see alpha particles as flashes on a fluorescent screen, except that the astronauts can dispense with the fluorescent screen. Their screen is their own retina.

The enormous heat required to induce fusion reactions is the major reason we don't have any fusion-based power plants. It is currently not feasible to generate enough power to sustain temperatures required for a fusion reaction, and there are no materials on Earth that could resist the heat needed to sustain them. So, even though the net output of a fusion reaction is more than the input energy, the obstacles to realizing that energy are substantial.

This is why claims of cold fusion attract so much attention. Cold fusion claims imply that fusion has been achieved by some novel mechanism that doesn't require input heat. This would mean that we could get fusion energy out without putting any heat energy in—a major technological advance that would promise easy and unlimited power forever. So far, all claims of achieving cold fusion have been debunked.[46]

In contrast, nuclear weapons based on hot fusion do exist. They are often called hydrogen bombs, or H-bombs, because they are based on the fusion of hydrogen nuclei;[47] but they are also sometimes referred to as thermonuclear bombs. Each bomb can release 1,000 times, or more, energy than the atomic bombs detonated on Hiroshima and Nagasaki. Due to the input energy required to initiate fusion, a fission bomb is needed to trigger a fusion bomb. So each fusion bomb is actually a combined fission/fusion bomb. Fortunately, a fusion bomb has not yet been detonated during warfare, although some have been test detonated by several countries. The United States test detonated the first fusion bomb, named Ivy Mike, on a coral atoll island in the South Pacific in 1952. The number of fusion bombs currently in existence worldwide is not known exactly, but it likely numbers in the thousands.

CODA TO PART ONE

This brings us to the close of Part One. We've learned the distinction between radiation and radioactivity and how each was discovered. And we've learned how Rutherford and his colleagues coaxed the atomic nucleus into revealing its structural secrets, thus toppling "plum pudding" as the standard model of nuclear architecture. In the process, we've seen that radiation comes in the form of either electromagnetic waves of different wavelengths, or in the form of different types of high-speed subatomic particles. Amazingly, both forms of radiation can carry their energy through solid objects, a phenomenon that Roentgen was the first to observe. But a portion of that energy is left within the object that the radiation traverses, thereby delivering to that object a radiation dose. But our knowledge of dose at this point in our story is still vague. We need to explore dose further if we are to understand radiation's health risks. Still, we now know enough nuclear physics concepts to engage in a meaningful conversation about the health effects of radiation.

So far, the health aspects of our story have been merely anecdotal and limited to the small group of scientists that worked with radiation shortly after its discovery at the beginning of the twentieth century. Although they had little knowledge of dose and scant understanding of the potential dangers of radiation, most of these scientists took some limited precautions to protect themselves from exposure, at least as much as was practicable for the work they conducted. If radiation work had remained confined to this small cadre of researchers, the effects of radiation on health would likely never have become a major public health issue. But radioactive materials soon became a commodity and were put to many uses in various consumer products. Public exposure to concentrated sources of radiation became increasingly prevalent. As such, there were bound to be health consequences to the public and particularly to the workers who handled radiation-producing devices and radioactivity. Before long, those health consequences became very evident.

PART TWO

THE HEALTH EFFECTS OF RADIATION

CHAPTER 5

PAINTED INTO A CORNER: RADIATION AND OCCUPATIONAL ILLNESS

No wonder you're late. Why, this watch
is exactly two days slow.

—*The Mad Hatter in* Alice's Adventures in Wonderland
(Lewis Carroll)

THE MYSTERIOUS ILLNESS OF SCHNEEBERG'S MINERS

Today we would call H. E. Müller a victims' advocate. Although not a medical doctor, he was a respected man of integrity, and the mining village of Schneeberg, Germany, hired him around 1900 to track down the source of a mysterious lung ailment that plagued the miners of the community.[1] He would not let them down.[2]

Lung disease among miners had been known for many centuries, and the term *bergsucht* (disease of the mountains) had entered the vernacular through the medical writings of the Swiss Renaissance physician Paracelsus (1493–1541). Bergsucht encompassed a collection of miners' lung ailments that modern physicians would collectively call chronic obstructive pulmonary diseases (COPD); but they also included lung cancers, as well as infectious lung diseases, such as tuberculosis and pneumonia, that were endemic among mine workers. The most prevalent form of bergsucht that Schneeberg's miners suffered seemed to be clinically different from that endemic to other mining regions. It didn't have as much variation in its symptoms, it had an earlier age of onset, and it killed more quickly.

In 1879 two physicians, F. H. Härting and W. Hesse—a family practitioner in Schneeberg and a public health official from nearby Schwartzenberg, respectively—decided to systematically study the disease. They performed autopsies on more than 20 bergsucht victims from Schneeberg and found that 75% of them had died in an identical manner. It turned out that the clinical symptoms of Schneeberg's bergsucht were better defined than elsewhere because the underlying disease was predominantly of a single deadly type—lung cancer. This discovery was made at a time when smoking was rare and lung cancer rarer. The autopsy findings were astounding. But what was the cause?

Two known toxins, arsenic and nickel, were considered likely causal agents for the lung cancer. Both were present in high concentrations in Schneeberg's ore. Cobalt and silica dust were also suspected. Unfortunately, none of these agents could be conclusively linked to the disease. Perhaps some specific combination of these candidates was required to produce illness. Decades passed, yet a definitive conclusion on the cause of Schneeberg's lung cancers remained elusive.

It is a sad truism of public health practice that the first indication of toxicity from any hazardous agent will always appear among those who are occupationally exposed. This is because workers are typically exposed to higher levels for longer periods of time than the general public, and higher and longer exposures often shorten the time to onset of disease and increase the severity of symptoms.

A classic example is mercury poisoning among hat workers. Mercury solutions were used in the hat-making industry during the nineteenth century to preserve the animal pelts used in hat production (e.g., beaver-fur top hats). Workers, called hatters, breathed the fumes, and the mercury accumulated in their bodies. They developed a wide range of neurological symptoms that were misunderstood as signs of insanity. It soon became generally known that hatters often suffered mental difficulties, and this developed into the colloquialism "mad as a hatter" to describe anyone who seemed a little daft. We now know that the hatters' apparent madness was due to the neurotoxicity resulting from their mercury exposure, and we have since adopted strict regulations limiting the general public's exposure to mercury.[3]

Mercury was by no means the only chemical to cause a particular illness in workers. As early as the eighteenth century, doctors were well aware of other occupational diseases, such as painter's colic (lead in paint), brass founder's ague (metal oxide fumes), and miner's asthma (rock dust).[4]

Even cancer was then known to be an occupational illness. In 1779, the surgeon Percivall Pott (1714–1788) reported that scrotal cancer (euphemistically called soot warts) was endemic among the chimney sweeps of England.[5] He claimed that their unusual affliction was a woeful consequence of prolonged exposure to chimney soot. It was thought that body sweat ran down their torsos and caused soot to accumulate around their scrotums, putting them at risk of scrotal cancer. Finding the cancer even among teenage workers brought attention to the fact that the chimney sweep industry was employing boys as young as four years old; they were hired because their small size allowed them access to even the narrowest chimneys. The British Chimney Sweep Act of 1788 raised the minimum age for chimney sweep apprenticeships to eight years old and required employers to provide clean clothing; otherwise it did nothing to protect the workers from soot exposure. It took more than a century to experimentally confirm Pott's contention that soot contains cancer-causing compounds (i.e., *carcinogens*).

Thus, workers like hatters and chimney sweeps often served as the canaries in the coal mine for the general public.[6] They warn the rest of us of otherwise hidden hazards so we can take precautions.[7] In the early years, the general public had a romantic view of radiation. Even though people didn't understand it and didn't fully appreciate its technological potential, they did appreciate its novelty and entertainment value. Unlike the radiation researchers, however, the general public was not fully cognizant of the fact that radiation exposure could pose significant health problems. That would soon change. What was needed was radiation's equivalent of a miner's canary to warn the public to proceed with caution. Ironically, the miners' canaries for radiation's hazards turned out to be actual miners themselves. And the major health hazard of radiation had been documented and characterized well before radiation had even been discovered. Still, someone needed to put all the salient facts together and find the cause of the miners' ailments.

Time plodded on, and the miners continued to suffer from their mysterious plague. Then a fortuitous finding in the field of geology broke the case. Professor Carl Schiffner (1865–1950) of the Mining College of Freiberg was intrigued by the discovery of radon-222 by the German chemist Freidrich Dorn (1848–1916).[8] Schiffner decided to perform a geological survey of the radon concentrations of all natural waters of the Saxon region, including Schneeberg. His survey revealed very high concentrations of radon in Schneeberg's waters and even in the air of its mines.

Learning of Schiffner's findings, Müller, who was still working on the Schneeberg mine problem, immediately suspected that radon was the cause of the miners' lung cancers. Müller knew that x-ray exposure had previously been associated with cancer in those who worked with x-ray machines. He also knew that radioactive substances, such as radon, emitted x-ray-like radiation. Putting these two observations together, it was no great leap to suppose that radon caused lung cancer by virtue of its emitted radiation. And that is exactly what Müller proposed in 1913. He wrote, "I consider Schneeberg lung carcinoma [i.e., lung cancer] to be a particular kind of occupational disease, acquired in the mines having rock strata containing radium, and thus showing high [radon] levels."

Radon is produced when radium decays. A radon atom is born whenever a radium atom dies as part of a long decay chain, the lineage of which starts with uranium-238 and ultimately ends with stable lead-206. Radon is called the *progeny* and radium is its *parent*.[9] All of the radioactive isotopes in this decay chain are solids except for radon, which is a gas. That is why radon alone was a problem for the Schneeberg miners; only radon could escape from the soil and be inhaled.

Müller's hypothesis about the association between lung cancer and mining had many skeptics, but it also could not be entirely dismissed. Besides, even if not radon, it was likely that some type of airborne toxin was involved. Regardless, the remedy seemed to be the same. This amounted to improved mine ventilation and other air containment precautions, in order to reduce miners' exposure to radon or other airborne mine contaminants. With these occupational hygiene changes in place the problem was thought to be resolved. Moreover, World War I was beginning, and the burdens of war made the threat posed by radon seem trivial. People stopped worrying about radon.

After the war, physician Margaret Uhlig revisited the issue of radon in the Schneeberg mines. She showed that the limited remediation efforts that had been deployed were inadequate and that miners were still dying of lung cancer in large proportions. Many studies followed, and any remaining skepticism was put to rest. A major review of 57 radon studies published through 1944 concluded, "There is a growing conviction in the United States and abroad that radiation emitted by radon inhaled over a long period of time . . . will cause cancer of the lung in man."[10] In the years since then, the fundamental truth of this statement has never come under any serious doubt.

Although the fact that radon causes lung cancer is unquestionable, what has come into question is whether the levels of radon often found in typical homes pose a significant lung cancer risk to residents. As you can imagine, in addition to the fact that home radon levels are typically much lower than mine levels, homes and their residents differ from mines and miners in many ways that can affect cancer risk. We will visit the issue of radon in homes in Part Three. For now, just know that radon was the first radioactive isotope to be associated with cancer in humans. Its ability to cause cancer has been a generally accepted fact since as early as 1944.

BAD TO THE BONE

Frances Splettstocher (1904–1925) never thought of herself as a radiation worker although she knew her work involved the use of radioactive material. She was a factory worker in the watch-making industry, and her employer was the Waterbury Clock Company in Waterbury, Connecticut.[11] She loved her job, as did all her female coworkers. For a 17-year-old girl without higher education, it was one of the best jobs available in Waterbury. In addition, the factory was less than a mile from her home on Oak Street, where she lived with her parents, three sisters, and three brothers. Her father also worked at the factory, so she likely walked to and from work with him every day. Thus, Frances perfectly typified the women working at the plant, who were mostly in their midteens to early twenties and came from upper middle-class working families. Life was good for Frances Splettstocher in 1921. Who could have imagined that by 1925 she would be dead?

Frances's job was to paint the numbers on watch dials with a fluorescent paint that contained radium. This paint had been formulated in 1915 by an Austrian immigrant Sabin von Sochocky (1883–1928), and the active ingredients in the formula were radium and the fluorescent compound zinc sulfide.[12] The radium emitted radiation that made the fluorescent paint glow in the dark and allowed the watch wearer to tell the time even in total darkness. Quite a novelty indeed!

At the time, wristwatches themselves were actually novelties for men. They were originally invented by the Swiss watchmaker Patek

Philippe & Company in the late 1800s, and were designed specifically for women. Men typically used the larger, more masculine pocket watches. Nevertheless, wristwatches proved to be highly practical for soldiers waging war in the trenches during World War I. Coupled with a glow-in-the-dark dial, they were extremely valuable for night maneuvers. Since there was nothing effeminate about a trench soldier, wristwatches with glowing dials suddenly became associated with machismo.

By 1921, the war was over and everyone wanted to own this new high-tech time device that was so popular among the soldiers. The watches were in great demand, and Frances and her coworkers were kept very busy. They were paid by the dial. The faster they painted, the more money they made. Good workers could paint as many as 300 dials a day, making $24 per week ($317 per week in 2015 dollars)—an excellent daily wage for women in 1921. (The average wage for women at this time was about $15 per week.[13])

The radium dial watches were profitable for the Waterbury company, but they faced stiff competition from other radium watch companies in Orange, New Jersey, and Ottawa, Illinois.[14] The demand for the watches was high and growing. While fewer than 10 thousand radium watches were purchased in 1913, by 1919, over 2 million were sold. One year later, Americans purchased 4 million radium dial watches, an astounding number considering that the population of the United States in 1920 was just a little over 100 million. It seemed as if soon every American would be wearing a radium watch.

Then, in January 1925, the good times came to an abrupt end for Frances. She became weak and anemic. Her face became painful to the touch, and her teeth began to hurt. She saw a dentist who pulled a suspect tooth. A piece of her jaw came out with the tooth. Soon the soft tissues of her mouth began to deteriorate, and she ended up with a hole in her cheek. By February, she was dead.

What few in Waterbury knew at the time was that Frances's counterparts in watch factories in New Jersey and Illinois were also suffering the same symptoms. In New Jersey, four dial painters had died, and eight were ill with symptoms similar to Frances's. Shortly after Frances died, one of her coworkers, Elizabeth Dunn,

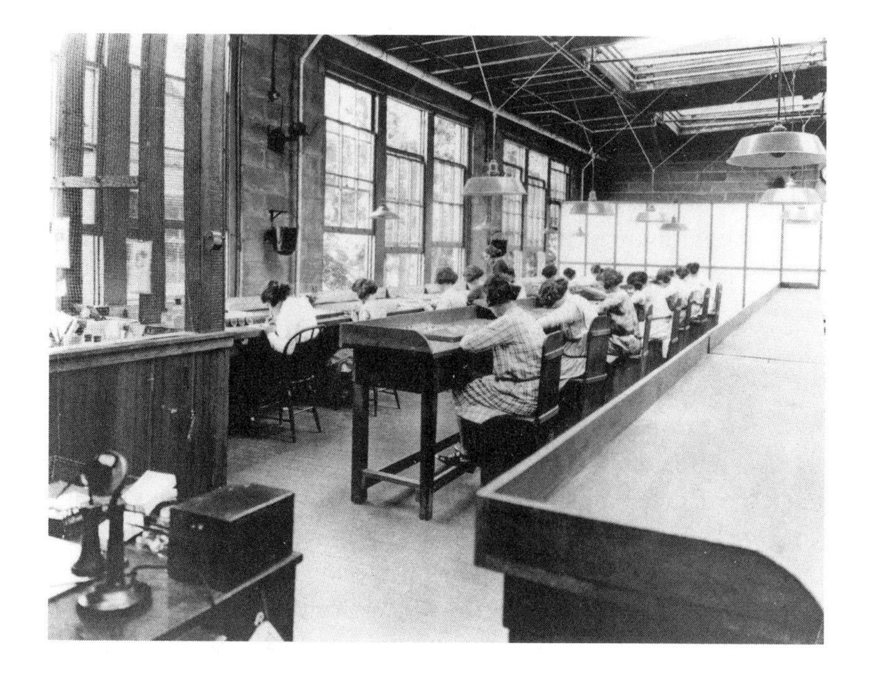

FIGURE 5.1. RADIUM DIAL PAINTERS AT WORK. Young middle-class women found good-paying jobs painting watch dials with glow-in-the-dark radioactive paint. But the painting procedure they used led to inadvertent ingestion of small amounts of the paint. Since the paint contained radium, ingestion resulted in large radiation doses to the women's bones. They suffered bone diseases and cancer, and many died. This was the first time that the general public became aware that ingesting radioactivity could be hazardous.

broke her leg for no apparent reason while dancing. She then developed mouth symptoms similar to Frances's and ultimately died in 1927. After her, coworker Helen Wall contracted similar symptoms and died.

One by one, more and more dial painters were falling sick in Connecticut, New Jersey, and Illinois. It is estimated that, at its peak, the industry employed over 2,000 radium dial painters, most of them centered in northern New Jersey; a good number of them fell ill. Dentists in New Jersey saw so many dial painters with bone necrosis that they dubbed the condition *radium jaw*, thus defining a new occupational disease. And the popular press tagged these women dial painters with the unflattering name *radium girls*, a term by which they became commonly known to the public.

The major nuisance of painting watch dials by hand was that the paintbrush tip often became blunted, making fine and detailed painting of numbers difficult. For this reason, the women often restored their brush's point by twisting the bristles in the corner of their lips. In fact, they actually received company training in how to do this properly. Each time they "lip pointed," they ingested a small amount of paint. Over time they ingested large amounts of paint, all of which contained radium. It has been estimated that a typical worker might have lip pointed her brush four times per dial, consuming about 0.001 grams of paint each time. At 300 dials per day for a five-day week, this could have resulted in about six grams (about one teaspoon) of paint being consumed each week. Over a year's time the worker would have consumed 300 grams of paint (about one coffee cup).[15] Although radium concentrations varied between companies and exact paint formulas were proprietary, it is estimated that this amount of paint would have contained about 375 micrograms of radium (about the same mass as a mosquito's head). It's a very small amount in terms of its mass, but a very large amount to have ingested because the radioactivity had been highly concentrated during purification from its natural state in mineral ore.

Radium is relatively insoluble, and most of it would have passed through the dial painter's gut without being absorbed. Nonetheless, approximately 20% would have been absorbed, and that would be enough to cause trouble. It would accumulate in bones and remain there. Day after day, year after year, the radium would continuously irradiate bones.[16] Because bones are lined with living tissue and filled with blood-producing bone marrow, irradiation from radium permanently lodged within bone can have serious health consequences. All this happened to bones because of a very unfortunate chemical property of radium. Radium is a *bone seeker*.

☢

What exactly is a bone seeker? Apart from their physical properties, radioisotopes are still chemical elements and retain their distinct chemical properties. As we have previously seen, an element's chemical identity is determined by its number of protons.[17] Thus, the proton number determines not only the name of the chemical but also its reactions. It was radium's special chemistry that was responsible for the dial painters' unique health problems. Radium belongs to the chemical family of alkaline earth metals (column 2 in the periodic table of elements). This family of elements includes calcium, the major mineral constituent of bone.

Unfortunately for the radium girls, elements within the same family often mimic each other's chemistries. Radium mimics calcium. So, when radium enters the body, it goes to bone, the same place calcium normally goes. The radium gets incorporated into the mineral component of the bone and becomes a permanent constituent. Though the radium itself doesn't hurt the mineral component of bone directly, the entire skeleton becomes radioactive, and the living tissues lining the mineral component of the bone are thus continuously exposed to radiation at high levels. This was the cause of the workers' health ailments.

Although lung cancer was not involved, the radium girls' ailments had a cause that was closely linked to the Schneeberg miners' sickness. The women's problems, however, were not due to radon. The radium paint couldn't have released enough radon gas to be a health problem. The radium girls' problem was the radium itself, which they were ingesting.

Radium and radon are both alpha-particle emitters and thus produce the same types of damage to cells. So why were the health ailments so different? Simply because different tissues were exposed to the radiation. The tissue that is exposed is determined by both the chemistry of the element (radium vs. radon) as well as the route of exposure (ingestion vs. breathing). The miners suffered lung ailments because the radioisotope was in the air, thereby exposing their lungs but not their other tissues. The dial painters mostly experienced bone problems—bone cancers and blood dis-

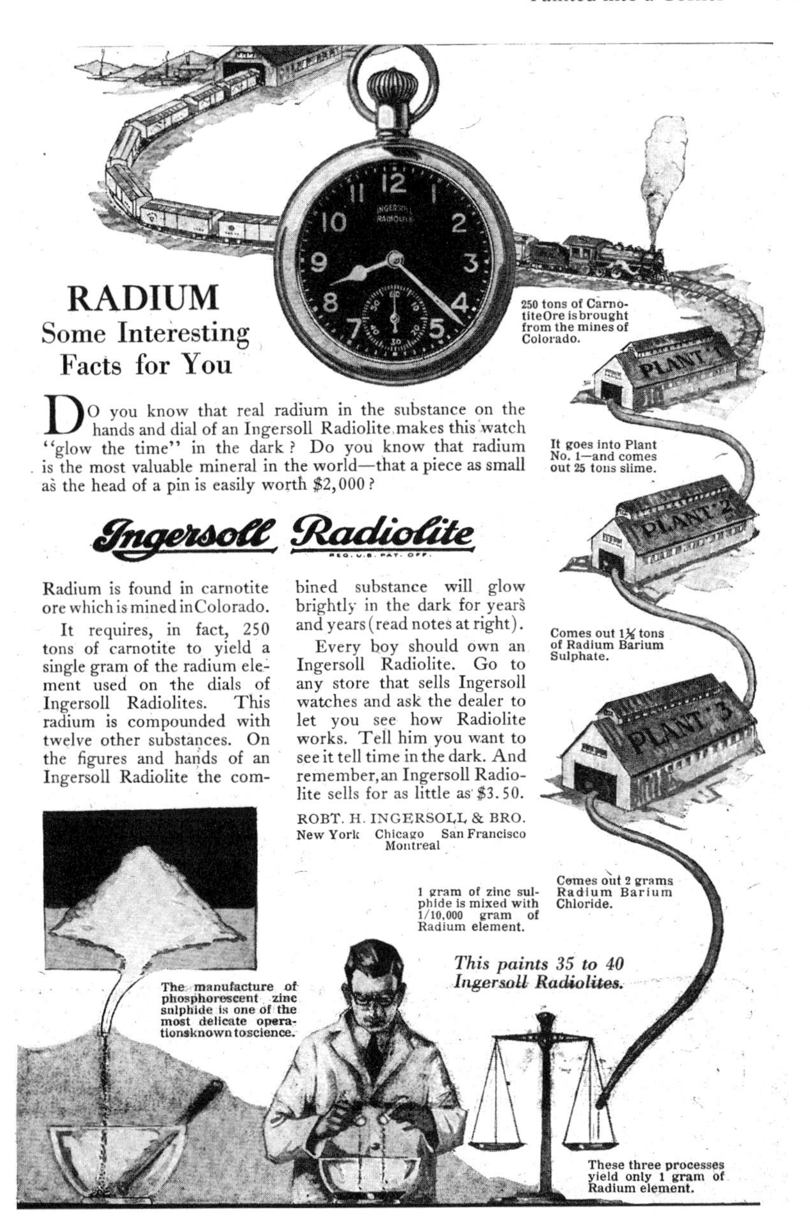

FIGURE 5.2. INGERSOLL RADIOLITE WATCH ADVERTISEMENT. Men and boys were the primary target market for radium watch sales. This advertisement appeared in the October 7, 1920, issue of *The Youth's Companion*, a very popular boys' magazine at the time. It includes a brief description of the entire radium refining process.

orders—because the ingested radium ended up in bone. We will see later that when the alpha emitter strontium was found to be a hazardous byproduct of nuclear fission reactions, it was possible to accurately predict its effects on health, because strontium's atomic number placed it in the exact same family on the periodic table as radium, making it another bone seeker.[18]

After much denial and resistance, but under pressure from the courts, the watch companies finally accepted the blame for the radium workers' ailments. From 1926 to 1936, Waterbury Clock Company paid out $90,000 ($1.5 million in 2015 dollars) in settlements to 16 women who had taken ill and put aside a reserve of $10,000 ($170,000 in 2015 dollars) to cover any future claims.[19] New Jersey and Illinois watch companies made similar settlements.

Radium dial painting continued, but by 1927, the painters were working in fume hoods, and wore hairnets and rubber gloves. Lip pointing of brushes wasn't permitted any longer. No cancers or other health effects were detected in workers hired after these precautions were implemented, although they continued to accumulate radium in their bodies at a rate of approximately one quarter of a microgram per year.[20] It seemed that in the case of radium, like x-rays, the body could tolerate a certain level of ingestion before health effects would become apparent. But what exactly was that tolerance level? More studies would be needed.

Frances died on February 21, 1925. Her funeral was held at the Church of St. Stanislaus in Waterbury and her body interred at Calvary Cemetery. By now, all that is left of her remains are her bones. Being that the half-life of radium is 1,600 years, they still contain nearly the same amount of radium as they did on the day that she died.

Marie Curie learned about the health afflictions of the radium girls in June of 1925 when her friend Marie "Missy" Mattingly Meloney (1878–1943)—a feminist, *Washington Post* editor, and strong supporter of Curie's scientific work—wrote to her from America.[21] Although Curie was concerned for others, she did not think that she herself had incurred any significant risk since she had not actually ingested any radium. Also, her laboratory had taken some precautions to limit contamination. Curie was more concerned that she had overexposed herself during the x-ray work she had done during World War I. Shortly after Roentgen had discovered x-rays, Curie had developed a portable x-ray machine that she helped deploy to field hospitals on the war front. During that time, she learned of radiologists and x-ray technicians who had suffered severe damage from overexposure to x-rays. They had lost limbs and eyesight as a consequence and likely faced more long-term health problems. In fact, in her book *Radiology and the War* Curie had railed against overexposure to x-rays, writing that "a person who receives the rays feels no pain to warn that he has been exposed to a noxious dose."[22] No, Curie was not worried about internal contamination from radium so much as external exposures from gamma and x-rays, which she saw as a bigger threat.

RADIOACTIVE SNAKE OIL

Although the radium girls' story was publicized in American newspapers as early as 1925 and protective standards for workers were put in place by 1927, at least one person seems to have missed the memo. Eben McBurney Byers (1880–1932) was a prominent industrialist from Pittsburgh, Pennsylvania, but he was even more famous for his golfing prowess. He competed in many amateur tournaments and was the US Amateur Golf Champion of 1906. While returning to Pittsburgh after attending a Harvard-Yale football game in the fall of 1927, he fell from his train berth and broke his arm. A doctor prescribed Radithor to aid healing. Radithor was a popular radium-containing patent medicine sold by William J. A. Bailey (1884–1949), a Harvard dropout who falsely claimed to be a medical doctor.[23]

Radithor was simply radium dissolved in distilled water. It was sold by the case in one-ounce bottles.[24] Radithor sales and Bailey's profits were huge. Each case of 30 bottles contained about $3.60 worth of radium, and the cases sold for $30.00 each, an 800% markup.[25]

Apparently, Mr. Byers really liked the stuff. He estimated that over the next three years he drank about 1,400 bottles of Radithor, and he even shared his supply with his friend, Mary Hill, who also became a devotee of the drink.[26] By 1930, Byers had lost most of his jaw and had holes in his skull. He died a gruesome death on March 31, 1932. Six months later, Mary Hill was also dead from similar causes. Byers's body now rests in a mausoleum in Allegheny Cemetery in Pittsburgh, in a coffin lined with lead to protect cemetery visitors from radiation exposure.

The shame of all this was that radium had been known to be a bone seeker since 1913, when British scientist Walter Sydney Lazarus-Barlow (1865–1950) published a research paper that clearly demonstrated ingested radium ends up in bone.[27] Paradoxically, far from drawing concern, it was widely interpreted to be beneficial, due to the popular assumption that radioactivity fostered a variety of health benefits by providing the body with added energy. Why would people think otherwise when charlatans were claiming that science had shown that radium water could enhance the sexual passion of water newts?[28] Why should newts have all the fun?

Nonetheless, in 1914 Professor Ernst Zueblin of the University of Maryland Medical School published a review of more than 700 scientific publications in the international literature that reported on the internal use of radioactivity for therapeutic purposes.[29] Although he concluded that there should be guarded enthusiasm regarding the therapeutic potential of radium, he also cautioned that ingested and injected radium could result in problems with necrosis and ulcerations. Unfortunately, his advice went unheeded.

Thomas Edison had even made public warnings about the health risks of radium as early as 1903. Edison had received a sample of radium from France by way of a former employee, William J. Hammer (1858–1934), who had visited the Curies while traveling in Europe in 1902. Wary because of his prior bad experience with eye injuries from x-rays, Edison kept his distance while experimenting with radium and even attempted to test its biological effects on insects.[30] Despite his precautions, one time he inadvertently put a vial of radium in his vest pocket and carried it for days, alarming both his staff and his family but apparently suffering no ill consequences. When interviewed about radium he urged people to exercise caution: "It took centuries to develop electricity upon its present scale, and it may take years for us to get any definitive ideas about radium." Another time he publicly remarked, "There may be a condition into which radium has not yet entered that would produce dire results, and everyone handling it should take care."[31] Unfortunately, by this time, the public had likely tuned out all health warnings coming from Edison, due to his incessant harangue about the dangers of AC electricity during the prior decade. Sadly, this time he was correct.

Luckily, there were two factors that kept radium health potions from ever becoming a major public health epidemic. First, most products were absolute frauds. It has been estimated that 95% of all radium solutions on the market at the time contained no radium at all.[32] They were a total scam and had no health effect, good or bad. Second, the potions that actually did contain radium, like Radithor, were too expensive for people of average means to use as a daily elixir. Only the wealthy could afford this indulgence. Although the actual body count was low in terms of a public health problem,[33] radium poisoning had struck near both ends of the socioeconomic spectrum–young working-class women and middle-aged, aristocratic men. Such a demographically broad scourge on society could not be allowed to stand for long.

At the time, the US Food and Drug Administration was in its infancy and lacked the regulatory teeth it has today. Even so, the Federal Trade Commission was able to close shady businesses. Following Byers's death and the publicity it generated, the Federal Trade Commission came down on "Doctor" Bailey, and he was forced to close up shop and stop producing Radithor. Still, Bailey remained resourceful. Shortly afterward, he started a new enterprise, the Adrenoray Company, which marketed a belt that held radium sources over the adrenal glands—the hormone-producing organs located on top of the kidneys—to increase vitality and "treat sexual strength."[34]

Remarkably, Bailey's supplier of radium was United States Radium (US Radium) in Orange, New Jersey; this was the same supplier that the watchmakers used for their paint. Bailey began selling his Radithor in 1925, the same year as the peak of the radium girls' health problems. By this time, both US Radium and Bailey well knew of the radium workers' plight, but they denied that the health problems were due to radium ingestion. Bailey claimed he would even be willing to swallow in one single dose the equivalent of all the radium that was used in a watch dial factory over an entire month of production. It is not clear that Bailey ever made good on this offer. It's unlikely, since such a dose would probably have been lethal, and Bailey lived another 24 years, dying in 1949 at the age of 64, and outliving poor Mr. Byers by 17 years.

Von Sochocky, the producer of the radium paint, did not fare as well as Bailey did. His body contained enough radium to interfere with his laboratory radioactivity measurements, and his fingers were badly damaged. It was too late for Von Sochocky to benefit from the workplace radium protections instituted in 1927. By then, Von Sochocky and his laboratory assistant Victor Roth were suffering from chronic anemia that had begun around 1924. In 1927, just as workplace radium standards came into effect, Von Sochocky and Roth both died within a few months of each other.[35]

The legacy of the radium dial painting and Radithor lasted for years. In the 1980s, some homes around Orange and Montclair, New Jersey—the center of US Radium activities—were found to have elevated levels of radon. Radon can be detected only by measuring radioactivity levels in the air. Widespread testing has shown that radon contamination in homes typically results from radon gas emanating naturally from radium decay in the soil, just as in the Schneeberg mines, and such radon contamination is not uncommon in New Jersey. The source of radon in these particular homes, however, turned out to be sand fill used at the housing sites, indicating that the sand used in construction was contaminated with radium. The sand's source was traced to the byproduct residue from the old radium extraction processing facility of US Radium in Orange. The homeowners filed suit against the successor company, Safety Light Corporation. In the end, the courts ruled that Safety Light executives should have known the dangers of radium contamination and their failure to disclose the presence of radium to the homeowners constituted negligence. The New Jersey Supreme Court held the company liable since "radium has always been and continues to be an extraordinarily dangerous substance."[36]

NONE TOO SOON: WORKER PROTECTION ARRIVES

The Schneeberg miners and the radium dial painters were the most spectacular early examples of many people being exposed to radiation in the workplace with disastrous consequences. Still, there had been a long chain of anecdotal reports preceding them: dermatitis of the hands (1896), eye irritation (1896), hair loss (1896), digestive symptoms (1897), blood vessel degeneration (1899), inhibition of bone growth (1903), sterilization in men (1903), bone marrow damage (1906), leukemia in five radiation workers (1911), and anemia in two x-ray workers (1912).[37] Various cancers among medical radiologists would soon be added to this list.[38]

The above reports, coupled with the findings related to the Schneeberg miners and the radium dial painters, underlined the need for radiation protection standards. The Roentgen Society, whose membership was primarily made up of medical x-ray workers, began pushing for safety standards as early as 1916.[39] In 1921, the British X-ray and Radium Protection Committee endeavored to establish an occupational dose limit for radiation, which they ideally sought to define in terms of "a reproducible biological effect, expressed as much as possible as a measurable physical unit of radiation dose."

Considering the list of known biological effects of radiation, choosing a relevant endpoint was not easy. Even so, the health effect of most concern was cancer, and the medical community's thoughts on the origins of cancer at that time were greatly influenced by the work of Rudolph Virchow (1821–1902), a physician and scientist who correctly postulated that cancerous cells originated from normal cells. This was a major advance in the understanding of cancer biology. Furthermore, Virchow had witnessed in his medical practice that nonmalignant lesions often turned into cancerous ones. Based on this observation, he had proposed that cancer arose from chronic irritation of normal tissue. And, in fact, there was much circumstantial evidence for this in the case of radiation. It was well known that the cancers radiation workers commonly suffered on their hands from exposure to x-ray beams typically were preceded by years of skin irritation (i.e., radiation dermatitis).

Because irritation was thought to precede cancer, it followed that radiation doses insufficient to cause irritation would also be insufficient to cause cancer. Based on this principle, skin reddening (e.g., erythema) was adopted as a highly relevant biological endpoint for radiation protection purposes. If radiation exposures were kept low enough to prevent skin reddening, they should also be too low to cause cancers of the skin or any other tissues. Thus, skin erythema was chosen as the most relevant measurable and reproducible biological effect on which to base radiation protection standards.

With the first part of their charge complete, the committee turned to the second issue. The biological effect needed to be "expressed as much as possible as a measurable physical unit of radiation dose." This task seemed to be even more straightforward than the first. Since x-rays are a type of ionizing radiation, and the number of ionizations produced is directly dependent upon the amount of energy deposited by the radiation, it follows that one should be able to determine the dose simply by measuring the number of ionizations produced. The problem was simply how to measure the number of ionizations accurately.

Here, too, that answer seemed straightforward. Since the ions produced are mostly electrons, they can produce an electric current. Measuring the electric current would thus be a good index of radiation dose. In fact, this method of measuring radiation dose was already under development. The device invented to perform the task was called a *gas ionization chamber*. When such a chamber of gas is placed in a radiation beam, the amount of gas ionization, measured as an increased flow of electrical current through the gas, reveals the radiation exposure level. And to honor the discoverer of x-rays, the unit for this radiation exposure measurement was named the *roentgen*.[40]

With proper tools now in hand, scientists sought to determine the *maximum tolerable dose* (*MTD*)—that is, the permissible level of exposure—as expressed in roentgens.[41] Studies were performed and data meticulously analyzed. Based largely on studies of people who had worked with radiation for many years without ill effects, a conclusion was finally drawn. In 1931, an MTD for x-rays was announced by the US Advisory Committee on X-Ray and Radium Protection: on any given day, workers were not to be exposed to more than 0.2 roentgen (i.e., one fifth of a roentgen unit).[42] This dose limit, however, was based on a relatively small number of workers with rather uncertain exposure levels. It was a "best guess" estimate, but it would have to do until better data arrived.

Unfortunately, the MTD based on skin erythema arrived too late to save Marie Curie's group. Her laboratory had chosen another biological effect by which to limit their exposure levels, and their biological endpoint allowed much higher whole body doses. The effect they monitored was anemia, the suppression of circulating blood cells. Since 1921, everyone in her laboratory had been having regular blood work done to monitor their exposure levels. When workers' blood counts dropped too low, they were pulled off the job for a while and sent out into the country to rest and recuperate until their blood counts returned to normal. Their blood counts typically did return to normal, and this reinforced the erroneous belief that radiation effects on health were always reversible.[43]

The use of anemia to monitor worker radiation exposures had a tragic effect of double jeopardy. First, it allowed whole body doses much higher than would have been permissible based on the skin erythema MTD. In fact, the Curies and their staff were resigned to their constant skin dermatitis as the price that they had to pay to be in the forefront of science. Pierre even boasted: "I am happy with all my injuries. My wife is as pleased as I. You see, these are the little accidents of the laboratory. They shouldn't frighten people who live their lives [purifying radioactivity]."[44] Second, because the Curies assumed that since their blood counts fully recovered with sufficient rest from work, so would all other health effects. They further supposed that there would be no long-term consequences from living a life with rollercoaster blood counts. In this they were sadly mistaken.

The anemia threat came to a head for Marie Curie in 1925 when a terrible radiation accident killed two of her former students—Marcel Demenitroux and Maurice Demalander. The two engineers were constructing an irradiation machine for medical use in a small factory outside of Paris. They sustained high radiation doses from the radioactive thorium they were using in the machine. The two lingered in an anemic state for months and ultimately died within four days of each other. Remarkably, it wasn't until their last days of life that the pair realized their coincidental illnesses were due to their high radiation exposures. The deaths shook Marie Curie, but she was in denial. Although she accepted that the radiation had killed the engineers, she attributed their deaths to the fact that they were living in quarters next to the laboratory because of their heavy workload and thus "didn't have a chance to get out and take the air."[45] In fact, Curie found a whole host of explanations for illnesses among her workers that were, in retrospect, obviously related to radiation.[46]

With time Curie developed her own illnesses, which she tried to keep secret. Ultimately, she admitted to herself that they were due to radiation. Increasingly frail and virtually blind from cataracts, she tried to continue working, but fell seriously ill in the summer of 1937. She traveled to an infirmary in the Savoy Alps to recuperate, but it was too late. Her doctor examined her blood work and diagnosed "anemia in its most extreme form." Her fever diminished some, but her fate was already sealed. Soon, she was dead. Her doctor later pronounced that her bone marrow never rallied to restore her blood counts "probably because it had been injured by a long accumulation of radiations."[47]

Based on the experience of the radium dial workers, it had been generally assumed that Marie Curie, too, had been poisoned from her work with radium—the radioisotope that had won her a Nobel Prize and started the whole commercial radioactivity craze. Then, in 1995, her coffin was exhumed and transferred to the Pantheon, the national mausoleum of France. To be interred in the Pantheon is France's highest burial honor, and the event was to be attended by many French dignitaries.[48] There was some concern about the safety of the process, however, because Curie's skeleton was suspected to contain radium. So the disinterment was carried out by the French Office of Protection Against Ionizing Radiation (OPRI).

When her gravesite was opened, Curie was found to have been buried in a multilayered nested coffin. The inner coffin was made of wood, the next of lead, and the outer one again of wood. To estimate the radioactivity within, the coffins were punctured and an internal air sample taken to measure its radon content. Since radium decays to produce its progeny, radon, a radon measurement allows estimation of the parent radium level that produced it. Surprisingly, the radon levels of the coffin air suggested that the radium content of Curie's body was not high enough to have been the cause of her death. Apparently, Marie Curie had been right about her radium exposure. She never ingested or breathed radium anywhere near the levels that the radium girls had. Instead, the facts suggested that Curie's radiation ailments were likely the result of gamma and x-ray exposures coming from radiation sources outside of her body, rather than from any internal exposures due to ingestion of radioactivity, just as she herself had suspected.[49]

WOMEN AND CHILDREN FIRST

Despite the fact that a daily dose limit for x-rays based on skin reddening had already been set, by 1931 there was still no dose limit set for radium ingestion. Then, in June 1931, a preliminary report on the health of radium dial painters employed after January 1, 1927— the date that lip pointing had been banned—was presented by a team of US Public Health Service scientists at an American Medical Association (AMA) conference.[50] The scientists presented all the exposure and health data that were collected from these radium dial painters who had never lip pointed. They made estimates of the amount of radium radioactivity in each worker's body and then looked for a biological change that could be used as a relevant endpoint for health effects, similar to how skin reddening had been used as a basis for setting x-ray dose limits. Despite the cessation of lip pointing, the scientists found that workers continued to accumulate radium in their bodies, albeit at a much slower rate. This was presumably due to inhalation and ingestion of the radium dust with which workplace environments were still contaminated.[51]

X-ray images of their jaw bones showed detectable changes in workers with as little as one microgram of radium in their systems; but no apparent adverse health effects were associated with these changes, and it was not clear whether dental procedures may have contributed to them. There were also slightly lower than average red blood cell and hemoglobin levels in the workers, but the differences were not statistically significant.

No radiation injuries could be detected in any radium dial painter with fewer than 10 micrograms of internalized radium. Borrowing the concept of MTD from the x-ray protection community, some scientists proposed to define a maximum permissible *body burden* for internalized radium as 10 micrograms.[52] Others thought it should be lower than 10, and some thought it should be zero. Consequently, the committee published its final report in 1933 without recommending a maximum body burden. Apparently, there was too much controversy within the committee on whether 10 micrograms or something lower should be the appropriate limit.

Remarkably, it took the threat of another world war to jostle the radiation protection community enough to finally arrive at an MTD standard for radium ingestion. In 1941, on the eve of the United States' entry into World War II, military preparedness required the production of large numbers of fluorescent dials for various pieces of military equipment. Frustrated with the lack of progress in setting a safety limit for radium, the US Navy insisted on having an MTD that they could use as a protection standard for radium dial production.[53]

Lauriston Taylor (1902–2004), a scientist who would later become the preeminent advocate for the establishment of scientifically based radiation protection standards, assembled a committee of experts under the auspices of the National Bureau of Standards (now called the National Institute on Standards and Technology, or NIST) and gave it the charge of establishing an MTD for radium.

Taylor's interest in radiation protection was personal. While calibrating x-ray exposure meters in a laboratory at the National Bureau of Standards in 1928, he was involved in a radiation accident of his own doing. He was calibrating the meters using a high intensity x-ray beam. While swapping instruments in and out of the beam, he forgot to replace a protective lead shield. He didn't notice that the shield was missing until the beam had been on for several minutes, and it wasn't clear whether or not he had been exposed to a lethal radiation dose. It was later estimated that he received about half of a lethal dose, and he ended up having no long-term health consequences from his exposure.[54] Nevertheless, Taylor was changed by the experience. He became a pioneer in the field of radiation protection, later founding the National Council on Radiation Protection and Measurements (NCRP), which exists to this day in Bethesda, Maryland, just a few miles away from NIST.[55]

One of the radium committee experts, Robley D. Evans (1908–1996)—a scientist from the Massachusetts Institute of Technology (MIT) who was well experienced with measuring radium body burdens—suggested using a "wife or daughter" standard to arrive at an acceptable number.[56] He asked his all-male committee to review all available data on the effects of ingested radium on dial painter workers and recommend an MTD they would feel comfortable letting their wife or daughter accumulate in their bodies. Starting with the 10 microgram limit that the earlier Public Health Service team had tossed around, the committee worked downward until 0.1 micrograms (i.e., 1/100th of the 10 microgram starting value) was unanimously agreed upon to be safe for people to ingest or breathe. In this way, an MTD standard for radium was born.

Thus, as World War II approached, the radiation protection field was in full stride, and the quantitative means to provide safety standards for radiation exposures had been established. Limits for x-ray exposure in the workplace were defined, and an MTD for radium was in place.[57] Progress! Or so it seemed.

MOTHER OF INVENTION: RADIATION PROTECTION STANDARDS FOR THE MANHATTAN PROJECT

Although the chosen approach for radiation protection was ostensibly logical and practical, it was not without its critics. As mentioned, the radiation MTD value was based on a relatively small number of workers with rather uncertain exposure levels. More worrisome, it relied on two major assumptions that had not been confirmed.

The first assumption was that there actually existed a maximum "tolerable" dose (MTD) for radiation. In other words, were there really radiation dose levels that the human body could tolerate with no risk of adverse health effects? The MTD was a concept that had been borrowed from chemical studies. For chemical toxins, it had long been established that there are dose levels below which the body can biologically inactivate a foreign chemical, by either metabolizing it or excreting it, and no significant health hazard should thus exist. Nonetheless, at some higher dose level, the organ systems become overwhelmed and can no longer adequately handle the burden of dealing with the chemical. Doses that exceed the body's capacity to neutralize a chemical's adverse effects are toxic, making the otherwise benign chemical a poison. This is a fundamental principle of toxicology that was first expounded by Paracelsus in the sixteenth century. (The very same Paracelsus who first described the miners' lung disease, bergsucht.) Paracelsus's most famous quote is: "The dose makes the poison." By this he meant that all chemicals become poisons at some level of dose. The trick for toxicologists is to determine the dose below which the body can safety tolerate the chemical. Unlike for chemicals, however, no proof of tolerance existed for radiation.

The second major assumption was that all radiation doses are equal in terms of biological effects. This, too, had not been verified, and there were good reasons to believe that it was not true in all cases. Evidence was already accumulating that identical doses from different types of radiation could result in different levels of biological effects. That is, the biological effects were dependent upon the type of radiation. And it was also beginning to be appreciated that factors such as dose rate, time between exposures, and other conditions of irradiation could influence biological effects. Apparently, all doses were not created equal.

The bottom line was that very little research on the biological effects of radiation had been done to this point. Although radiation was by then in widespread use in both medicine and commerce, virtually nothing was known about the mechanisms through which it produced its effects on health. Many scientists found that current state of knowledge completely unsatisfactory. Among them were the leading nuclear physicists of the day, most of whom were former students of the first radiation scientists and who now represented the next generation of atomic geniuses. They knew about the health consequences from radiation exposure suffered by their old professors, and were concerned about their own health.

The discovery of fission just prior to World War II had sent Allied and Axis nations alike on a race to create a nuclear bomb (i.e., atomic bomb). Production of a nuclear bomb would rely upon amassing an amount of uranium sufficient to reach criticality and unleash an uncontrolled chain reaction of nuclear fission that would be capable of instantaneously releasing a massive amount of energy. In response to a secret letter from Albert Einstein warning of the prospect of Axis forces obtaining this technology,[58] President Franklin D. Roosevelt initiated an enormous government project in 1939 to investigate the possibility of producing such a bomb. This secret project became known by its code name, the Manhattan Project.

The Manhattan Project recruited the best and brightest nuclear physicists of the day, sequestered them at secret locations (mostly Los Alamos, New Mexico), and sent out secret agents around the world to buy up every available quantity of uranium they could get their hands on, much of it from Africa.[59] The accumulation and concentration of radioactive materials at these quantities was unprecedented and worrisome. It made the radium levels that killed the dial painters look small. Nuclear physicist Arthur Holly Compton (1892–1962) recalled the mindset of the Manhattan Project scientists in 1942: "Our physicists became worried. They knew what had happened to the early experimenters with radioactive material. Not many of them had lived very long. They were themselves to work with materials millions of times more active than those of the earlier experimenters. What was their own life expectancy?"[60] No one could answer that question.

The United States' entry into World War II was looming.[61] The Manhattan Project couldn't wait until more radiation biology data accumulated to refine the project's safety procedures. The physicists decided to press on using the current radiation safety standards of the day, even modeling their fume hood and ventilation systems after designs that had been employed to protect the radium dial painters.[62] To cover all bases, they also planned a crash program in radiation biology research within the Manhattan Project itself, to run concurrently with the nuclear physics program. So as not to reveal its mission, the radiation biology research program was cryptically named the Chicago Health Division (CHD).

The CHD immediately recognized the inadequacy of existing radiation protection standards, which had been designed purely for routine radiation work, in safeguarding against the enormously diverse and novel exposures that workers would experience in the Manhattan Project. For one thing, there wasn't much known about acceptable tolerance levels for the known particulate radiations, let alone the highly novel particles produced through fission reactions. Furthermore, the experience of the radium girls had shown that ingested or inhaled radioisotopes could concentrate in tissues and create their own unique problems. Virtually nothing was known about whether other internalized radioisotopes would also seek out particular organs. And some of the radioisotopes that were being produced had no counterparts in nature.

Another big problem was the actual measurement of the amount of radiation to which individual personnel were exposed. What was needed was some type of instrument that was small enough to be attached to a worker's body, or placed in a pocket, to measure cumulative radiation doses throughout a normal workday. Such compact personal ("pocket") ionization chambers had become commercially available from the Victoreen Company as early as 1940, but they tended to give unreliably high readings if dropped or exposed to static electricity. Because of this problem, it was routine to carry two of them and only record the lower reading of the two. Dental x-ray films, carried in a pocket, also allowed for crude monitoring of the radiation exposures of personnel, but they were even worse when it came to reliability.[63] The CHD thus considered improvements in the technology of dose measurement (i.e., *dosimetry*) as essential to their mission, and the research team made major advances in this area.

The CHD scientists also had qualms about the concept of maximum tolerable dose, because it was still an unverified assumption. They further recognized that exposure to radiation, as measured by gas ionization, was only an imperfect estimate of the amount of energy deposited in body tissue (i.e., the dose).[64] They understood that dose, not exposure, was driving biological effects. The same exposures from different types of radiation could result in different doses. They were more interested in dose than exposure, and they worked to find better ways of measuring it.

Since human cells are approximately 98% water, CHD scientists found the amount of energy deposited in water was a much better estimate of the radiation dose to human tissues than could be achieved by measurement of energy deposited in a gas (which was the way roentgens were quantified). They defined a new dose unit called a *rad*, an acronym of *radiation absorbed dose*, which represented the amount of energy deposited in a specific mass of tissue. The rad soon replaced the roentgen as a radiation protection unit.

But they also soon found that rads from different types of radiation could differ in their tissue-damaging efficiency. To account for this, they applied weighting factors to doses from different radiation types. This allowed them to standardize the radiation doses so that they would all produce the same level of biological effect regardless of the radiation type. These new weighted units were called *rems* (an acronym for *rad equivalence in man*), and were referred to as *dose equivalent* units.

THE NEW GUY IN TOWN: MILLISIEVERTS

If you find all this business about roentgens, rads, and rems confusing, you are not alone. Even the experts have a hard time juggling them. This ever-changing metric for measuring radiation risk has been a major obstacle to conveying risk levels to an increasingly risk-attuned public. As stated previously, the dose equivalent is the only valid measure of conveying health risk information because it accounts for the different biological effects of the various types of ionizing radiations and puts them all on a level playing field. For example, the health risk of neutrons can be directly compared to the risk of x-rays using the dose equivalent unit. So the dose equivalent unit is really the only unit needed to measure health risks from ionizing radiations. The exposure and dose units can be entirely dispensed with, if the question is solely health risk.

This insight about the utility of dose equivalent units was largely the contribution of the Swedish medical physicist Rolf Maximilian Sievert (1896–1966), who pioneered the development of biologically relevant radiation measurements. In 1979, a standard international (SI) unit for dose equivalent was defined and deployed by the General Conference on Weights and Measures, the international body that sets the standards for scientific measurement units. To honor Sievert, the SI unit for dose equivalents was named after him.

The sievert (Sv) and its more popular offspring the millisievert (mSv)—1/1,000th of a sievert—are very practical units for the field of radiation protection practice because they are tied directly to health effects. Although the mSv is really just a derivative of the older working unit, the millirem (mrem),[65] it has some practical advantages over the mrem.[66] Namely, within the range of typical human radiation experience, from background to lethal levels, dose equivalents can be expressed in mSv without resorting to using any decimals. For example, the lowest whole-body dose equivalents are around 3 mSv per year for background radiation, while lethal dose equivalents are slightly in excess of 5,000 mSv. So there's no need to wander too much out of this dose equivalent range if your interest is strictly human health risks.[67]

It is hard to overstate the importance the concept of dose equivalence has had on the radiation protection field. Unfortunately, the term "dose equivalent" is cumbersome to use in everyday speech. No wonder it often gets truncated to simply "dose." In fact, we will do the same throughout the rest of this book for that very reason. Just remember, when a radiation quantity is expressed in mSv, it's a dose equivalent measurement and nothing else. And remember, you don't need to know which type of ionizing radiation it is in order to evaluate the health risks from radiation doses expressed in mSv. All those considerations are worked into the mSv unit.[68]

WANTED: MECHANISTIC INSIGHT

The early deployment of the dose equivalence concept to radiation protection practice is particularly impressive given that the scientists who developed the approach had no basic understanding of the underlying radiation biology principles involved. In fact, Robert S. Stone (1895–1966), the scientist who spearheaded the approach within the Manhattan Project's radiation protection group, lamented: "Beneath all observable effects was the mechanism by which radiations, no matter their origin, caused changes in biological material. If this mechanism could have been discovered, many of the problems would have been simpler."[69]

Although they lacked knowledge of the mechanism by which radiation damaged living tissue, the radiation protection group's conclusions and approach to the problem were exactly correct. More than 70 years later, we now know the mechanistic details of how radiation damages tissues (which we will explore in later chapters). And those mechanisms actually explain the empirical findings of the Manhattan Project scientists. Over the years, the dose equivalence concept has been refined and recalibrated. Nevertheless, the general approach has stood the test of time, supporting the fundamental principle that radiation doses, when adjusted by a biological-effectiveness weighting factor for the radiation type, are accurate predictors of human health outcomes.[70] Or more simply, as Peracelsus might have put it: "The [radiation] dose makes the poison."

The Manhattan Project scientists were also prescient in their distrust of the MTD. They decided to put aside the concept of MTD. They reasoned that until the MTD was validated as applicable to radiation effects, they would work under the premise that the lower the radiation dose the better, and they promoted the notion that no worker should receive any higher dose than was absolutely necessary to complete his or her job.[71] And for ingested and inhaled radioisotopes, their goal was zero exposure. They believed that enforcing strict industrial hygiene practices in the workplace could prevent workers from being contaminated with radioactivity from their work.

MURPHY'S LAW

Although the Manhattan Project scientists had largely solved the most pressing radiation protection problems for the world's largest and most complex radiation job sites, which included thousands of radiation workers, their protection solutions primarily addressed only the radiation hazards of routine work, in which all workers complied with safety regulations. Unfortunately, the physicists working on the bomb were pushing the envelope on what occupational activities could be considered routine. Sometimes the physicists pushed too hard, creating unique radiation hazards with consequences for themselves that no radiation protection program could have easily foreseen.

On the evening of August 21, 1945, Harry K. Daghlian Jr. (1921–1945), a Manhattan Project physicist, was working alone after hours; this was a major violation of safety protocol. He was experimenting with neutron-reflector bricks, composed of a material (tungsten carbide) that could reflect neutrons back toward their source. Reflecting neutrons was considered one possible means of inducing criticality in an otherwise subcritical mass of plutonium. Unfortunately for Daghlian, the reflection approach turned out to be correct, and he had slippery fingers. He was handling one of these bricks and accidentally dropped one on top of the plutonium core. There was a flash of blue light in the room, and Daghlian instantly knew what that meant: the core had gone critical.[72] He also knew what he needed to do and what it would mean to his life. He reached into the pile with his hand and removed the brick, thereby returning the core to subcritical. Then he called in authorities and awaited his fate.[73]

This criticality accident resulted in a burst of neutrons, as well as gamma rays, which irradiated Daghlian's entire body. It is estimated that he received a whole body dose of 5,100 mSv—more than 1,000 times a typical annual background dose. This dose resulted in severe anemia that would ultimately kill him. His hands, which had actually reached into the core to remove the brick, received a much higher dose. They soon became blistered and gangrenous, causing him excruciating pain. His team supervisor, Louis Slotin (1910–1946), sat by his bedside in the hospital every day until Daghlian finally died on September 15.[74]

Not only had Daghlian broken a standard safely rule by working alone, he had also broken a cardinal rule of core assembly: never use a method that would result in criticality if some component were dropped on it. Rather, criticality should only be tested by raising that component; then, if it slips, gravity will pull it away and not toward a critical mass situation.[75]

Regrettably, it seems that Slotin, who should have known better, did not learn from Daghlian's experience. One year later, Slotin was lowering a hemisphere of beryllium over the exact same plutonium core with the help of a screwdriver.[76] The screwdriver slipped, the beryllium dropped, and the core went critical, just like it had for Daghlian. Within just a few minutes, Slotin began vomiting—an indication that he had received a fatal dose (i.e., much higher than 5,000 mSv). He was taken to the hospital to face his destiny. He died nine days later in the same hospital room that Daghlian had.

Seven other men were working in the laboratory at the time of the accident, but only Slotin received a fatal dose. One of the seven, Alvin C. Graves (1909–1965), was also hospitalized with symptoms of radiation sickness, but he recovered and was released from hospital care after a few weeks, suggesting that he had received a whole-body dose somewhere between 2,000 and 5,000 mSv. Unfortunately, this would not be the last time that Graves was involved in a fatal nuclear accident.

FIGURE 5.3. THE LOUIS SLOTIN ACCIDENT. This photograph depicts a reenactment of the accident that killed Louis Slotin. It accurately shows the spatial arrangement of every item in the room when the accident took place. Ignoring safety rules, Slotin lowered a beryllium hemisphere over a uranium core using a screwdriver for leverage. Unfortunately, the screwdriver slipped, the beryllium hemisphere fell, and the uranium went supercritical, causing a burst of radiation to be emitted. By reenacting the circumstances of the accident, scientists determined that the whole-body radiation dose that Slotin received must have exceeded 5,000 mSv (a lethal dose). Slotin died 9 days after the accident.

HOT CHOCOLATE

Few know that the Manhattan Project was not the only secret radiation project going on in the United States during World War II. There was another one, of nearly as significant military importance, being conducted in the Radiation Laboratory at MIT. The focus of this project was on improving the deployment of radar, but it would ultimately have much more widespread influence.

Radar is an acronym for "radio detection and ranging." It is a technology that uses radio waves to detect planes and ships. It can also determine their altitude, speed, and direction of movement. It is indispensable to modern warfare for monitoring movements of enemy forces, as well as for civilian activities, such as aviation.

Radar exploits the same property of radio waves that allowed Marconi to transmit his signals across the Atlantic Ocean; that is, their tendency to bounce. Marconi's radio waves bounced off the inside layer of the atmosphere, skipping their way around the globe as a pebble skips across the surface of a pond. In addition to this skipping phenomenon, when radio waves directly hit large objects in their path, some bounce back to where they came from. Measuring the time it takes for radio waves to return to their source is the basis of radar. If you can detect the radio waves bouncing back, you can estimate the size of the object from which they bounced and, since they travel at a constant speed (i.e., the speed of light), you can calculate how far away the object is by how fast the signal returns to its source.[77]

Radar was developed simultaneously and in secrecy by different nations in the years just prior to World War II. At the core of the technology was a *magnetron* that generated microwave signals. Microwaves are radio waves with wavelengths ranging from about one meter (about one yard) to one millimeter (about 1/25 of an inch). They are "micro"—the Latin word for small or short—only with respect to other radio waves, which can be as long as an American football field. They have mega wavelengths compared to x-rays and gamma rays. The military required a lot of magnetrons to satisfy their radar needs, and they had contracted with the Raytheon Company to produce magnetrons for the Radiation Laboratory at MIT. Raytheon was charged with building as many magnetrons as possible, as fast as they could, and delivering them to MIT. But "as fast as they could" turned out to be just 17 per day, which was woefully inadequate. Enter the hero of this story, Percy Spencer.

Percy Spencer (1894–1970) was a Raytheon engineer and a world expert on radar tube design. Spencer had originally become interested in radio technology at the age of 18 when the Titanic sunk and he heard how a Marconi shipboard radio had broadcast the distress signals that brought other ships to the rescue. Shortly thereafter, he entered the US Navy to obtain training in marine radio and was sent to the Navy Radio School. When he finished his tour of duty, he got a job in the private sector, working for a company that made radio equipment for the military. In the 1920s he joined Raytheon.[78]

It was largely on the reputation of Spencer that Raytheon had gotten the magnetron contract in the first place. Spencer soon developed a way to mass-produce the magnetrons. Instead of using machined parts, he punched out smaller parts from sheet metal, just as a cookie cutter cuts cookies, and then soldered the punched out parts together to make the required magnetron components, which were then assembled to make a magnetron. This change eliminated the need for skilled machinists, who then were in short supply, and allowed unskilled workers to take over magnetron production. Eventually, 5,000 Raytheon employees were working in production, and Raytheon's magnetron output increased to 2,600 per day.

There were few concerns about health risks from radar equipment at the time, since radar tubes emitted only radio waves and not ionizing radiation. Consequently, there were no protective procedures to guard against the microwaves emitted by the radar. One day, while Spencer was working with a live radar apparatus, he noticed that a candy bar he had in his pocket had completely melted. Intrigued, he set up a contraption to do a little experiment on an egg. He focused a high intensity microwave beam on a raw egg to see what would happen. The egg exploded due to rapid heating. He extended the experiment to popcorn kernels and found he could make popcorn with the microwaves. (By now, you should be able to see where this story is going, so we'll dispense with further details.) On October 8, 1945, just two months after the atomic bombs were dropped on Japan, Raytheon lawyers filed a patent for a microwave oven. They appropriately named their first commercial product the Radarange (a contraction of Radar Range).

Fortunately for Spencer, chocolate candy bars melt at temperatures just below human body temperature, and he noticed that the candy bar had melted well before any of his tissues were burned by the microwave heat. As it turned out, the inventor of the microwave oven died of natural causes at the age of 76, apparently never suffering any burns or other known health ailments from his years of work with microwaves.

While all these developments were taking place with consumer products in occupational settings, the medical community was simultaneously developing radiation technology for its own purposes. And there was great excitement. At the same time that occupational exposures were revealing the cancer risks associated with high doses of ionizing radiation, physicians were showing, ironically, that high doses of ionizing radiation, when focused on tumors, could actually *cure* cancer. Curing cancer is something that had never been achieved before without radical and disfiguring surgery. This was the first inkling of the highly mercurial nature of radiation, and it would complicate people's views about radiation and health forevermore.

CHAPTER 6

THE HIPPOCRATIC PARADOX: RADIATION CURES CANCER

Physician, heal thyself.

—*Luke 4:23*

FOOLS RUSH IN

No one was better witness to the schizophrenic nature of radiation than Chicago physician Emil Herman Grubbe (1875–1960). He was the first to recognize that radiation might cure cancer, as well as cause it. But he learned this the hard way.

Grubbe came to the world of radiation the way many had—through the electric light bulb. At the tender age of seven, he was taken to McVicker's Theatre in Chicago to see a public demonstration of Edison's newly invented light bulb.[1] As fate would have it, by the age of 20, he was actually employed in the business of making light bulbs in partnership with an itinerant German glass blower, Albert Schmidt.

Grubbe had entered the light bulb business because of his earlier entrepreneurial interests in platinum mining. Platinum had not yet made its debut in the jewelry business, but it was used in the electronics industry, primarily for carrying electrical circuits through the walls of glass containers. Since this was precisely the circuitry requirement of a light bulb, the major market for platinum was in electric light bulb manufacturing. Seeking to expand the Chicago market for platinum, Grubbe decided to establish his own electric light bulb company.[2]

It wasn't long before Grubbe started thinking of new products that his light bulb company could manufacture, and he began, under Schmidt's urging, to subscribe to the journal *Annalen der Physiks und Chemie* (*Annals of Physics and Chemistry*) to mine for ideas. It was in one of the issues of this journal that he learned of Crookes tubes, and thought they might be a potential new product for his company. He allegedly even corresponded with William Crookes about the design, and soon Grubbe and Schmidt were making Crookes tubes.

In preparation for entering the commercial market, Grubbe and Schmidt produced a number of prototype Crookes tubes. Their approach was to empirically test different types of electrode shapes to see which gave the best performance in a trial and error process similar to what Edison had done to identify the best light bulb filament. In doing this work, Grubbe's hands became itchy, swollen, and blistered. It was about this same time (January 6 to January 19, 1896) that Grubbe first learned of Roentgen's discovery that x-rays were emitted from Crookes tubes, and he suspected these to be the cause of his hand problems.

Now it just so happened that Grubbe was a very busy guy. Not only was he running his own light bulb business, he was also studying part time to be a physician at the Hahnemann College of Medicine. When he attended school with his hands bandaged, his professors inquired about his health troubles. He told them about his work with the Crookes tubes and how he assumed that x-rays were to blame. One of the professors, Dr. John Ellis Gilman (1841–1916), remarked that if x-rays were so damaging to normal tissue, they might be effective in destroying diseased tissues such as cancerous tumors. With that remark, the field of radiation oncology was born.[3] The date was January 27, 1896—just one month after

Roentgen's publication of the discovery of x-rays (December 28, 1895).

As remarkable as it may seem, Grubbe started treating patients just two days later! Dr. Reuben Ludlam (1831–1899), another one of Grubbe's professors, referred one of his difficult breast cancer cases to his student.[4] Mrs. Rose Lee had advanced breast cancer that had returned after two surgeries. Desperate, she arrived at Grubbe's light bulb factory at 10:00 a.m. on January 29 for treatment with x-rays. Grubbe administered to her the first of what would ultimately become a total of 18 x-ray treatments. The treatments did reduce her pain. Nevertheless, she died a month later.

Other late-stage patients came and were treated and though most died shortly afterward, Grubbe was not deterred. He knew that doctors were sending him only their worst cases; those who had advanced disease and were very close to death. He hoped that by demonstrating some improvement even in these patients doctors would start sending him patients with earlier-stage disease—ones in which a significant therapeutic benefit was more likely to occur. Grubbe later recalled: "I continued to make use of x-rays for treatment purposes for several years. Most of the . . . patients referred to me were moribund and . . . many died . . . soon after I began to make x-ray applications. [Later], patients exhibiting more favorable . . . conditions arrived for treatment [and in some cases, the results] were so striking as to create quite a sensation."[5]

To put this in perspective, in 1896 there were virtually no effective medical treatments for most diseases, let alone cancer. Furthermore, there was no mechanistic understanding of disease beyond germ theory, and even for the diseases caused by germs, there were still no antibiotics to treat them.

On top of this, the physicians of the day had divided into two warring camps, the allopaths and the homeopaths, each with diametrically opposed philosophies about how disease should be treated. Of the two battling groups, the homeopaths seemed to be winning the war.

The allopaths used noxious drugs, such as arsenic and mercury, and other aggressive treatments, such as bloodletting, to drive out disease. Patients often became sicker and died because of the treatment. The basic underlying treatment philosophy was "what doesn't kill them, makes them stronger." Many did not get stronger. They simply died.

Homeopaths on the other hand were disciples of Christian Friedrich Samuel Hahnemann (1755–1843), a German physician. At one point Hahnemann was experimenting on himself to determine what effect, if any, overdoses of cinchona bark would have on a patient. The bark of the cinchona tree (actually a large shrub native to the tropical Andes) was first used in herbal medication for fever by Quechua tribes of South America. It eventually became widely used in Western medicine as the only effective remedy for malaria.[6] Hahnemann slowly increased his dose of the bark and found that he experienced symptoms that he thought mimicked the symptoms of a malarial patient. From this anecdotal experience, he made a universal conclusion about how medications worked to control disease, and invented a new branch of medicine grounded largely on two principles. The primary principle was that agents that cause specific symptoms in healthy people could be used as treatments for the same symptoms in ill patients. This contrarian approach was highly counterintuitive and, in fact, is complete nonsense. Regardless, homeopathic patients often improved, while allopathic patients typically deteriorated.

The comparative success of homeopathy is likely explained not by its first principle, but by its second: the therapeutic benefit of any agent is enhanced through dilution. Thus, many homeopaths diluted their chemical remedies to such an extent that there were only trace quantities of the agent in the doses they delivered to patients. So technically speaking, they were not treating with anything at all. Their better results over the allopaths are, therefore, best explained by the fact that while the allopaths were poisoning their sick patients, the homeopaths were just letting nature take its course, and their patients were getting better on their own.

Grubbe was being trained at a homeopathic medical school, so it is no surprise that his professors would look at a toxic agent that damages normal tissue in an otherwise healthy individual to be a potential therapeutic agent for diseased tissue. That was their mindset, given their homeopathic philosophy on therapy. It seemed obvious, to them, that x-rays should be therapeutic.

In light of the overall ineffectiveness of most therapeutics of the time, whether allopathic or homeopathic, the ability of x-rays to combat the most formidable disease—cancer—was nothing short of miraculous. No wonder Grubbe said his x-ray treatments created "quite a sensation."

Although the therapeutic effectiveness of radiation in treating cancer soon became widely accepted, the underlying biological mechanism remained a complete mystery. Some proposed that the radiation transformed the cancer cells back into normal cells. Others held that cancerous tumors resulted from bacterial or parasitic infections, and that the radiation killed the parasite or bacteria. But Grubbe had his own hypothesis.

Grubbe believed that x-ray exposures caused high levels of irritation to the tumor, which, in turn, resulted in an increase in its blood volume. The increased blood then brought large number of leukocytes (white blood cells) that then choked circulation through the tumor. Deprived of circulatory nutrition, the tumor starved to death. Remarkably, this explanation, based on an irritation hypothesis, is reminiscent of the mechanism for cancer causation proposed by Virchow. You may recall Virchow's hypothesis was that tissues suffering prolonged irritation were at risk of becoming cancerous. Thus, according to Grubbe, irritation was supposedly the underlying mechanism for both the cause and cure of cancer. Although this was a unifying mechanistic hypothesis, it still didn't adequately explain how radiation could produce two opposite biological effects. That explanation would be a long time coming.

With a lack of any validated biological mechanism, x-ray treatment of cancer needed to advance empirically, with trial and error refinements to doses, numbers of treatments, and intervals between treatments. These treatment parameters needed to be determined by each physician, based on his own experience with patients when using his particular equipment, because, as Grubbe had warned, Crookes tube x-ray outputs were not standardized. There could be very different x-ray doses between Crookes tubes even at the same voltage and current settings. Grubbe advised his fellow practitioners that they must proceed with caution when treating patients because treatment results could vary tremendously, and perhaps catastrophically, between Crookes tubes. Not only that, different tissues seemed to have very different sensitivities to radiation, although why that should be was not yet clear.

Grubbe downplayed the significance of the radiation burns that patients sustained on their skin as a result of x-ray treatment, saying that these effects were always reversible with time (just as the Curies had believed for both skin ulcers and anemia) and, in any event, severe skin burns could be effectively treated with petroleum jelly. Nevertheless, he recommended the use of a mask made of lead foil, with an opening just over the tumor, to minimize the dose to surrounding normal tissues. Yet, he made no recommendations to the physicians about protecting themselves during patient irradiation, and presumably he took none himself.

Why Grubbe decided to deliver his radiation doses to tumors a little at a time is unclear. For his very first patient he spread the dose out over 18 separate treatments spaced days apart. It may be that he was worried about overdosing the patient. By delivering the dose a little at a time (i.e., fractionating the dose), while closely monitoring the patient's response, he lessened the chance of a catastrophic overdose. But even after he had empirically determined appropriate dose levels for treatment, he still used highly fractionated doses. Possibly, he simply wanted to increase profits by charging patients by the treatment. (Grubbe was highly mercenary.) Still, it is equally likely he believed that by diluting the radiation doses he was increasing their potency for cure, since this was one of the tenets of his homeopathic training. In any event, his decision to fractionate the doses no doubt contributed immensely to his treatment successes, but it had nothing to do with homeopathy philosophy.

We now know that fractionated doses enhance a phenomenon that Grubbe had observed but couldn't explain; tumors are more sensitive than normal tissues to x-rays. The relatively high tumor sensitivity was critical to successful treatment, but nothing was known about why it occurred. An answer to this question wouldn't be found for another twenty years. When it finally arrived, it came from an unexpected source: ram testicles.

Scientists in France in the 1920s were searching for a quick and effective alternative to surgical castration (which had high morbidity and mortality rates) for sterilizing rams. They thought that radiation might do the trick, because even as early as 1903 radiation exposure of the genitals was associated with infertility in male radiation workers.[7] They found exposure of the ram testicles to x-rays worked. But when the sterilizing dose was delivered all at one time, severe irritation of the scrotal skin occurred. However, if the same dose was delivered a little at a time over several days, sterilization could be still be achieved, yet without the scrotal skin complications.[8] But why?

It was later found that rapidly dividing cells, such as the cells that produce sperm (spermatogonium), are relatively sensitive to the killing effects of radiation, and are spared just slightly by the dose being spread out over time. In contrast, slowly dividing cells, such as skin cells (keratinocytes), are less sensitive to the radiation to begin with, and their sensitivity can be even further reduced if the radiation dose is spread out over a protracted period. The net result for testicle irradiation is that fractionating the dose preferentially spares slower growing cells to the detriment of the faster growing ones. Thus, if you fractionate the dose, you can kill off sperm production without critically damaging scrotal skin.

This was also the long-sought explanation for the effectiveness of fractionated dose delivery in tumor radiation therapy. Although men may take issue with the comparison, the situation of testicles surrounded by a scrotum is not unlike the situation of a tumor surrounded by normal tissue. Tumors, like testicles, contain fast-growing cells within, and killing those cells with radiation necessarily involves delivering a dose to the surrounding normal tissue, which can be equated to the scrotum. So by fractionating the dose delivery, tumors can be "sterilized" with radiation while the normal tissue is relatively spared. This is the underlying cellular mechanism that enabled Grubbe to successfully treat tumors with fractionated radiation therapy.[9]

Grubbe consistently practiced radiation therapy throughout his professional career, while experiencing ups and downs in his health and personal life. A tumultuous marriage to an unfaithful wife ended in divorce in 1911, producing no children, and an engagement to another woman was ended by her shortly before the wedding. He carried on alone, with no family. By 1929 his poor radiation protection practices had finally caught up with him. He had multiple surgeries for a tumor on his upper lip that severely disfigured him. (He had already had his left hand amputated at the wrist earlier that year as the result of a hit-and-run car accident.) Over the following years he had more and more surgeries and amputations over various parts of his body, and he even removed 15 lesions from his own body by himself by means of electrical cauterization. (It is not clear whether he ever treated himself with radiation.) During this time his radiation therapy practice slowly dwindled. It's likely that his patients had second thoughts about subjecting themselves to radiation therapy treatments when they saw their grossly disfigured doctor. In 1948 he officially retired. By 1951, he was so badly disfigured by his multiple surgeries that his landlord asked him to vacate his apartment because his grotesque appearance was scaring away tenants. He lingered on in agony for nine more years, sometimes contemplating suicide. He finally died in 1960. According to his death certificate, he died from pneumonia while harboring multiple squamous cell carcinomas (skin cancers) with regional metastases.[10]

Grubbe's contribution to radiation therapy went unnoticed for many years, because he failed to publish his original findings. He was also a flamboyant and clownish figure with a grandiose image of himself and a colorful imagination, well known to embellish, exaggerate, and even lie when promoting his own interests. Very few people were, therefore, inclined to take his many stories about his life's escapades seriously. His claims of priority in radiation therapy, made many years later, understandably met with much skepticism. But at least with regard to the essential facts regarding his initial radiation therapy activities, independent investigation of historical records has confirmed his claims.[11] He was indeed the first to treat cancer with x-rays, and he did meet with some remarkable successes. More remarkably, he achieved success despite his adherence to flawed homeopathic concepts. His therapeutic approach was right, but for the wrong reasons. The right reasons wouldn't become apparent until radiation's target for cell killing was discovered, and that was still a long way off.

HOLD THE PHONE: BRACHYTHERAPY IS BORN

Grubbe didn't have a monopoly on cancer radiation therapy for very long. Others independently found the same things that he had, and many physicians demonstrated favorable results with both x-rays and radium exposures. (The first "cure" reported in the scientific literature was for skin cancer in 1899.) But the best results always involved tumors on the body surface. Tumors deep in the body were another matter. For these, penetration of the radiation deep into the tumor tissue was a major problem. X-rays produced by Crookes tubes were not very penetrating because they had relatively low energies. They deposited most of their dose in the overlying surface tissues while the deeper tumors were spared. Even dose fractionation could not overcome this problem.

One potential solution to the depth problem was to increase the energy of the x-rays used in therapy (i.e., shorten their wavelengths; see chapter 2) so that they would be more penetrating, and some investigators were pursuing that approach. This strategy, however, was largely dependent upon technological progress by physicists and engineers in producing the next generation of x-ray machines, and that progress would occur slowly. Current cancer patients would require an alternative, shorter-term strategy, not dependent on future advances in physics. The best alternative turned out to be the employment of radioactive sources, typically radium or radon, to irradiate tumors. The gamma rays from these radioactive sources had higher energies and were thus more penetrating than Crookes tube x-rays, thereby allowing treatment of even the deepest tumors.

Another advantage of radioactive sources was that they were quite small. They could, therefore, be used either externally, by holding the source over the area of the body containing the tumor, or internally, by placing it directly within or on the surface of the tumor.

One of the first recorded suggestions to use radium sources internally came from an unlikely person: Alexander Graham Bell (1847–1923), the inventor of the telephone. In a 1903 letter to a physician in New York City, Bell proposed his idea:

> I understand . . . that [x-rays], and the rays emitted by radium, have been found to have a marked curative effect upon external cancers, but the effects upon deep-seated cancers have not thus far proved satisfactory. It has occurred to me that one reason for the unsatisfactory nature of these latter experiments arises from the fact that the rays have been applied externally, thus having to pass through healthy tissues of various depths in order to reach the cancerous matter. The Crookes tube, from which the [x-rays] are emitted is, of course, too bulky to be admitted into the mass of a cancer, but there is no reason why a tiny fragment of radium sealed upon a fine glass ampule should not be inserted into the very heart of the cancer, thus acting directly upon the diseased material. Would it not be worthwhile making experiments along these lines?[12]

Bell was indeed on to something. Why not bring the radiation source directly to the tumor? (This therapeutic strategy is now called brachytherapy—from the Greek word *brachys*, meaning "short distance"—and is in widespread use in radiation therapy to this day.[13]) But this idea of Bell's wasn't the brainstorm that it might seem. Others had thought of it, but there was a huge obstacle that Bell failed to appreciate. Purified radium, one of the rarest materials on Earth, was extremely expensive even when obtainable. It simply wasn't an option to routinely treat patients with radium unless there was a ready, and affordable, supply. In 1904, the dean of New York Medical College (another homeopathic medical school) lamented:

> Further progress in the use of radium for curing disease will be practically impossible. . . . When Prof. Curie and other eminent European scientists are totally unable to procure desirable specimens of the substance, there is small chance of anyone else doing so. The Austrian government has positively refused to allow any more of it to leave that country for the present,[14] and there is as yet no other known source of what may be called a working supply of the element.[15]

With time, several hospitals in Europe were able to secure small amounts of radium and use it with some success in cancer treatment, but no hospitals in America had any therapeutic quantities. This was still the situation in 1908—the fateful year that Eleanor Flannery Murphy was diagnosed with uterine cancer.

Eleanor Flannery Murphy (1856–1910) was the beloved sister of prominent Pittsburgh industrialists James J. Flannery (1855–1920) and Joseph M. Flannery (1867–1920). The Flannery brothers were among the wealthiest men in the United States. Starting out in a highly successful undertaking business, they later wanted to diversify their holdings into other business ventures. So they used some of their fortune to buy the patent rights for a flexible stay bolt that was used in the manufacture of railroad locomotives at a time when the railroad industry was in its prime.[16] At some point their bolt manufacturing business was in search of better metal alloys, and they soon identified vanadium alloy steel as superior to all others. But vanadium was in short supply. This led them into the vanadium mining business. Soon they were supplying vanadium for the steel industry. Vanadium alloy steel was soon in high demand for a variety of industrial applications, not just bolts. (It was used in the lock gates of the Panama Canal, and in Ford automobile parts.) By 1908, it seemed that the Flannery brothers had the Midas touch. They were millionaires three times over, thanks to successive fortunes made in undertaking, bolt manufacturing, and vanadium mining.

When doctors told them their sister's condition was terminal, the brothers did not take the news with resignation. The only glimmer of hope offered by physicians was treatment with radiation from radium. Since purified radium for medical treatment was unavailable in the United States, Joseph Flannery immediately set sail for Europe to find some, buy it, and bring it back home for his sister's treatment. He spent months combing Europe, desperately trying to purchase radium from anyone, at any price, to save his sister; but he found little, and no one who had any was willing to sell. Disheartened, he returned home to be with his sister at her death.

After Eleanor died in 1910, Joseph vowed he would cure cancer by commercially producing radium for radiation therapy. He and his brother incorporated the Standard Chemical Company of Pittsburgh for the sole purpose of commercial radium production for medical use. Radium had recently been discovered in the carnotite ore of the Paradox Valley in Colorado and Utah.[17] The plan was to purchase this ore and extract the radium.

But the brothers found it hard to recruit financial backers. Carnotite ore contained relatively little radium, and there was no known method to commercially extract it. The brothers were, therefore, forced to invest all of their personal wealth into the project, hoping that success would restore their fortunes, and Joseph devoted his entire attention to that success. He bought mining claims, mining equipment, and a stove factory in Canonsburg, Pennsylvania, 18 miles southwest of Pittsburgh, which he converted into a radium extraction plant.[18]

By 1913 the company produced its first purified radium, but at tremendous cost. To manufacture one gram (about three aspirin tablets) of purified radium required 500 tons of carnotite ore, 500 tons of chemicals, 10 million liters (about 2.5 million gallons) of water, 1,000 tons of coal, and the labor of 150 men.[19]

Unfortunately, production costs were so great that Standard Chemical's radium prices were still too high for most American hospitals to buy any. Sadly, Standard Chemical ended up selling almost its entire radium supply to different European countries, and American hospitals continued to do without.

But one American physician, Dr. Howard Atwood Kelly (1858–1943), a gynecologic cancer surgeon at the Johns Hopkins School of Medicine, was able to obtain some of Standard Chemical's radium with the help of James S. Douglas (1837–1918).[20] A medical philanthropist, Douglas bought some and donated it to Johns Hopkins for medical research. Douglas's motivation was similar to the Flannery brothers'. He had also suffered a personal loss—a daughter to breast cancer—and was also on a mission to cure cancer with radium. Kelly used that radium with some success to treat gynecologic cancers, and he wanted more, a lot more, for his private gynecology clinic. But he wanted it at substantially reduced prices. Working with Douglas, who was also a mining engineer, Kelly lobbied the federal government to nationalize domestic sources of carnotite ore to keep it out of the hands of private entrepreneurs, who could manipulate prices. But the US Congress refused to go along.

After being rebuffed by Congress, Kelly and Douglas decided to purify their own radium. They created the not-for-profit National Radium Institute, which soon cut a deal with the US Bureau of Mines. The institute would buy the carnotite ore and the bureau would provide the technical expertise to extract the radium.[21] Whatever technology the bureau developed for radium extraction would be freely disseminated to others. Since no secrets would be kept, the free availability of the information about radium extraction technology would presumably encourage others to enter the market, thus further increasing the radium supply and driving down the cost. Moreover, the institute agreed to transfer its entire radium-processing facilities over to the government once Kelly and Douglas had satisfied their own radium needs.

Employing the bureau to develop the extraction technique proved to be a shrewd move. Its scientists developed a methodology that turned out to be much more efficient than Standard Chemical's. This allowed the National Radium Institute to produce a gram of radium from 200 tons of ore, as opposed to the 500 tons required for Standard Chemical's procedure.[22] By 1916, the insti-

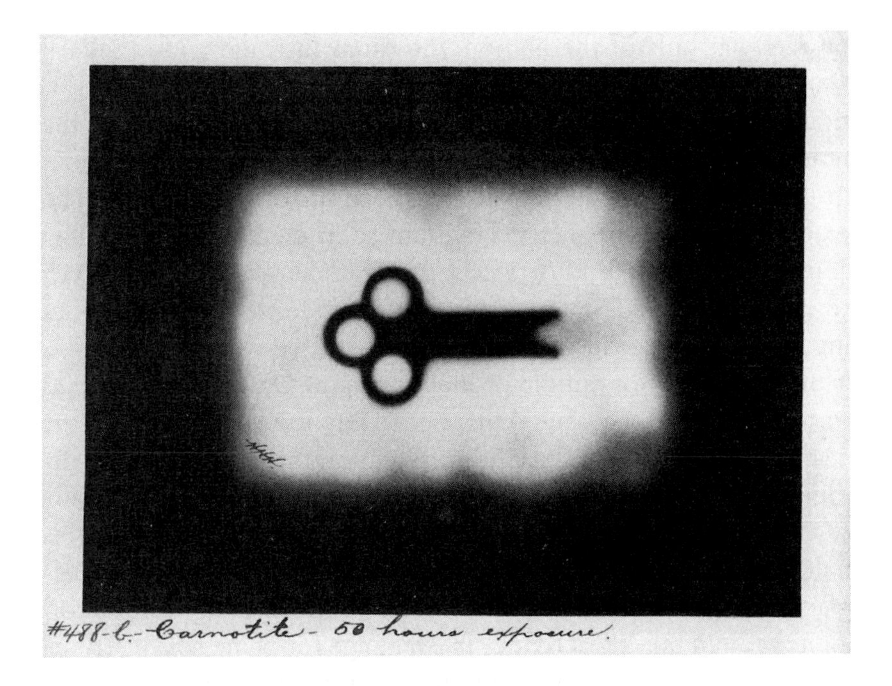

FIGURE 6.1. AUTORADIOGRAPH OF RADIUM-CONTAINING ORE. Howard Atwood Kelly used Becquerel's photographic film procedure (termed *autoradiography*) to detect and measure the presence of radium in raw ore. By comparing the ability of ores from different locations to expose photographic film over a fixed time period, he could judge the relative radium contents of different deposits. This 50-hour exposure of film to carnotite ore from the Paradox Valley of Colorado includes a clearly silhouetted image of an overlying key. The metal key blocks the ore's emitted radiation and serves as an internal negative control for the exposure. (*Source*: Photograph provided courtesy of the Alan Mason Chesney Medical Archives, The Johns Hopkins Medical Institutions)

tute had produced several grams of radium at a cost of only $40,000 ($869,000 in 2015 US dollars) per gram; less than one-third of radium's price on the world market. Once Kelly and Douglas had all the radium they needed (8.5 grams; about the mass of two US Jefferson nickels) they dissolved the institute and handed the facilities over to the US Bureau of Mines as promised. The joint venture between the federal government and the private sector had been an unqualified success, and others jumped into the business on the coat tails of the National Radium Institute and Standard Chemical. Soon, newer and more efficient extraction and purifications methods were developed. Prices then dropped further, to the point that it even became feasible to use small quantities of radium in consumer merchandise, such as in paint for watch dials. The ready availability of purified radium soon created a whole new consumer products industry based on radioactivity.[23] From 1913 to 1922 the United States dominated the radium market, producing 80% of the world's supply.

But the heyday of radium production in the United States started to wane in 1920. For one thing, both Flannery brothers died that year from the Spanish flu. But more importantly, higher-grade radium ore was discovered in the Belgian Congo, and the inferior carnotite ore from Paradox Valley simply couldn't compete. Finally, in 1922, Standard Chemical signed a contract with the Belgian producer, Radium Belge, in which the former agreed to stop all its radium mining activities in exchange for being sole distributor of Radium Belge's radium in the Western Hemisphere.[24]

Back in Baltimore, Kelly's radiation therapy practice was growing in leaps and bounds. His private clinic, at 1418 Eutaw Place, expanded to include the neighboring dwellings (1412–1420) and became known as the Kelly Hospital.[25] It had both diagnostic and therapeutic radiology equipment for diagnosis and treatment with x-rays, and a stockpile of five grams of radium—the largest stash of purified radium in the world.[26] His hospital was the largest radiation therapy operation in the nation at the time and performed virtually all of the radiation therapy in the state of Maryland.[27]

In 1916, Kelly presented his hospital's data on radium treatment of 347 women with cancer of the uterus or vagina to the American Gynecological Society.[28] Kelly described the impressive results: "The most remarkable fact about the radium treatment . . . was that it often cleared up cases which had [spread all the way] to the pelvic wall . . . great massive cancers choking the pelvis . . . Over 20% of this remarkable group had been apparently cured." These findings by the most respected gynecological surgeon of the age made brachytherapy an overnight sensation and the treatment of choice for most gynecological cancers.

Although all of Kelly's brachytherapy procedures were commonly referred to simply as radium therapy, in reality much of his radium therapy was actually done with radon that was "milked" from his radium stockpile. Solid radium continually leaches off its progeny, radon, and this radioactive gas was collected using an apparatus given to Kelly by Ernest Rutherford.[29] The collected radon was encased in glass ampules that were then placed within brass capsules. It was these brass capsules (sometimes called seeds) of radon, not the parent radium itself, that were typically positioned in and around the tumors.[30] When the radon in the ampules had decayed away to the point they were no longer useful for treatment, the ampules were discarded.[31]

Despite the fact that Kelly was well aware of fractionated radiation therapy, and he readily admitted that smaller quantities of radium introduced over a longer period of time might prove advantageous, he was not a big fan of it. His reasons are not clear. Perhaps he felt he could treat more patients overall if he used a single treatment for each patient, or maybe his surgical training predisposed him to single-intervention procedures. Whatever his reasons, we might expect that his therapeutic results with brachytherapy may have been even more spectacular had he adopted fractionated brachytherapy as the norm, just as Grubbe had for x-rays treatments.

Even though Kelly worked almost exclusively on gynecological cancers, he by no means thought that radiation therapy's utility was limited to such tumors. He was prescient in his prediction that brachytherapy of prostate cancer might prove to be a particularly promising field. (In 2012, about half of all prostate cancers in the United States were treated with brachytherapy.) He also recommended irradiating neighboring nondiseased lymph nodes when treating Hodgkin's lymphoma, to quell the spread of the disease. As we'll see, this recommendation would prove to be visionary.

Notwithstanding the spectacular successes in his professional career, all was not smooth sailing for Kelly. In particular, his use of single high-dose treatments rather than fractionated treatments ended up getting him in some trouble. In late 1913, New Jersey Congressman Robert Gunn Bremner (1874–1914) came to him for treatment of a rapidly growing tumor in his shoulder. Congressman Bremner had been referred to Kelly by fellow New Jerseyan President Woodrow Wilson, who had heard about Kelly's work with radiation therapy. By the time Bremner ultimately arrived at Kelly's clinic, however, the tumor had reached a massive size, simultaneously growing in both directions, down the front as well as the back of the shoulder, and nearly meeting under the armpit. On Christmas Day, 1913, Kelly surgically inserted 11 radioactive ampules into the tumor. High doses of cocaine were also administered to reduce the surgical pain, and the ampules were left in place for 12 hours. The *New York Times* reported on the treatment the next day and noted it was "the largest number of ampules and the most expensive quantity of radium ever used in a single operation." It touted the treatment as "one of the most important that has been performed in this country and, if successful in producing improvement or cure, will mark a notable [advancement] in the treatment of cancer."[32]

But the accolades were premature. The congressman quickly took a turn for the worse, having apparently been overdosed with radiation. When Bremner died in Kelly Hospital a few weeks later, Kelly was accused of being a quack and summoned to appear before the Maryland State Medical Society to explain himself.[33] Kelly left for Europe to escape testifying in the kangaroo court, and made the rounds of various European radium experts while he waited for the fury to die down at home. He would later claim that the medical profession "wreaked its vengeance" upon him as "an innovator."[34]

Kelly soon resumed practicing radiation therapy at his hospital. But then, in 1919, The Johns Hopkins University instituted a new policy under which all of the medical school's faculty members were required to be full-time employees of the Johns Hopkins Hospital. Kelly reluctantly resigned his faculty position in favor of maintaining his own private hospital, which he continued to operate for another 20 years.

During all his years of radiation therapy, Kelly was never known to have suffered any health consequences from radiation exposure. He was well versed in the fundamental concepts of radiation physics, and he had consulted with many of the early European nuclear physicists, including Rutherford, about how best to handle radioactivity. Kelly always kept his personal radiation doses to a minimum by working quickly, using long forceps to pick up radium sources, and protecting himself behind a lead barrier as much as possible. Today we recognize these safety practices as the fundamental tools of radiation protection practitioners (known as *health physicists*), who routinely minimize radiation doses to people by means of (1) reducing exposure time to the radiation source, (2) increasing the distance from the source, and (3) shielding the source from the body.

Kelly enjoyed good health and continued working in radiation therapy into his 80s. Then, in 1943, he contracted pneumonia and was hospitalized at Union Memorial Hospital in Baltimore. He died on January 12, at the age of 84. His wife, Laetitia, to whom he had been married for 53 years, and with whom he fathered nine children, was hospitalized at the same time. She died in the next room, six hours later.

ROLLING OUT THE BIG GUNS

During the heyday of radium therapy, x-ray therapy for cancer had taken a back seat. The cumbersome Crookes tube, with its poorly penetrating x-rays, was no match for radium's penetrating gamma rays. But all that was about to change. The physicists had a new toy, and the physicians wanted to play with it.

As we have already seen, the Cavendish scientists had produced a "machine gun" that shot streams of highly accelerated atomic particles, and they used it to split the atom in 1932. This instrument was called a *linear accelerator*, or simply a *linac* (pronounced: *'li-ˌnak*),[35] because it accelerated the particles in a straight line toward their target. Conceivably, a similar instrument could be used to accelerate electrons at high voltages and, thereby, produce x-rays with higher energies than those that came from Crookes tubes (by this time, more commonly called *x-ray tubes*). But a linear accelerator was not an instrument that could routinely be used in a medical setting to treat patients, largely because of the difficulty in maintaining stable high voltages.

All this changed in the late 1930s when physicist Charles Christian Lauritsen (1892–1968) of the California Institute of Technology (Cal Tech) built a sophisticated 750,000-volt transformer for use in his linac research. Lauritsen recognized that the high-voltage transformer might allow medical applications for linacs, so he contacted acclaimed Californian radiation oncologist Albert Soiland (1873–1946), who had a clinic nearby.[36] In a scene reminiscent of Grubbe's first x-ray treatment of advanced breast cancer, Soiland brought one of his patients over to Lauristen's laboratory for treatment. The man had advanced rectal cancer. The cancer was not amenable to treatment with standard x-ray therapy because the lower energy x-rays would have burned his anus and scrotum, so the man's prospects were quite grim. Unlike Grubbe's first patient who died despite treatment, Soiland's patient was apparently cured. Shortly following treatment with linac x-rays, the tumor shrank away and the patient sustained only minor skin reactions. Two years later, the patient had just a small lesion where the tumor had been.[37] The medical community took note.

Unlike the Crookes tube, however, which was cheap and readily available worldwide when its medical applications were first reported, a linac was an exotic instrument with an enormous price tag. Also, at the time, the radioisotope cobalt-60 had largely replaced radium as the radioisotope of choice for external beam radiation therapy.[38] It had many physical and practical advantages over radium, including affordability. Most physicians were quite satisfied with its clinical performance, and they were very comfortable using it. These old dogs were not much interested in the linac's new tricks, unless linacs became available at prices comparable to cobalt-60 machines.

In the case of radium, we saw that medical needs drove radium-refining technology to the point that prices dropped substantially, making it financially feasible to even use radium in military equipment such as watch and instrument dials. In the case of linacs, the opposite was the case. Linac-related technology found utility in World War II, and military research ultimately drove down manufacturing costs to the point where it was feasible to use linacs for medical purposes.

Specifically, it was radar research that ultimately produced the technological bridge that made medical linacs feasible. As we've seen, production of radar equipment was critical to the United States military during World War II, and there was a major need to produce a small radar device that could produce microwaves of very high power. Initially, a microwave radio transmitter called a *klystron* was invented for this purpose. (It was ultimately replaced by the magnetron that Percy was instrumental in mass producing.) After the war, the klystron technology was applied to linacs. The approach was to use a waveguide—a copper pipe into which electrons were injected—and push the electrons down with microwaves emitted from the klystron. This is analogous to a surfer being pushed along by an ocean wave. The result was a compact and relatively inexpensive linac with electrons traveling near the speed of light, with energies equivalent to 6 million electron volts. At a price tag of around $150,000 ($1,500,000 in 2015 dollars), it would still be a long time before most radiation oncologists would feel the need to have one in their clinic. The radiation oncologists would need to be impressed by what these fancy new toys could do for patients before they were going to invest that kind of money.

Just as Grubbe knew that the true value of x-ray therapy for cancer would not be appreciated until it was specifically deployed in the patients most likely to benefit (i.e., the early-stage breast cancer patients), the linac needed a patient population in which to show its stuff. Radiologist Henry Kaplan (1918–1984) thought that Hodgkin's disease victims might represent such a population. Kaplan understood better than most that the most critical factor determining success of cancer therapy was accurately matching the patient to the appropriate treatment. That is, only by understanding the underlying biology of a disease can you most effectively prescribe the treatment. And he thought he understood Hodgkin's disease biology well enough to match it with a perfectly suited treatment, and that happened to be linac therapy.

Hodgkin's disease is a form of cancer of the lymph nodes that most commonly strikes young men.[39] It was discovered in 1832 by British pathologist Thomas Hodgkin (1798–1866). He had collected a series of cadavers of young men who had died of a strange disease characterized by enlargement of the lymph nodes in the chest. He did not know it was a type of cancer. Hodgkin presented his finding of this "new" disease to his colleagues and published a paper, thereby recording his primacy in its discovery. But since the disease was uniformly lethal and Hodgkin did not recommend a treatment, his report was largely ignored until 1898. Then, Austrian pathologist Carl Sternberg (1872–1935) was able to show that the affected nodes actually contained cancerous lymphocytes (a lymphocyte is a type of white blood cell), thus identifying the disease as another form of lymphatic cancer.

Hodgkin's disease has an unusual and distinct disease progression pattern.[40] Instead of randomly metastasizing to distant lymph nodes, as cancer so often does, it progressively spreads along a chain of lymph nodes in linear sequence. This makes it easier for physicians to precisely define the limits of the disease in any particular patient. Yet, the Hodgkin's disease lymph nodes are often deep in the chest; this meant that lower-energy x-rays, which are poorly penetrating, were ineffective. It was these combined characteristics of Hodgkin's disease—deep tumors with well-defined locations—that would prove to be the game changer for the linac.

Kaplan felt that Hodgkin's disease was highly amenable to linac radiation therapy because it is largely a localized disease and radiation therapy is a local treatment. That is, radiation can only eliminate the cancer localized within the treatment beam. If the cancer has spread to regions outside the radiation beam, the therapy would produce temporary local control, but metastases (i.e., sites of cancer that have spread far from the primary tumor) would ultimately cause treatment failure. Since Hodgkin's disease is usually localized, Kaplan reasoned that it should be curable. But it would require the higher-energy penetrating x-rays from a linac to do the trick.

Kaplan capitalized on prior research by Swiss radiologist René Gilbert (1892–1962) and Canadian surgeon Mildred Vera Peters (1911–1993) that suggested Hodgkin's disease responded well to radiation therapy if the treatment field was extended to include both diseased and neighboring lymph nodes. Complete cures, however, were still elusive due to the limited penetration of conventional x-rays. Kaplan reasoned that if he combined penetrating linac x-rays with Gilbert and Peters' treatment approach, he would be able to produce cures, provided he chose patients with highly localized disease.

Kaplan understood that poor staging—the practice of assigning patients to treatment groups based on the stage of their disease—would result in the erroneous inclusion of more advanced-stage patients among the early-stage patients that he intended to treat with the linac. Since advance-stage patients with widespread disease had no hope of a curative benefit from localized radiation therapy, their inclusion would decrease the apparent cure rate of the linac. Consequently, Kaplan took special pains in meticulously staging his patients. He went as far as doing surgical explorations with node biopsies before concluding that the patient's disease was truly localized and, therefore, amenable to curative linac radiation therapy.

Although this may sound like Kaplan was simply stacking the deck in favor of his chosen treatment, the fact is that cancer is a very diverse disease, and treatment options are highly variable as well. Mismatching patients with treatments does no one any good. Since the curative potential of radiation is restricted to the location of the radiation beam, the real issue was not whether a particular treatment was good or bad, but whether that treatment had been tried in the right patients. No cancer treatment could be all things to all patients. Kaplan understood this well, but it would take years for many of his clinical colleagues to come around to this realization.

Kaplan began a clinical trial with his well-defined early-stage Hodgkin's disease patients and soon found that the linac x-rays were capable of effecting nearly miraculous cures. Within a relatively short time, Kaplan demonstrated that Hodgkin's disease could be consistently cured with linac x-rays. Soon, 50% of Hodgkin's disease patients were being cured of their disease, thanks to the linac. As of 2010, 90% are being cured. (The more advanced cases are, of course, less curable, but even for advanced disease, current cure rates now stand at about 65%.[41])

The concept of matching subsets of cancer patients with distinct treatments that target their specific form of disease is now fully accepted as a powerful therapeutic approach, but Kaplan was among the first to fully appreciate this. As such, he was able to reveal the true value of linac x-rays in curing cancer.

Today, new technologies for defining cancer patient subsets, based on genetic traits and other molecular and cellular factors, are becoming available at a rapid pace. These advances will continue to improve medicine's ability to define and characterize individual cancers in a highly sophisticated fashion that goes well beyond simply identifying the organ or tissue from which it arose. Significant improvement in outcomes can be expected if such efforts result in better matches between patients and treatments, even if the repertoire of current treatments doesn't substantially change. But radiation treatment options are continuing to expand.

Linacs are just the tip of the iceberg. The modern radiation oncologist is armed with a whole host of radiation-producing machines that can be selectively used to suit the circumstances of the disease and the patient being treated. Regardless of the nature of the specific machine, all are designed to better place a radiation dose directly within the tumor, and kill as many tumor cells as possible while sparing normal tissue. The challenge now is how best to use these powerful tools and in exactly which patients.

Physicians Grubbe, Kelly, Soiland, and Kaplan, as well as the physicists that helped them, were true pioneers of radiation therapy for cancer, and they met with remarkable successes, including some of the first cures for multiple types of cancer.[42] From their time to the present, there have been great advancements in both our understanding of tumor radiation biology and the technology used to deliver therapeutic radiation exactly where it needs to be. Radiation therapy for cancer has progressed precisely because radiation physics and medicine have advanced hand in hand.

Why aren't all cancers curable with radiation? If cancers were always localized, radiation would be able to cure most of them. Unfortunately, cancer often spreads around the body, making radiation therapy an endless task of finding the cancer and irradiating it before it does damage. The more likely a cancer will spread, the less likely that radiation can produce a cure. In the case of cancer that has spread, radiation therapy needs to be combined with some type of chemotherapy that can circulate through the body and kill distant metastases before they even become clinically apparent.[43] Modern cancer therapy amounts to orchestrating an intricate dance between radiation therapy, chemotherapy, and surgery to find the optimal combined-treatment strategy for each individual patient; this treatment strategy is based on both the extent of the disease and the biology of the tumor.

But even for the cancers that radiation can't cure, radiation therapy often has an important function. It can shrink the mass of tumors, thus stalling the disease, and, as Grubbe even saw with his very first patient, radiation can also relieve pain.[44] So radiation therapy often plays a major role in cancer treatment even when it can't cure. Currently, nearly two-thirds of all cancer patients receive radiation therapy at some point during their treatment.

Cancer patients receiving radiation to cure their cancer can experience side effects due to unavoidable irradiation of their normal tissue. Sometimes the side effects are mild, such as skin irritation.

Sometimes the side effects are more severe, such as nerve damage. These side effects occur because some normal cells are innocent bystanders that are killed along with the cancer cells. When radiation doses are high enough to kill cells, there is always some risk of such complications. But most often these complications of treatment are localized because the radiation dose is localized, and they usually can be mitigated with medication. Frequently, the side effects go away completely with time. Unfortunately, some linger permanently and are the regrettable price that patients must sometimes pay for the cure of their cancer.

Most practicing radiation oncologists, however, will never see any of the severe systemic illnesses that radiation can cause (collectively called radiation sickness) because this result is typically produced only when the entire body is irradiated to doses high enough to kill cells (i.e., more than 1,000 mSv to the whole body). On rare occasion, radiation sickness occurs because of some catastrophic accident, such as Daghlian suffered when he dropped the reflector brick on the uranium core during the Manhattan Project, and irradiated himself to a fatal dose. But it wasn't until the atomic bombs were dropped on Japan, in 1945, that the medical community saw large numbers of people with whole body doses high enough to cause radiation sickness. And then there would be even more lessons learned about how radiation affects health.

CHAPTER 7

LOCATION, LOCATION, LOCATION: RADIATION SICKNESS

We will now discuss in a little more detail
the struggle for existence.

—*Charles Darwin,* On the Origin of Species

155 DEGREES

It's no small feat to drop an atomic bomb from an airplane and not fry your own ass in the process. Yet this was the charge that Lieutenant Colonel Paul Warfield Tibbets (1915–2007) was given. He was to pilot a B-29 atomic bombing mission over a yet to be determined Japanese city, and return the plane and crew safely home.[1]

Tibbets's task was to design and fly a safe mission. But what is considered safe? The US airmen were not kamikaze pilots. The odds needed to be in the favor of the plane and its crew for a mission to get the command to go ahead. How much in their favor? During World War II, a mission with a risk level of no worse than one in ten (i.e., no more than a 10% risk of plane loss) was thought to be safe.[2] This would be considered a high level of risk under most circumstances, but not during a war operation.[3] Of course, a precise risk calculation for the very first atomic bombing mission was not possible. But Tibbets had many successful bombing missions under his belt from his prior service in the European and North African war theaters, so any flight plan in his hands typically fell well within the one-in-ten risk line. Although he couldn't precisely measure risk, Tibbets knew how to minimize it.

Tibbets, an expert on B-29s, knew them to be mercurial beasts. At lower altitudes their performance was spectacular. At high altitudes, however, their engines tended to overheat during long trips, and their steering suffered when overloaded. To deal with this, he planned to fly most of the distance between his base on Tinian Island in the South Pacific and the bombsite in Japan at the relatively low altitude of 9,000 feet (2,700 meters), and then climb to 30,000 feet (9,000 meters) for the final approach.[4] Thirty-thousand feet was a compromise altitude—high enough to minimize the probability of getting shot down by ground artillery, but not so high that the thinness of the air interfered with control of the plane's flight during the bomb drop. Tibbets also planned to strip the plane of all armaments but its tail guns, reducing the plane's weight by about 7,000 pounds (3,000 kg), thereby allowing the plane to better accommodate the added weight of the 9,000-pound (4,000 kg) atomic bomb. The weight reduction would also increase the plane's flight speed. Tibbets similarly knew that the lighter weight would improve the plane's maneuverability and, most importantly, tighten its turning radius. Since there was no way an overweight B-29 could win an all-out dogfight anyway, they might as well dispense with the heavy guns to increase the flight speed and let the tail gunner cover the plane's rear as the pilot tried to outrun its attackers.

Still missing from Tibbets's bombing plan was a strategy for a safe return to base. Scientists had told him that a minimum safe distance would be eight miles from the bomb's hypocenter.[5] The plane's altitude would provide some of that distance in the form of vertical height. (30,000 feet is nearly six miles of altitude.) But the rest would need to be provided by horizontally fleeing the scene as fast as possible. A little trigonometry told Tibbets that, at 30,000 feet, he needed five miles of horizontal distance (ground distance) to produce a slant-line distance of eight miles between the plane and the target site. So Tibbets had to get the plane five miles (of ground distance) away from the target site as fast as possible. Allowing that the jettison of the bomb would lighten the plane by 9,000 pounds, thereby further increasing the plane's flight speed, and given the limited time he had to flee the area (83 seconds between bomb drop and detonation), he calculated the maximum ground distance he could achieve as six miles. Safe by war standards, and with a mile to spare!

Even still, attaining the required distance was going to necessitate some precision piloting, and one more calculation. Tibbets had just recently learned about $E = mc^2$ and what it meant for atomic bombs, but he wasn't too much concerned with that equation. He left $E = mc^2$ to the physicists. He had another equation to grapple with:

$$\text{Optimal Angle of Departure} = 180° - \text{cotangent}(r/d) \times 2$$

In this equation, r stands for the plane's turning radius, and d is the lateral distance of the bomb's movement once dropped; that is, d is the forward distance that the bomb would be expected to move after it was released from the plane.[6]

It might appear at first that this math should be unnecessary. Intuition suggests that the ideal piloting strategy would simply be to approach the target from upwind, using the tailwind to maximize air speed, then drop the bomb and just keep going straight at top speed. But in reality that strategy has two problems. First, by moving with the wind at high speed, there would be a price to pay in terms of accuracy in hitting the target. To maximize bombing accuracy, approaching into the wind is best. It slows the plane and gives the bombardier time enough to adjust his sites before releasing the payload. Second, dropping a bomb without changing course means that the bomb's forward momentum is carrying it directly underneath the plane for a time, literally chasing the plane and its crew as they scramble to get out of town.

The best bombing strategy would actually be to approach from downwind, drop the bomb, make a 180-degree turn on a dime, and accelerate out with the wind at your tail. The problem with this plan is that planes don't turn on a dime. For any given airspeed, they have a minimum turning radius specific to the plane model. The reality is that, after a plane releases a bomb, the bomb moves forward as the plane starts to make its turn. At some point the turning plane is broadside to the forward-moving bomb, and then the nose of the plane starts facing away from the bomb. Somewhere during the turn the plane is facing directly away from the bomb, and that's the point when the pilot needs to stop turning and hit the gas. But what precisely is that optimal turning angle? That's where the equation above comes in. It told Tibbets that the needed angle was exactly 155 degrees. So that's what he would do: make a 155-degree turn and then rev those engines.

And what was it exactly that Tibbets's plane was running from? Was it the bomb's burst of radiation, moving toward the plane at the speed of light? No, the real foe was moving much more slowly. Tibbets was outrunning the bomb's shock wave, which would be following the plane at approximately the speed of sound. It was the shock wave and not the radiation that would potentially bring the plane down, and that fact would buy him even more time to save it. The speed of sound is 768 miles per hour (1,236 kilometers per hour) not too much faster than a B-29, which had a top speed nearly half of that. So even after the blast occurred, it would take some time for the shock wave to overtake the plane, allowing Tibbets to put even more distance between himself and ground zero.

The angle of departure was critical to safe return. Everything hinged on Tibbets's ability to make a perfect 155 degree turn in a B-29 and bring the plane to top exit speed as quickly as possible. It was something that Tibbets would practice again, and again, and again, knowing his life and the lives of his crew depended on it.

Although release of radiation is an added hazard of atomic bombs, their destructive power is still mostly through percussive effects (shock waves) and incendiary action (fire). This makes them similar to conventional bombs in the way they kill people and damage property. Most buildings and other man-made structures cannot withstand a shock wave greater than 5 psi. (The term "psi" stands for pounds per square inch and is a measure of the strength of the shock wave.) This fact led to the 5 psi rule, which allows prediction of the destructive area of a bomb based on its known psi profile. Everything within the 5 psi radius from ground zero will probably be destroyed, while structures beyond 5 psi are usually spared.[7] Since collapse of buildings and flying debris are responsible for a large proportion of human deaths, the 5 psi radius typically demarcates the area of greatest human fatality as well. Fire produced by bombs is less predictable and can, of course, spread well beyond the 5 psi radius, but it typically moves downwind and is slowed by firebreaks. Radiation is a threat only for those who survive both the shock wave and the fire. But radiation's killing, like shock waves and fire, behaves according to some rules.

90 DEGREES

Terufumi Sasaki would also be making a turn that would save his life, but he didn't know this and it required no calculation on his part.[8] He made the same turn every day. Sasaki was a 25-year-old physician, recently trained at the Japanese Eastern Medical University and then assigned to the Red Cross Hospital, located on the western side of Hiroshima. To get to the hospital each morning, he first had to take a commuter train from his home, situated thirty miles northeast of Hiroshima, into the city's center. Then, he switched to a streetcar headed westward toward the outskirts of the city, where the hospital was located. This transfer between transport vehicles amounted to making approximately a 90-degree turn in his direction of travel, and it took him directly away from downtown Hiroshima.

On August 6, 1945, Sasaki completed his commute as usual and arrived at his hospital at 7:40 a.m. The hospital was located 1,650 yards (1,500 meters), or just under one mile, from the Aioi Bridge in the center of Hiroshima. The bridge had a ramp midway that gave it a distinctive T shape, making it highly distinguishable on a map (and from the air) from the many other river bridges that crisscrossed a city built across the highly pronged delta of the Ota

River. It's been said that the city's shape resembles that of a left hand.[9] With palm down and figures spread, the back of the hand would correspond to the main stem of the Ota River, which divides into the fingered waterways of the delta, with fingertips reaching out toward the sea. The T-shaped bridge lay where a wedding ring would be worn.

Sasaki quickly shuffled around the hospital as his workday began. He was headed down a hallway toward the hospital's testing laboratory with a blood specimen from a patient he had just examined. Sasaki suspected that the patient was suffering from syphilis, a disease with a poor prognosis at the time, but he wanted a lab test to confirm his diagnosis before delivering the disturbing news to the frightened man. At 8:15 a.m., when Sasaki was halfway down a windowed brick hallway, Tibbets completed his 155-degree turn and headed directly away from Hiroshima. All Sasaki would later recall of the moment was that the hallway was instantly illuminated by a brilliant flash of light. If there was any sound involved, it did not make a memorable impression.

7,200 DEGREES

As Sasaki walked down the hospital hallway, 53-year-old office worker Shigeyoshi Ikeda was just sitting down at his desk to begin his day's work at the Kansai Oil Company. His desk was located within the Hiroshima Prefecture Industrial Hall, in downtown Hiroshima. This concrete building happened to be just 175 yards (160 meters) from the Aioi Bridge. Before the sound of the bomb's explosion even had time to reach his ears, Ikeda's flesh was completely vaporized.

Ikeda's wife and his 11-year-old son went to the building shortly after the bombing in search of his body. All they found was Ikeda's skeleton, still sitting in his office chair and identifiable only by a few fragments from his pants and his wristwatch. Ikeda's wife said to the boy, "This is your father." They packed up his remains, including the watch, and took them home.[10]

Tens of thousands in downtown Hiroshima met a similar fate that morning, but few of them left any remains, unless they happened to be within a stone or concrete building—the only structures that didn't completely vanish along with the victims.

The heat from the blast was intense. The fireball at ground zero was estimated to have a temperature of 7,200°F (3,980°C), exceeding the surface temperature of even the sun (about 5,500°F; 3,037°C). It was as though a piece of a star had suddenly appeared on the surface of Earth. This fact would not escape the notice of nuclear bomb physicists.

Even at 380 yards (329 meters) from ground zero, mica was found fused to granite gravestones in a cemetery.[11] Since mica has a melting temperature of 1,650°F (900°C), the temperature must have been at least that high at 380 yards out. Being that Ikeda was located at less than half of this distance, his body must have been exposed to temperatures well in excess of 1,650°F.

The heat from the bomb directly produced fires that combined with secondary conflagrations caused by debris falling on cooking stoves and electrical wires. Fanning of flames by the high winds that followed the blast produced a firestorm that engulfed about four square miles of the city center,[12] an area that included three-quarters of the city's wartime population of about 245,000 people.[13] This roughly circular, burned area had an average radius of 1,500 yards (1,371 meters) from ground zero.

BAD NEWS COMES IN THREES

Sasaki was the only Red Cross Hospital doctor who was unhurt. He was violently thrown down. He broke his glasses and lost his slippers, but was uninjured. Not so for the rest of the hospital's staff and patients, including the syphilis patient Sasaki had left behind down the hall. He was already dead.

Of the 30 doctors at the hospital, only 6 were able to function. The nursing staff fared even worse; only 10 out of 200 were able to work.[14] This small cadre was unprepared for the 10,000 bomb victims that would descend upon the 600-bed hospital before the day was out.[15]

FIGURE 7.1. THE PEACE DOME. The Hiroshima Prefecture Industrial Hall was one of the tallest concrete buildings in downtown Hiroshima at the time of the atomic bombing. It was also one of the few buildings that had a superstructure strong enough to survive the devastation. The top photograph shows what

As Sakai and the other doctors labored among the wounded, who were quickly spilling over onto the hospital grounds, they found some very odd symptoms. Although most victims had the lacerations, contusions, and abrasions typical for a percussion bombing, there were also blindness and severe burns, even among those who had not been in the firestorm. For some patients, silhouettes of flowers or other decorative patterns on their shirts had actually been etched onto their burned skin.

It had not even occurred to the doctors that radiation might be involved, but on the second day after the bombing, the staff discovered that all stores of x-ray film in the hospital had strangely been exposed, and the reality of the situation then began to dawn on them.[16] Sasaki knew little about radiation sickness, although he did know that x-rays had caused such illness among overexposed x-ray workers. But he had never been trained to deal with this illness, and the medical skills he did have were of little use as he tried to treat radiation sickness over the next days and weeks. He simply moved from patient to patient, swabbing burns with antiseptics and bandaging what he could. But no matter what he did, people continued to die.

Sasaki was unable to interpret the underlying disease mechanisms for the strange symptoms he was seeing. We now know that the unusual eye symptoms were not caused by fire or heat of the bomb, but rather by the intense light that the bomb emitted. Although the bomb's ground zero temperature equaled the sun's, the light emitted from the atomic bomb detonation was even brighter than the sun.[17] This had been anticipated by the physicists. Tibbets's crew

(*Figure 7.1 cont.*) remained of the building just after the bombing; it towered over the surrounding rubble of buildings that did not survive. The bottom picture shows the same building as it looks today, dwarfed by the surrounding reconstruction of the city. The building has been preserved as a memorial to the atom bomb victims and is currently referred to as the Hiroshima Peace Memorial, or simply the Peace Dome.

was issued welding goggles to protect their eyes from the blast,[18] but only the tail gunner actually witnessed the flash of light because the plane, as correctly calculated by Tibbets, was facing directly away from the blast when it occurred.[19]

Although the intense visible light was blinding, it was also accompanied by UV radiation, not visible but just as damaging to eyes and skin. This UV radiation, with its wavelengths sandwiched between invisible ionizing radiation and visible light, has properties of both (see chapter 2). Like ionizing radiation, it is damaging to tissues; however, similar to visible light, it penetrates poorly, leaving all its destructive energy in surface tissues.

The story of one bomb victim vividly demonstrates the nature of UV radiation casualties. It was said that two men were riding a bus at the time of the blast, seated one behind the other. One man had his window closed and flying shards of glass cut his body badly when the shock wave hit the bus and broke the window glass. The man behind him escaped the glass injuries because his window was open. So the unhurt man assisted the bleeding one and started to carry him toward the nearest hospital, but on the way their roles reversed. The uncut man's skin started to become so badly burned that he couldn't go any further, and the bleeding man then carried the burned man the rest of the way.[20]

Although it is impossible to verify the authenticity of this story, the details ring true. Windowpane glass is quite effective in filtering out UV radiation. It is likely that the UV radiation, traveling at the speed of light, exposed the unprotected man rather than the one behind glass, but the shock wave, traveling more slowly at the speed of sound, arrived and broke the glass of the bus window seconds after the ultraviolet light had passed. The rapid appearance of the ultraviolet radiation burns suggests that the man's ultraviolet radiation dose was extremely high. And this detail seems credible as well.

Within a short time, Sasaki's hospital was filled with such burn victims, some of whom were also vomiting for no apparent reason. Most suspected that the vomiting was caused by a chemical toxin that was incorporated into the bomb, since many people reported smelling a strong "sickening" odor immediately after the blast.

The odor was actually just ozone (O_3), the product of the ionization of oxygen gas (O_2) in the atmosphere by the radiation. Ozone is produced whenever oxygen is ionized, either by radiation or electricity, and is responsible for the pungent odor one smells after electrical storms. Although some may find a strong smell of ozone nauseating, the majority of the vomiting that people experienced was actually the result of internal radiation injuries, and was simply a prelude to what would come.

For all of Sasaki's labors, most of his burn patients died. The UV radiation burns were a sure indication of full exposure to all types of bomb radiation—radiation that included not just the visible and UV type, but also the highly penetrating ionizing radiations as well. These ionizing radiations were mostly gamma rays and neutrons, and their health effects would be more insidious and delayed, but no less severe, than the surface burns. The internal radiation injuries were harder to see. Out of sight, out of mind. But just as deadly.

Over the following days, people at the hospital started dying in great numbers, and removing so many bodies became a serious problem. Workers threw the bodies of the dead onto a pyre that had been built outside the hospital. The ashes of the cremated victims were collected and saved for their families. Having no use for x-ray envelopes due to the exposure of all the x-ray films, the envelopes were used to hold the ashes of the dead. The envelopes were labeled with the person's name and stacked in piles in the main office of the hospital until relatives could be notified.

Even as the radiation deaths continued, the causes of death changed over time, and previously healthy individuals inexplicably sickened even weeks after the bombing.[21] As delayed bouts of illness overtook the surviving population, the rumor began to spread among the patients that the atomic bomb had spread some kind of slow-acting poison that would make Hiroshima unlivable for seven years.[22] But this was an idea ahead of its time.

Though it was difficult to see at the time of the bombing, Sasaki and his staff later came to realize that radiation sicknesses came in three successive waves, each with its own symptoms. Likewise, each killed in a different manner.

We now know the three waves of radiation sickness represented three subsets of victims suffering from different ranges of radiation dose.[23] The first radiation syndrome was hardest to recognize because it was mixed with the severe physical trauma and skin burns experienced by those closest to ground zero. Nevertheless, against the background of physical injuries, there was a characteristic combination of nausea, vomiting, and headache. Patients with these symptoms expired within the first three days. The exact cause of death was hard to determine because it usually was the result of a combination of both trauma and radiation injury. Still, there were surely some who, one way or another, escaped trauma and burns yet still died within hours or days of the blast. These patients were likely killed exclusively by internal radiation damage. Their location within some building at the time of the blast protected them from flying debris and burns, but only the strongest stone, brick, or concrete buildings could have shielded someone from the penetrating radiation, and few of Hiroshima's buildings were constructed of those materials.

This first radiation syndrome occurred among those who were closest to the blast and who thus received the highest radiation doses (greater than 20,000 mSv).[24] When a human body experiences whole-body radiation doses at this level, all of the body's cells, including the brain's nerve cells (neurons), begin to die. Ironically, neurons are among the most radiation resistant of human cells because they never divide; yet, when doses get high enough for even neurons to succumb, it is these resistant cells that precipitate the victim's rapid death. This is because the brain is critical in controlling all of the body's physiological functions. As the brain's neurons begin to die, the brain swells, and coma and death cannot be far behind, as all systems begin to shut down. This first syndrome, found exclusively among the very highly dosed victims, is known as the *central nervous system* (*CNS*) *syndrome*. It is a type of radiation sickness from which none recover, and death comes mercifully soon.

After the first wave of fatalities there was a lull in the storm of death at the hospital. Once the early deaths had culled out the most severely injured, a second wave of death arrived a week later. These patients suffered primarily from gastrointestinal (GI) problems, including severe diarrhea, but also hair loss and high fever. Some of the worst cases included combined symptoms of malnutrition due to poor absorption of nutrients from the intestines, abdominal bloating, dehydration, internal bleeding, and infection from bacteria entering the body through the damaged intestinal lining. There was little Sasaki could do to help these patients. By two weeks after the bombing, all were dead.

These GI symptoms can all be traced to complete shutdown of the small intestines. Although it may not be apparent at first, the small intestines share something in common with testicles. They both contain rapidly dividing cells; crypt cells and spermatogonia, respectively. We have seen how spermatogonia are preferentially sensitive to radiation damage, due to their rapid cell division. The same situation exists with the small intestine's crypt cells. But while spermatogonia give rise to sperm, crypt cells give rise to structures called *villi*.

Villi are microscopic, fingerlike projections that line the inner surface of the small intestine and give it the appearance of a long-piled carpet. This carpet of villi increases the surface area of the intestines for the absorption of food, thereby facilitating the uptake of nutrients into the bloodstream. Since these villi wear down quickly as moving food rubs away their tops, cells need to be constantly replaced from below. At the base of the villi (between the fingers) are rapidly dividing crypt cells that constantly add newly made cells; these compensate for those cells lost at the top. In this way, the height of the villi remains constant as a growing cohort of recruits replaces the older cells that are swept away. The transit time for individual cells, from the base to the top of a villus, is about eight days. Thus, the maximum lifetime of a villus cell is eight days. So when crypt cells are killed by radiation and stop replenishing the villi, the villi waste away over this eight-day period. At its end, there are no villi left, so it's not surprising that the average time to death from this syndrome is eight days.[25]

Patients suffering from villi failure are said to have the *GI syndrome* of radiation sickness. These patients received a whole-body radiation dose too low to severely damage the nondividing and, therefore, resistant brain cells. Consequently, they don't develop the CNS syndrome. Still, their doses were high enough to kill the rapidly dividing crypt cells in their GI tract. It is important to appreciate that all tissues with rapidly dividing cells are affected the same way, and lose their population of dividing cells. GI symptoms appear first, however, because of the (1) short, eight-day lifetime of the cells of the villi, and (2) critical importance of the small intestine to survival due to absorption of nutrients and provision of a barrier between fecal matter in the intestines and the body's internal organs.

As with CNS syndrome, there was no hope of recovery from GI syndrome. The GI syndrome patients soon succumbed to all of the medical problems one would expect to encounter in people who no longer had functioning small intestines. Patients who die from the GI syndrome have usually received whole-body doses less than 20,000, but more than 10,000 mSv.

After the GI syndrome patients had all passed away, Sasaki noticed another lull in deaths that lasted a couple of weeks. But then at around thirty days after the bombing, some of the surviving patients had a nearly complete loss of all types of blood cells (pancytopenia) and started to die from all of the medical complications associated with this condition. Sasaki and his colleagues soon learned that two important factors were critical to the prognosis of these patients. Either a very high and sustained fever, or a white-blood-cell count that dropped below 1,000 per microliter of blood, was bad news. One of these two symptoms meant death was highly likely, but having both spelled certain demise.

By the time the third syndrome arrived the doctors were well aware that they were dealing with radiation sickness. They treated

FIGURE 7.2. ATOMIC BOMB VICTIM. A man, suffering the symptoms of radiation sickness, rests on a straw mat at a makeshift hospital in a Hiroshima bank building.

patients with liver extracts, blood transfusions, and vitamins, all of which were thought to boost blood counts. They also found that penicillin was effective in these patients because loss of blood components made them vulnerable to infections which penicillin helped to fight.

Nevertheless, these doctors had some crude ideas about the mechanism by which radiation was unleashing its health effects on the blood. They thought that gamma rays had penetrated the victims' bodies at the time of the blast and made the phosphorus in their bones radioactive.[26] The radioactive phosphorus supposedly started irradiating the bone marrow—the blood-forming tissue of the body—by emitting beta particles, which slowly started to break the marrow down. The doctors apparently had some knowledge of the radium girls and the problems they experienced due to radium substituting for calcium in their bones, and they also knew that the health effects of overexposed x-ray workers included delayed anemia. So they likely conflated these two entirely different circumstances into a unifying hypothesis, with a common causal mechanism involving radioactive bones. Although phosphorus is a major constituent of bone, and radioactive phosphorus-32 does, in fact, emit beta particles, gamma irradiation of phosphorus does not make it radioactive. A bone biopsy and a Geiger counter—a handheld instrument that identifies radioactive contamination and produces an audible "click" for each atomic decay (disintegration) detected—could have easily shown that their hypothesis was flawed. But, apparently, the wrongheaded idea went unchallenged.

We now have the correct explanation of why these blood problems appeared so late. Bone marrow cells—the cells that make all of the blood components—divide rapidly, just like the crypt cells. Additionally, the bone marrow cells are prone to an accelerated type of cell death,[27] making them even more sensitive to radiation than the crypt cells. Nevertheless, this syndrome doesn't precede the GI syndrome because the mature blood cells have longer average lifetimes in the blood's circulation than do the villi cells; thirty versus eight days, respectively. Thus, the blood can continue its functions much longer before the lack of new recruits is noticed. Nevertheless, by thirty days, the mature cells are old and starting to die off, and the absence of replacements results in severe anemia. The symptoms Sasaki witnessed included everything that would be expected for patients with no blood cells in their circulation. Without medical intervention, death usually followed. This scenario of acute anemia caused by radiation is called the *hematopoietic* (Latin for "blood forming") *syndrome* of radiation sickness. People who suffer from this syndrome received whole-body doses insufficient to wipe out crypt cells but high enough to kill some or all of blood-forming cells residing in bone marrow (1,000 to 10,000 mSv).

The probability of dying from hematopoietic syndrome is decidedly dose dependent. Those with whole-body doses greater than 5,000 mSv will likely die within thirty days. Those with doses near, but still below, 5,000 mSv will probably recover because they usually have enough surviving cells squirreled away somewhere in their marrow to ultimately grow back and start making blood cells again. Recall that Marie Curie and her staff frequently suffered bouts of anemia, followed by apparent full recovery, suggesting that their whole-body doses were somewhere in the 1,000 to 4,000 mSv range.

Unlike CNS and GI syndromes, hematopoietic syndrome is sometimes amenable to treatment, including blood transfusions and antibiotics. The blood transfusions, in particular, may buy these patients the time they need for their own blood-producing capabilities to grow back, in about 60 to 90 days. But the prospects are again highly linked to dose. Patients receiving 4,000 to 8,000 mSv may be helped by such transfusions, but those above 8,000 mSv are not likely to recover because at the higher end of the range, the more lethal GI syndrome starts to predominate.[28] In contrast, those receiving substantially less than 4,000 mSv are highly likely to survive even without medical interventions, because normal individuals are generally able to tolerate a modest level of cell loss, and are fully capable of regenerating all or most of the lost cells.

What about Sasaki's own health? His location at 1,650 yards (1,509 meters) from ground zero, plus the fortuitous fact that he was shielded by the hallway's brick walls rather than being in front of a hallway window, likely resulted in minimizing his dose. He probably received a radiation dose well below 1,000 mSv. At doses less than 1,000 mSv, the potential for developing radiation sickness drops drastically, and below 500 mSv radiation sickness is not possible. This is because doses below 500 mSv are typically not very toxic to cells, whether rapidly dividing or not. Radiation sickness is a disease caused by the killing of large numbers of cells in vital body tissues. In the absence of cell killing, radiation sickness cannot occur. In fact, most tissues can tolerate significant losses of cells before health problems ensue. For this reason, radiation sickness has a practical starting point at about 1,000 mSv.[29]

Fortunately, it takes a lot of radiation to kill significant numbers of body cells. In addition, partial body irradiations, even in great excess of these threshold doses, are not likely to produce radiation sickness. This is because surviving cells in the nonexposed body areas can come to the rescue of the overdosed region. For example, dividing blood stem cells can enter the circulation and recolonize radiation-depleted bone marrow. Thus, the survivable doses for partial body irradiations are much higher than for a whole-body irradiation.

In summary, for any whole-body irradiation incident, whether from an atomic bombing or an accident, those coming down with radiation sickness experience one of three possible syndromes, depending upon the dose they receive. And the dose that they receive is highly dependent upon their exact location relative to the radiation's source (i.e., ground zero). Radiation dose drops off rapidly with distance.[30] For this reason, nearly all of the radiation sickness patients in Sasaki's hospital were likely to have been between 1,000 yards (915 meters) and 1,800 yards (1,645 meters) from ground zero when the bomb was dropped. Any closer than 1,000 yards, they probably would have died from the percussive effects (i.e., the 5 psi rule), and further than 1,800 yards away, they wouldn't have experienced enough radiation-induced cell death to have any radiation symptoms at all. Thus, at 1,650 yards (1,509 meters), the hospital was ideally located to witness large numbers of patients suffering from the complete spectrum of radiation syndromes.

WAGES OF WAR

By November 1945, the radiation syndromes in Hiroshima had run their course. Those destined to die had done so, and those fated to survive were on their way to recovery. It isn't clear how much influence medical intervention had in determining these outcomes, but it definitely wasn't much. There were very few medical services available to the victims since only 3 of the city's civilian hospitals out of 45 were seeing patients and both military hospitals were destroyed. It is estimated that 90% of the city's doctors and nurses had been killed or injured.[31]

Recovery from radiation sickness would not be complete for every survivor, as there were lingering side effects, including chronic and debilitating fatigue. Still, these side effects paled in comparison to the horrific disfigurements suffered by the burn survivors. So those who had survived radiation sickness without any obvious physical deformities tended to suffer in silence, thankful that they were not among the burned or dead. Nevertheless, their problems weren't over. Some of them would go on to experience radiation's late health effects, as we shall soon see. But, for the moment, they were simply glad to be alive.

No one is really sure how many people were killed in the atomic bombing of Hiroshima. The exact casualty numbers for the city are elusive, because many people left no corpses and it is always hard to account for the missing, particularly in wartime. Estimates range from 90,000 to 165,000 deaths. Although the numbers are fuzzy, it's thought that of those killed by the bomb, about 75% died from fire and trauma and 25% from radiation effects.[32] Not all were Japanese. About 50,000 inhabitants of the city (20% of the population) were Korean conscripts performing forced labor,[33] and there were even a couple of dozen American prisoners of war held in downtown Hiroshima.[34] None of these Americans survived.[35]

THE PROBLEM WITH NEUTRONS

When the Japanese government failed to immediately surrender in the wake of the Hiroshima bombing, the United Stated dropped another atomic bomb, this time on the city of Nagasaki. While the Hiroshima bomb employed uranium-235 as the fissionable material, and had a yield of 15 kilotons (equivalent to the detonation of 15,000 tons of TNT), the Nagasaki weapon used plutonium-239 and had a yield of 21 kilotons. Plutonium-239, a man-made radioactive element, was included in the bomb production plans of the Manhattan Project when it became apparent that it would not be possible to purify enough uranium-235 to satisfy the United States' atomic bomb requirements. The plutonium bomb used a different detonation device than the uranium bomb (implosion rather than gun-assembly triggering[36]), but otherwise they worked on the same fundamental principle. In each case, detonation of a conventional explosive device within a confined space propels subcritical masses of fissionable material together with sufficient force and speed to instantaneously create a single supercritical mass. This supercriticality of fissionable material results in nearly simultaneous fission of large numbers of atomic nuclei, thereby releasing massive amounts of energy all at once.

Although the Nagasaki bomb had 40% greater explosive power, the Hiroshima bomb's radiation was potentially more lethal because a higher proportion of its radiation energy was emitted in the form of neutrons, rather than gamma rays. This was due to differences in the underlying physical properties between their nuclear fuels—plutonium-239 fission (Nagasaki) and uranium-235 fission (Hiroshima).[37] Although these differences are of no consequence to the shock waves or the firestorm generated, neutrons can enhance the lethality of the bomb's radiation to humans, since fast neutrons are ten to twenty times more damaging to human tissue than gamma rays. Thus, bombs with high neutron yields can be particularly lethal.

At first it might seem strange that neutrons—those "ghost particles" that cause no direct ionization themselves and, thus, eluded discovery by ionizing radiation detectors for so many years—might be among the most lethal types of ionizing radiation. But a fast neutron's biological effects are delivered through an emissary—a fast moving proton. Neutrons are highly penetrating to most matter because they have no electric charge and, therefore, plow blindly through all the vacant space that constitutes atoms, unperturbed by the electrical forces that push and pull other types of particulate radiation. (Recall the empty baseball stadium analogy in chapter 4.) When neutrons ultimately do happen to collide with the much larger mass of a typical atomic nucleus, most bounce back; an example is how Rutherford's alpha particles bounced back from gold nuclei, leaving the neutron-impacted nucleus unfazed by the experience. Still, not all nuclei are as large as gold's. In fact, hydrogen's nucleus consists of just a single proton. Remember, a proton has exactly the same mass as the neutron, albeit positively charged; herein lies the explanation for a neutron's lethal powers.

If a neutron were the size of a ping-pong ball (i.e., a table tennis ball), the nucleus of a typical element would be about baseball size or larger. Try throwing a ping-pong ball at a baseball and see how much you can move it. Not much, eh? But hydrogen looks just like another ping-pong ball to the colliding neutron, and the high-speed neutron collision can send the sedentary proton off on its own high-speed journey; a journey that now involves a positive electrical charge ripping electrons from the molecules it encounters along the way. (Recall that it was the reported interaction of a yet unidentified type of radiation with the hydrogen atoms in paraffin that had betrayed the very existence of neutrons to Chadwick; see chapter 4.) Dislodged electrons cause ionization damage. So neutrons can indirectly ionize and damage molecules with the cooperation of hydrogen nuclei.

Biological tissues are loaded with hydrogen. Not only is the body composed mostly of water (about 55%), which has two hydrogen nuclei for every oxygen nuclei (H_2O), but also all biological molecules are comprised largely of hydrogen, and a few other elements. The bottom line is that about 65% of the atoms in the human body are hydrogen. What does this mean for the human body in terms of sensitivity to neutrons? Ironically, the human body, because of its high hydrogen content, is better at absorbing neutrons than a slab of lead. And those absorbed neutrons conspire with the proton nuclei of the hydrogen atoms to produce a tremendous amount of cellular damage. For living tissue, neutrons are bad news indeed. This was another salient fact that did not go unnoticed by nuclear bomb physicists.

After suffering two atomic bombs, and potentially facing more, the Japanese had had enough. On August 15, 1945, in a recorded radio address broadcast simultaneously to the entire nation, Emperor Hirohito (1901–1989) announced unconditional surrender and officially ended World War II. He explained to his people:

> The enemy has begun to employ a new and most cruel bomb, the power of which to do damage is, indeed, incalculable, taking the toll of many innocent lives. Should we continue to fight, it

would not only result in an ultimate collapse and obliteration of the Japanese nation, but also it would lead to the total extinction of human civilization.

It was the first time the entire Japanese population had heard the broadcast voice of their emperor, and it gave them pause. They would not hear a national broadcast from their emperor for 64 more years, when still another nuclear catastrophe would once again necessitate an imperial address.

THE FIRST CASUALTY

It's been said that the first casualty of war is the truth.[38] This was exactly the case following the bombing of Hiroshima and Nagasaki. Japanese military officials ordered Tokyo newspapers to downplay the bombings. After the bombing of Hiroshima, the Tokyo daily *Asahi* ended a general story about recent American bombing raids with the single sentence regarding Hiroshima: "It seems that some damage was caused to the city and its vicinity."[39] Furthermore, the American occupation authority, under orders from General Douglas MacArthur (1880–1964), censored all information related to the nature of the bombs, leaving the Japanese population with no information.[40] Even John Hersey's book *Hiroshima*, which described the victims' plight in graphic detail, was not published in Japanese until 1949, three years after it was published in America, and four years after the bombing. This paucity of credible information spawned some fantastic rumors among the bomb victims about the nature of the bombs. Some initially thought, because of the bomb's brightness, that magnesium (the flash power once used by portrait photographers) had been sprayed over the city and then ignited. Others thought the bomb was an enormous "bread basket"—a self-scattering cluster of conventional bombs.

Despite all the false rumors and lack of official information, most of the bomb's survivors soon learned what had actually happened to them, even if they got the details a bit muddled. One woman offered a description: "The atom bomb is the size of a matchbox. The heat of it is six thousand times that of the sun. It explodes in the air. There is some radium in it. I don't know just how it works, but when the radium is put together, it explodes."[41] That about says it all.

THE SILENT CASUALTIES

It is an odd fact that, for all the focus on radiation emitted from atomic bombs, radiation constitutes their smallest energy output. In a typical atomic bomb, fully 50% of the energy goes into the shock wave and 35% into heat, leaving only 15% emitted as radiation.[42] It is the combined shock wave and fire that account for most of the lives lost. The rings of devastation around ground zero are primarily the overlapping of the shockwave damage with the firestorm damage, leaving only people who happened to be outside this area, in the minimally damaged fringe zone, vulnerable to radiation effects. In the case of Hiroshima victims, it was largely the irradiated people from this fringe zone who lived to tell their stories. Consequently, the human experiences of the atomic bombing were related through their perspective, with eyes that mainly saw the radiation effects. We have heard nothing from the shock wave and firestorm victims who didn't live to tell their tales. Doubtless, they would have told a completely different story about how atomic bombs affect health.

Ironically, radiation emitted from ground zero becomes even less important as the kilotonnage of atomic bombs increases. At bomb sizes even slightly larger than Hiroshima's and Nagasaki's (15 and 21 kilotons, respectively), the radii for shock wave and burn deaths increase to distances beyond the radius for radiation deaths.[43] This means that every person who receives a significant radiation dose is also simultaneously cremated. It's a sad fact that, for bombs greater than 50 kilotons, no one dies from the blast radiation. But as far as the health effects of the *radioactivity* released by these king-sized nuclear bombs go, that's quite another story.

CHAPTER 8

SNOW WARNING: RADIOACTIVE FALLOUT

[My body] is radium. If I should strip off my skin the world would vanish away in a flash of flame and a puff of smoke, and the remnants of the extinguished moon would sift down through space, a mere snow-shower of gray ashes!

—The Devil, in Mark Twain's short story Sold to Satan *(1904)*

I will show you fear in a handful of dust.

—T. S. Elliot, The Waste Land *(1922)*

NEPTUNE'S WRATH

The fishing vessel *Lucky Dragon No. 5* had left its homeport of Yaizu, Japan, in January 1954, with a crew of 23 fishermen onboard. It was the beginning of a very long offshore fishing voyage that placed the boat about 80 miles (129 kilometers) east of Bikini Atoll in the central Pacific Ocean, just north of the equator, in the early morning hours of March 1. As the men worked their baited lines, the sun rose over the vast expanse of ocean to the east. But before the sun had the chance to fully brighten the sky, a flash of light appeared over the western horizon, far outshining the sun. Minutes later there was a roar louder than any thunder the crewmen had ever heard. Shortly after that it began to snow. However, the "snow" turned out to be a coarse powder that sprinkled down on their boat for hours. "We had no sense that it was dangerous. It wasn't hot. It had no odor. I took a lick; it was gritty but had no taste," recalled crew member Matashichi Ōishi.[1]

By the time the *Lucky Dragon No. 5* returned to Yaizu with its catch two weeks later (March 14), the fishermen were already experiencing the symptoms of radiation sickness.[2] Nevertheless, they managed to unload their fish and take them to market.

Once the fish were sold, the fishermen sought medical help. After hearing their story and witnessing their symptoms, it wasn't hard for doctors to diagnose radiation sickness. The story quickly hit the news media, and the radioactive fishermen were the front-page news of every major newspaper in the world.

When word got out about the radioactive fishing boat, the catch had already reached the regional fish markets. Yashushi Nishiwaki, a young biophysicist living in Osaka, heard the news, grabbed his Geiger counter, and headed down to his local fish market. All the tuna in the market were heavily contaminated with radioactivity. The news about contaminated tuna in the outlying markets hit the press and widespread panic ensued, as public health officials tried to retrieve all the contaminated fish from the various markets.[3] Meanwhile, local fish sellers bought up all the available Geiger counters so that they could demonstrate to their customers that their particular catch was not contaminated and was safe to eat. But it was a hard sell. Then fish sales collapsed totally when it was reported that Emperor Hirohito had stopped eating fish. With no customers, the Tokyo fish market was forced to close until all the radioactive fish were recovered.[4]

What the fishing crew had inadvertently witnessed was the United States' second test of a new breed of nuclear weapon that was based on fusion of atomic nuclei, rather than fission. This new type of bomb was called a *hydrogen bomb*.[5] The name comes from the fusion reaction between isotopes of hydrogen, which is the source of the bomb's energy.[6] Problematically, hydrogen atoms don't like to fuse, so it takes a lot of input energy to force them together. Nevertheless, once they do fuse, the output energy they then release is enormous, much greater than the input amount. The net result is a massive explosion.

The energy barrier for fusing atoms is so great that fusion does not occur naturally on Earth. Only the sun has high enough temperatures for fusion to take place. But nuclear physicists had noticed that fission bombs, like the ones dropped on Hiroshima and Nagasaki, produced temperatures rivaling the sun's. In that fact, they saw an opportunity to use fission bombs to trigger fusion bombs. So a hydrogen bomb is really a hybrid explosive in which a fission explosion kindles a fusion explosion.

Fusion bombs release so much more blast energy than fission bombs that it's hard to comprehend. If the energy of the atomic (fission) bomb detonated over Hiroshima was likened to one page of this book, a hydrogen bomb's energy would equate to eight full books. For a damage comparison, if a fission bomb like Hiroshima's (13 kilotons) were dropped on the Capitol Building in Washington, DC, it's likely that the White House would be spared severe damage. In contrast, a large hydrogen bomb (50,000 kilotons) dropped on the Capitol Building would take out all of Washington, DC, and the neighboring city of Baltimore as well.[7] If accompanied by winds from the southwest, lethal radioactivity would blanket Philadelphia, New York, Boston, and possibly Bangor, Maine.

Far from being secret tests, much of the world's populace had been very familiar with the nuclear bomb testing occurring at Bikini Atoll ever since the first exhibitions of atomic bomb power took place there eight years earlier. Just as Edison had publicly felled poor Topsy the elephant to indict AC electrical current, and as Rutherford had showcased the dangers of electrical coronas by blowing up a hapless rat, the US military felt it also needed to publicly demonstrate the destructive power of its newly acquired nuclear weaponry. During the summer of 1946, before throngs of reporters, US politicians, and representatives from all the major governments of the world, a 95-vessel fleet of obsolete warships, including American, German, and Japanese vessels, was demolished with an airdropped atomic bomb (June 30), followed a few weeks later by an underwater detonation (July 24).[8]

Since 1946, there had been a total of 11 nuclear bomb tests at Bikini and its neighboring atoll Enewetak,[9] including ten atomic (fission) bombs and one previous hydrogen (fusion) bomb.[10] The March 1, 1954, bomb test that the fishermen had witnessed was the twelfth and largest test of any nuclear bomb to date—a massive hydrogen bomb facetiously nicknamed "Shrimp." Shrimp delivered a much grander spectacle than the elephant, the rat, or, for that matter, any prior nuclear detonation.[11]

Complicating the fishermen's medical treatment was the fact that, despite repeatedly bathing, they were still heavily contaminated with radioactivity. It was too much for the local hospital to handle, so a plan was hatched to send the patients to Tokyo hospitals, but no commercial trains or planes would transport them. To break the impasse, the United States military sent two C-54 warplanes to Japan to fly the fishermen to Tokyo, where Tokyo University Hospital doctors awaited their arrival.

The men were admitted to an isolation ward, where they would remain for one year. At first, their main problems were surface burns where the radioactivity had touched their skin. The doctors initially weren't sure what other health problems to expect since the radiation doses that the men had received were unknown. There were no significant gastrointestinal (GI) symptoms, so whole body doses over 5,000 mSv were ruled out. But then, after a couple of weeks, the men's blood cell counts started to drop precipitously, as would be the case for people receiving doses over 2,000 mSv. Some even had white cell counts below 1,000 per microliter—the threshold that Dr. Sasaki had found to be a portent of death for Hiroshima bomb victims.

Recognizing the sensitivity of the spermatogonia cells of the testicles to radiation, the doctors measured the men's sperm counts to get yet another inference as to the radiation dose levels they might have received. Sperm levels were very low, with some men having no detectible sperm at all, suggesting they received doses greater than 3,000 mSv.

It was apparent that the men were all suffering from the hematopoietic syndrome of radiation sickness, so treatment consisted largely of antibiotics, vitamins, and blood transfusions to ward off death until their bodies could recover their blood producing capabilities. Still, based on the clinical findings, Dr. Masao Tsuzuki of Tokyo University Hospital was not optimistic. He indiscreetly divulged to the inquisitive newspaper journalists his concern that as many as "10% of the 23 crew members may die."[12] As it turned out, all but one survived. On September 23, Aikichi Kuboyama, the boat's radio operator, died.[13] Although his immediate cause of death was ruled liver failure, most likely due to the hepatitis infection he had contracted from contaminated blood transfusions he was known to have received in the hospital, the media asserted it was radiation exposure that had destroyed his liver. But radiation-induced death was unlikely, since the liver is a fairly radiation resistant organ and does not contribute significantly to any of the three lethal radiation illnesses (i.e., GI, CNS, and hematopoietic syndromes, as discussed in chapter 7).

Although there were no direct measurements of the doses the fishermen received, the clinical findings alone told a fairly accurate story. The symptoms (anemia and low sperm counts, but no GI symptoms) suggested a whole body dose range from 3,000 to 5,000 mSv. In contrast, the skin burns the fishermen experienced were not a good indicator of internal doses, because radioactive fallout delivers much of its dose through beta particle emissions; these have limited ability to penetrate into internal organs. Thus, it would be a mistake to gauge a fallout victim's whole-body dose based sole on their skin burns, since exposed skin receives a disproportionately higher dose. This may not have been fully understood by the physicians. Radioactivity raining down from the sky was a new twist the doctors weren't familiar with. Medical experience with high levels of radioactivity thus far had been primarily limited to ingested radium, not nuclear bomb radioisotopes.

This Japanese fishermen incident was very different from the previous Japanese experience with atomic bombs. In Hiroshima and

Nagasaki, all the radiation sickness was caused by exposure to blast radiation coming from ground zero. Fallout hadn't been a significant health threat. Now, fallout was the main and only health issue. The fishermen had received no radiation dose from the blast itself.

Fallout is the term used to describe radioactivity that settles to Earth's surface from the sky. Most frequently it is caused by the detonation of a nuclear device that spews various radioactive isotopes up into the atmosphere where they combine with bomb debris, dust, or even just water vapor. Sooner or later, it all settles back down to Earth.

At Hiroshima and Nagasaki, the local fallout was minimal because both bombs were detonated above ground level (1,980 and 1,540 feet, respectively).[14] Had the bombs been detonated at ground level, they would have kicked up much larger dust clouds and fallout would have been a far greater problem. Instead, the high altitude detonation dispersed the radioactivity into the upper atmosphere where it lingered, thus allowing some time for it to decay, dilute, and dissipate before settling mostly over the Pacific Ocean and away from people. As it turned out, radioactivity readings taken near ground zero just three days following the Hiroshima bombing showed minimal fallout, and it was safe for rescue workers to immediately enter the city and for residents to return to the area. Had significant fallout been present, rescue and reentry might have been considerably postponed until the radioactivity decayed away to less dangerous levels.

BLAME IT ON THE WEATHER

"The wind failed to follow the predictions" is the way Lewis Lichtenstein Strauss (1896–1974), chairman of the Atomic Energy Commission, explained how the crew of a fishing vessel outside the restricted zone received toxic doses of radioactivity from a US nuclear bomb test. He neglected to mention that the bomb also released more than twice the energy that scientists had predicted: 15,000 KT (15 MT) of TNT, rather than the anticipated 6,000 KT (6 MT). The stronger than expected blast was caused by an error in the theoretical yield calculation made by physicists at Los Alamos National Laboratory. But how could these genius scientists make such a basic mistake as to miscalculate the explosive yield of the fuel? Had Einstein's $E = mc^2$ failed them?

The fuel of hydrogen bombs is lithium deuteride. (Deuterium is another name for the nonradioactive isotope, hydrogen-2, which has one proton and one neutron.[15]) The physicists believed, based on theoretical considerations, that only the lithium-6 isotope of lithium deuteride could support a fusion reaction,[16] and that the lithium-7 isotope (approximately 60% of the lithium content) was inert (i.e., an inactive ingredient). They were wrong. The lithium-7 also contributed to the fusion reaction making the bomb's explosive yield more than twice that expected.[17] These two factors—the unreliable wind and the truculent lithium-7—conspired to make Castle Bravo, the code name for the test, one of the most infamous blunders of the US nuclear bomb program.

At the center of the Castle Bravo debacle was Alvin C. Graves, no stranger to nuclear mishaps. He had been working with Louis Slotin when Slotin's infamous screwdriver slipped, thereby uniting two halves of a plutonium core and producing criticality. The massively overdosed Slotin and Graves were both afflicted with radiation sickness, but Graves alone recovered and continued working in nuclear bomb research. Graves, like Slotin, had very high tolerance for risk, and he was used to pushing the safety envelope.[18] Unfortunately, for Graves, once bitten did not mean twice shy.

It was Graves who gave the final go ahead for the detonation despite deteriorating weather conditions. A change in the prevailing winds had increased the risk of downwind fallout casualties, but mathematical weather models predicted the risk level to still be marginally acceptable. Rather than wait until the wind pattern improved, Graves was willing to take the risk and authorized the test to proceed as planned.

JUMBO SHRIMP

Shrimp was a 15,000-kiloton hydrogen bomb. It was, and remains, the largest nuclear weapon ever tested by the United States.[19] The bomb was detonated on a platform constructed on a coral reef off of Namu Island, of the Bikini Atoll.[20] This atoll is part of the Marshall Islands, a vast chain of remote atoll islands in the central Pacific Ocean. The Bikini Atoll is composed of 23 small islands arranged in circular formation, the largest of which is named Bikini Island. Originally a German possession, the Japanese gained control of the Bikini Atoll during World War I, and the United States took control from the Japanese after their defeat in World War II.

Because of its remoteness, the United States military had chosen Bikini as a site for nuclear weapons testing,[21] and the responsibility of delivering the news to the islanders fell to Commodore Ben H. Wyatt, military governor of the Marshall Islands. Wyatt intercepted the entire community of 167 Bikini islanders as they were leaving church services one Sunday and told them that their island was needed for a project to benefit mankind. Somewhat in awe that the United States would select their humble islands for such an important role in human history, the Bikinians readily acquiesced to the request. "If the United States government and the scientists of the world want to use our island and atoll for furthering development, which with God's blessing will result in kindness and benefit to all mankind, my people will be pleased to go elsewhere," was Bikini chieftain King Juda's response to Wyatt. The military governor thanked him, and likened their exodus to "the children of Israel whom the Lord saved from their enemy and led into the Promised Land."[22]

In early 1946, just before the first atomic bomb tests took place, the Bikinians were relocated to Rongerik Atoll, 120 miles (200 kilometers) to the east.[23] Rongerik, however, proved to be no promised land. The atoll had been uninhabited, partly due to its very

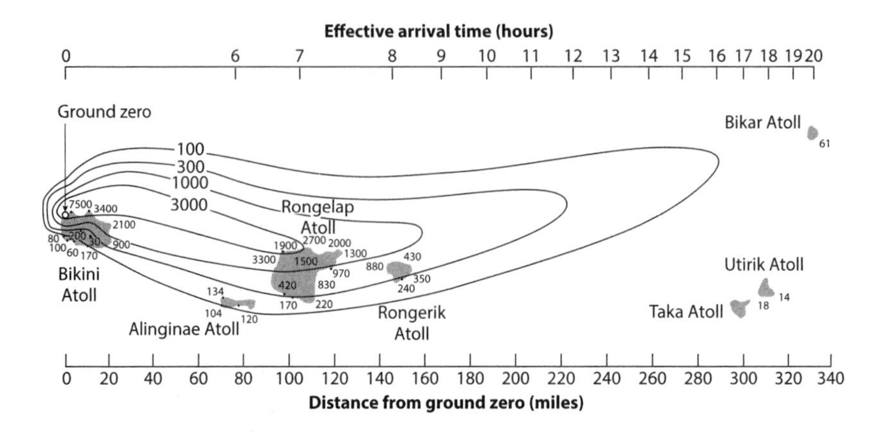

FIGURE 8.1. MAP OF HYDROGEN BOMB FALLOUT RANGE. The Bikini Atoll in the Marshall Islands was the site of Castle Bravo, a hydrogen bomb testing mishap that produced unexpectedly high quantities of radioactive fallout, which spread well beyond the anticipated range. Residents of other atolls, over a hundred miles east of Bikini, would require evacuation because of the high radiation doses produced by the fallout. The values shown on the isodose curves are exposure rates in roentgens per hour, at four days post blast. (*Source*: *The Effects of Nuclear Weapons* 3rd ed., S. Glasstone and P. J. Dolan, eds. Washington, DC: DOD and DOE, 1977)

small size (with a land area less than 0.65 square miles; 1.68 square kilometers), and partly because Marshall Islander folklore held that it was haunted with evil spirits. The move to Rongerik ended what had been an extremely long habitation for the natives of Bikini. Although there are no written records of Bikini's habitation prior to 1825, archeological discoveries and radiocarbon dating of human artifacts suggest that people had resided on the atoll as early as 2,000 BC (the approximate year that Stonehenge is believed to have been completed).[24]

Rongerik was not as lush as Bikini, and the islanders began to suffer malnutrition. They were then relocated to Kili Island, even further away from their homeland (425 miles south of Bikini), which turned out to be their greatest blessing in one sense. The Bikinians were already well settled on Kili Island when Shrimp was detonated on March 1, 1954. This was lucky for them, since

Rongerik Atoll was situated directly downwind of Bikini on that day. Not so lucky were 28 American servicemen, the only remaining inhabitants of the haunted atoll, who were based at the Rongerik weather station.

Shrimp was detonated at 6:45 a.m. It produced a fireball almost four and a half miles (7 kilometers) in diameter, which was visible over 250 miles (400 kilometers) away (the approximate distance between Washington, DC and New York City). It produced a crater over a mile wide and 250 feet deep (the height of a 25-story building). The mushroom cloud reached a height of 25 miles (40 kilometers) and its cap had a diameter of 62 miles (100 kilometers). The radioactivity-laden cloud contaminated more than 7,000 square miles of the Pacific Ocean (an area nearly the size of New Jersey). But the fallout was not evenly distributed. Areas downwind of the blast received the bulk of it, while locations upwind received virtually nothing. Given the prevailing winds from the west, potentially lethal levels of fallout stretched as far as 230 miles (370 kilometers) east of Bikini Atoll.

Soon after the blast, the US military became aware that radioactivity had escaped beyond their restricted area. Seven hours after the explosion, reports from the Rongerik weather station indicated high levels of radioactivity. The radiation recording instruments placed at the station by the Atomic Energy Commission were reading off scale. This meant that the servicemen at Rongerik (meaning "small lagoon" in Marshallese), and the islanders on nearby Rongelap ("large lagoon") and Utirik Atolls would need to be evacuated. Also, the off-limits area around Bikini had to be increased from 50,000 to 400,000 nautical square miles (nearly the land area of the state of Alaska) to keep ships and planes from being contaminated. The radius of this new danger zone around Bikini was 500 miles (the distance between Washington, DC, and Detroit).[25] Yet, the military did not know about the *Lucky Dragon No. 5*, so it was never intercepted. Thus, the fishing boat escaped the area and made it back to its homeport, bringing Shrimp's radioactive contamination with it.[26]

Three days after the bomb test, fallout had accumulated on some downwind atoll islands to depths of a half an inch (1.27 centimeters) or more. The native inhabitants of the contaminated atolls, as well as the 28 US servicemen from Rongerik, were evacuated to Kwajalein Atoll, the largest atoll in the world, and home to a US military base. The Rongelap natives received a particularly high dose (approximately 2,000 mSv) from the fallout. They had extensive burns on their skin and depressed white blood cell counts. The native inhabitants of other islands had no symptoms, and neither did the servicemen from Rongerik.

At Kwajalein, the contaminated islanders were treated by doctors and were immediately enrolled in a long-term health study led by physicians and scientists from Brookhaven National Laboratories. The health investigation was entitled "Study of Response of Human Beings Exposed to Significant Beta and Gamma Radiation Due to Fallout from High Yield Weapons" (commonly known as Project 4.1).[27] Radiation expert Dr. Eugene Cronkite (1914–2001) was assigned leadership of the project on March 8, barely one week after the blast, and before anyone yet knew of the contaminated Japanese fishermen.

With everyone evacuated to Kwajalein, the military thought that it had accounted for all the fallout victims. But then reports of the radioactive fishermen's arrival in Japan hit the newspapers on March 14. The Japanese government asked the US military to provide the fishermen's doctors with a breakdown of the fallout's radioactive composition. But military officials refused because of alleged concern that Soviet scientists would be able to use that information to deduce the mechanism of the hydrogen bomb.[28] This explanation does not seem credible, however, since that cat was already out of the bag. The Soviet Union had by then developed its own hydrogen bombs, and had even tested one on August 12, 1953 (six months before Shrimp's detonation) in northern Kazakhstan.[29] Nevertheless, the Americans would offer no information about the radioactive isotopes in Shrimp's fallout, so the Japanese scientists performed their own analysis from fallout collected from the fishing boat. They determined that it included, at the least, radioactive zirconium, niobium, tellurium, strontium, and iodine.[30] Of these, zirconium, niobium, and tellurium have no biological roles, and were, therefore, not of major concern. Strontium and iodine, however, were quite a different story. They were both biologically active. How would they affect health?

DÉJÀ VU ALL OVER AGAIN

Strontium was considered a problem by public health scientists because it is another one of the alkaline earth metals (column 2 on the periodic table), just like calcium and radium; therefore, it would be expected to incorporate into bone. Accordingly, these fishermen, who had actually breathed, ingested, and absorbed the strontium through their skin, ran the risk of becoming the strontium boys, the male equivalent of the unfortunate radium girls. Thus, radiation history appeared to be repeating itself. Suddenly, the experience of the radium girls, three decades earlier, seemed highly relevant to the new strontium problem. All of the radium health studies from that earlier time were revisited, and in order to assess long-term effects, a nationwide hunt was launched to find all surviving radium girls to enroll them in follow-up health studies.[31] In 1956, an Atomic Energy Commission official summed up the situation:

> Something that happened far in the past is going to give us a look into the future. Why, when these people took in their radium, there was no such thing as strontium-90, and yet they may help us determine today [how much is safe to ingest]. The way I see it, we're trying to follow up a wholly unintentional experiment that has taken on incalculable value.[32]

In the end, 520 surviving dial painters were located, and all their medical, dental, and laboratory records were collected and transferred for study to Argonne National Laboratory, a newly established national science and engineering research laboratory outside of Chicago.[33] After analysis, all of this data supported the concept that bone seekers, like radium and strontium, behave very similarly, both in their short-term and long-term health effects. One need only adjust for differences in their emission characteristics (half-lives, particle types, energies, etc.) to predict which health effects would be seen for any given exposure.

Iodine posed another tissue-specific threat. Iodine has an affinity for the thyroid gland, a butterfly-shaped organ that straddles the windpipe (trachea) just above the collarbone. The thyroid uses iodine to produce hormones that regulate body metabolism, an essential function. The problem arises because iodine is relatively uncommon in the natural environment and scarce in many human diets.[34] Consequently, whenever the thyroid gland detects some iodine in the bloodstream it snatches it up and stores it away for future use. In a normal individual, the concentration of iodine in the thyroid is 10,000 times the concentration in blood.[35]

The thyroid is essentially an iodine sponge. This had been known since 1895, the year before radioactivity was discovered by Becquerel, when German physiologist Eugen Baumann (1846–1896) reported that he found extremely high concentrations of iodine in thyroid tissue.[36] Understanding this, and knowing that radioactive elements behave chemically just like their nonradioactive counterparts, it isn't too much of a stretch to predict that radioactive iodine is not going to do good things to the thyroid gland.

Remarkably, the scientific community was slow to realize the unique danger that radioactive iodine posed to the thyroid, and surveillance of the gland was not part of the medical follow-up for any of the victims evacuated from the areas contaminated with Shrimp's fallout. Years later, Dr. Victor P. Bond (1919–2007), a member of the medical team that the United States sent to treat the fallout-exposed victims, recalled the following: "And quite frankly, I'm still a little embarrassed about the thyroid. [The] dogma at the time was that the thyroid was a radiation-resistant organ . . . [It] turned out that they had . . . very large doses of iodine . . . to the thyroid." Dr. Eugene Cronkite, Bond's supervisor and head of the medical team, concurred, noting "there was nothing in the medical literature . . . to predict that one would have a relatively high incidence of thyroid disorders."[37]

By the 1960s, many of the fallout-exposed islanders began to develop thyroid abnormalities, including three thyroid cancers on Rongelap and three on Utirik—a thyroid cancer incidence rate much higher than would normally be expected for a population of just a few hundred people. This caused distrust among the islanders of the medical team, because the team's leadership had predicted that there would be no thyroid cancers at all.[38] The thyroid gland is now understood to be very susceptible to cancer caused by radioactive iodine.

The composition of radioactivity in fallout is no longer considered a state secret since every nation in the world now knows the fundamentals of how nuclear bombs work. Radioactive fallout is composed largely of fission products—the isotopes that result when atoms are split. Although Shrimp was a fusion bomb, fusion bombs are always detonated with a fission bomb trigger, as previously discussed, and the energy of fusion can actually produce more fission reactions. So even fusion bombs emit a substantial quantity of fission fragments. In addition, fallout contains appreciable quantities of the fissionable materials uranium and plutonium, because the fission chain reaction does not reach completion before the core is blown apart.

Fissionable uranium and plutonium atoms can split apart in about 40 different ways, so about 80 different fission fragments (radioisotopes) are theoretically possible. In reality, some are rarely produced, while others are abundant. In addition, most have very short half-lives of less than one minute. These short-lived radioisotopes don't contribute anything to the fallout hazard. But a number of radioisotopes stay around long enough to cause problems. In addition, some, like the radioisotopes of strontium and iodine, have biochemical properties that cause them to be taken up in the body.

One major component of the fallout from Shrimp that the Japanese scientists overlooked was cesium-137. Cesium occupies column one of the periodic table, making it a chemical cousin of sodium and potassium, both important body electrolytes. Electrolytes are basically dissolved salts. In fact, seawater is an electrolyte solution, and human body fluids are not unlike seawater. The body is particularly dependent on sodium and potassium electrolytes, which are used for numerous physiological functions in all tissues and organs. As such, these electrolytes are quite evenly disseminated among soft tissues. Cesium, like sodium and potassium, is distributed uniformly throughout the body and doesn't seek out any particular tissues or concentrate in any one organ, although it is somewhat enriched in muscle. So that's not the basis for its health effects. Rather, the problem with cesium-137 is simply that it hangs around in the environment for a very long time (half-life = 30 years). The only silver lining to this cloud is that cesium, again like sodium and potassium, is highly soluble in water (hence, its electrolyte properties), and, therefore, is usually quickly dissolved by rainwater, swept into rivers, and deposited in the ocean, where currents mix and dilute it down to very low concentrations.[39]

The bottom line is that the radioisotopes in fallout with a significant potential to affect health are few: iodine-131 (8-day half-life), strontium-89 (52-day half-life), strontium-90 (28-year half-life), and cesium-137 (30-year half-life). As can readily be seen, the iodine-131 represents a relatively short-term risk, but strontium-90 and cesium-137 can persist for years.

For the Japanese fishermen, the treatment turned out to be as bad as the disease. Sadly, much of the blood they received during their transfusion therapy was contaminated with hepatitis virus, so most of them (17 out of 22) came down with hepatitis (i.e., liver disease).[40] To make matters worse, they were neither told of nor treated for their liver disease,[41] possibly because the doctors wanted to cover up their mistake, and many of the fishermen ultimately died from liver-related causes. A study initiated to follow their health over time, in order to learn more about radiation sickness from radioisotopes, eventually broke down because of the fishermen's distrust of medical authorities, and because their underlying liver disease would have compromised the validity of the findings anyway.[42] During the aftermath of the *Lucky Dragon No.*

5 episode, Japanese/American relationships reached a postwar low, especially since the American authorities insisted that the Japanese doctors share the culpability for the fishermen's health problems because they had administered the virus-contaminated blood.[43] After intense negotiations, each fisherman received $5,550 (about $49,000 in 2015 dollars) in compensation from the US government, in exchange for agreeing to push their legal claims no further.

LAND OF MILK AND HONEY . . . AND COCONUT CRABS

The evacuated islanders were eventually resettled on uncontaminated atolls. Rongelap's people were placed on the island of Ejit in Majuro Atoll. While they waited to return to their home atoll, the US government conducted environmental studies on Rongelap and monitored the gradual decrease in its environmental radioactivity levels. The natives were finally returned to Rongelap in 1957. They were told that the island itself was now safe, and the local foods were safe to eat.[44] But the local foods turned out not to be safe.

Besides breathing and absorbing fallout through the skin, another major means by which fallout enters the body is through food ingestion. The ingestion route is particularly important for those radioactive elements that can enter the food chain—the series of steps that lead from food production to consumption. Both iodine and strontium fall into this category.

There's some good news and bad news when it comes to iodine exposure through the food chain. As it turns out, we get most of our dietary iodine through dairy products. That's because cows get their iodine from eating grass and some of that iodine goes into their milk. Thus, nursing calves get the iodine they need through consumption of milk, until they start eating grass on their own. When we intercept the milk for dairy products we abscond with that iodine for our own bodies. Once we ingest those dairy products, the iodine goes to our thyroids. What this means is that if a cow is eating fallout-laden grass, we're consuming radioactive io-

FIGURE 8.2. COCONUT CRABS. This extremely large species of omnivorous terrestrial crab (*Birgus latro*) is native to the South Pacific. It has the ability to climb palm trees, crack coconuts with its very powerful claws, and eat the coconut meat. Radioactive strontium in the soil of the Marshall Islands from nuclear bomb testing fallout was taken up by palm trees and concentrated in their coconuts. Unfortunately, both the crabs and the coconuts formed a large part of the Marshall Islanders' diets.

dine (i.e., iodine-131) when we drink its milk (or eat dairy products). This is bad news if we're drinking a lot of iodine-131 contaminated milk right after a fallout event. The good news is that, after a brief period, the short-lived iodine-131 decays away so that the grass, the cow, and even the milk, no longer contain hazardous levels of iodine-131. (Since the half-life of iodine-131 is just eight days long, waiting three months before eating contaminated food reduces exposure to its radioactivity by 99.9%.) For iodine-131, it turns out that the best protection is temporary abstinence from locally produced food products, especially milk.

Strontium has its own unique food-chain story, reminiscent of the watch painter's unique problem with radium ingestion. In February 1958 (one year after the return of the islanders to Rongelap), US Navy scientists decided to test the Rongelap coconut crab population for radioactive contamination. They found boatloads of radioactivity.

Coconut crabs (*Birgus latro*) are very large terrestrial crabs, actually resembling lobsters more than crabs. They have claws powerful enough to crack coconuts, their major food source. The coconuts are rich in the calcium that the crabs need to sustain their large fast-growing shells (exoskeletons). Palm trees absorb the calcium from the ground and concentrate it in their coconuts. And strontium, as we already know, is a dead ringer for calcium as far as biochemistry is concerned. Consequently, the crabs of Rongelap had become highly radioactive from the strontium-90 contamination. To make matters worse, the islanders considered the coconut crabs a delicacy, and had been eating lots of them ever since their return to the atoll a year earlier. Then, in June 1958, the Rongelap Islanders were told by the navy not to eat their local coconut crabs anymore, yet another assault on their native lifestyle.[45]

TRIPLE JEOPARDY

The island of Bikini itself, the actual site of most of the bomb test detonations, wasn't resettled until 1969. This occurred once an AEC blue-ribbon panel had estimated the dose that returning islanders would receive from ingesting local food contaminated with radioactivity and deemed the levels to be safe. Unfortunately, the panel based its recommendation to resettle Bikini on a report with a misplaced decimal point—a typographical error that underestimated coconut consumption a hundredfold. Scientist Gordon Dunning, who authored the report, freely admitted, "We just plain goofed."[46]

Since coconut was the Bikinians' major foodstuff, the typographical error meant that the repatriated islanders had actually ingested massive amounts of radioactivity, perhaps more than any other known population. When the problem was discovered in 1978, they were evacuated from their home island once again.

This decimal point error was the third catastrophic error that the islanders suffered at the hands of the experts. As we've seen, the first error was when Los Alamos nuclear physicists erroneously assumed that lithium-7 would not contribute to a fusion reaction—a mistake that resulted in doubling the hydrogen bomb's yield, thus driving radioactive fallout farther than anyone expected and right onto the displaced islanders' laps. The second error was the failure of doctors to appreciate that the physiology of the human thyroid put it at high risk from radioactive iodine. Thus, danger to the islanders was denied or downplayed three times, and the islanders paid the price for mistaken expert opinions.

When all was said and done, the Marshall Islanders living on fallout-contaminated atolls had breathed, absorbed, drunk, and eaten considerable amounts of radioactivity. In the 1960s, thyroid cancers began to appear, and then other cancers were diagnosed, including leukemias (blood cancers). United States government physicians provided ongoing medical care. During the course of this treatment, Brookhaven National Laboratory scientists collected the islanders' health data for analysis. Health data was also collected from the natives of Kwajalein Atoll, which was not subject to the fallout, in order to have comparison data on an otherwise similar, but unexposed, group (i.e., a control population). This data set remains the largest study of the health effect of fallout radioactivity on humans, and future safety limits for fallout radioisotope exposures would be based largely on findings from the Marshall Island health study. The Marshall Islanders became to fallout what the radium girls were to radium. And both groups were used by the rest of us as miners' canaries.

The Brookhaven study of the Marshall Islanders officially ended in 1998. Much was learned about the risks posed by radioactivity absorbed into the body. Currently, the Department of Energy (DOE) continues to provide annual health screenings for radiation-exposed Marshallese at clinics located both in the Marshall Islands and the United States. The DOE includes information of the islanders' ongoing health conditions in its annual report to Congress.

In June 1983, the United States and the Republic of the Marshall Islands entered into an agreement in which the former recognized the contributions and sacrifices made by the Marshallese as a result of US nuclear testing. A Marshall Islands Nuclear Claims Tribunal was established to dispense compensation for personal injury and property damage to those deemed to have health problems caused by the testing. As of 2003, over $83 million in personal injury awards had been awarded to a total of 1,865 individuals who claimed a total of 2,027 radiation-related ailments.[47] By May 2009, all funds awarded by the US Congress were completely exhausted, with $45.8 million in unpaid personal injury claims still owed the Marshallese fallout victims.[48] Many have died while waiting to be paid. It is estimated that over 50% of the valid claimants died before realizing the full amount of their awarded compensation.[49] Currently, the tribunal exists in name only; that is, until Congress restores funding, which by all accounts is extremely unlikely to happen.

THE SALT OF THE EARTH

Biological and chemical weapons have never proved to be effective weapons of mass destruction largely because they are hard to control. Once biological and chemical weapons are unleashed, environmental conditions can throw the weaponized agents back in the faces of their users. Such was the case during the Battle of Loos, France, in World War I, when shifting wind blew chlorine gas released on German soldiers back into the trenches of the British soldiers. If a weapon's lethality can't be controlled, it isn't of much military use. In that regard, radioactive fallout presents a similar military problem; it can come back and get you. After seeing how fallout-contaminated battleships were as incapacitated by radioactivity as had they been sunk, Admiral William H. P. Blandy (1890–1954), the first commander of the Bikini Atoll nuclear bomb tests, remarked, "This is a form of poison warfare."[50]

Uncontrollable fallout is a major obstacle to effective use of nuclear weapons. But if unwanted atmospheric fallout were minimized, while radioactivity in the targeted blast area were maximized, it might provide yet another means of making nuclear weapons more lethal to enemy combatants. Furthermore, the targeted radioactivity could make the enemy's lands uninhabitable, thus thwarting recovery and deterring them from future aggression. This concept is reminiscent of the idea of the Roman general, Scipio Aemilianus (185–129 BC), who allegedly salted the earth of defeated Carthage at the end of the Third Punic War (146 BC) in order to avert a Fourth Punic War.[51]

Neutrons provide an opportunity to salt the earth in another manner due to one of their unique physical properties. As it turns out, any high-speed neutrons that don't happen to hit an atomic nuclei while traveling through matter eventually slow down to a stop and become "free" neutrons. Nature, however, doesn't like free neutrons. Neutrons are supposed to live in atomic nuclei, not float around on their own without companion protons. So, once these neutrons stop moving, nearby atomic nuclei suck them up. The nuclei of some elements are better at doing this than others; nevertheless, they all do it to some extent, and the result is always the same. The nucleus with an additional neutron is now neutron rich. That is, it has moved farther away from the one-to-one neutron-proton ratio that stable nuclei prefer. And yes, as you may have guessed, most of these neutron-rich nuclei are then radioactive. The extra neutron bloats their stomachs, so the nuclei want to convert the neutron to a proton, and spit out a beta particle, to relieve their indigestion. These brand-new beta emitters have a range of half-lives, mostly short. There are some, however, like cobalt-60, that have half-lives as long as a year or more. Quite long enough to salt the earth.

This unique property of neutrons to make elements radioactive led physicist Leó Szilárd in 1950 to propose a concept for a new weapon usually called the *cobalt bomb* (or sometimes *salted bomb*).[52] Such a nuclear bomb would have mixed within its warhead large quantities of cobalt-59, a nonradioactive metal that is very efficient at absorbing stopped neutrons. When this happens, the stable cobalt-59 becomes radioactive cobalt-60, with a half-life of five years. Not long enough to make land permanently uninhabitable, but likely long enough to put it out of commission for the remainder of the war.

No cobalt bombs are known to exist, but the concept has captured the public's imagination. Cobalt bombs, or other types of salted bombs, have made regular appearances throughout pop culture, including some novels (*On the Beach,* by Nevil Shute), movies (*Goldfinger* and *Dr. Strangelove*), and television programs (*Star Trek*).

A BITE WORSE THAN ITS BARK

Before we leave hydrogen bombs and the fallout problems they pose, one other thing should be mentioned. Their efficiency in killing enemy troops can be greatly enhanced. Hydrogen bombs are like fission bombs—they destroy mostly by shock waves and fire. And although this may be a moot point to the civilian population, enemy armies can make manned fortifications that are resistant to shock waves and fire. It is, therefore, conceivable that soldiers might survive even a hydrogen bombing and then carry on the battle in the resulting wasteland. Neutrons can provide a solution to this problem of combatants who refuse to die.

Neutrons, as we know, are both highly penetrating to lead (and other metallic shielding) and highly lethal compared to other types of radiation. Capitalizing on these two characteristics of neutrons, weapon scientists have been able to produce what are called *enhanced radiation bombs* with greatly increased yields of neutrons. This is achieved by using atomically inert shell casings, devices called x-ray mirrors, and other modifications to fusion bombs that allow a maximum number of neutrons to escape from the blast. These neutrons are able to penetrate armor and kill hiding soldiers by radiation sickness, even when the shock waves and fire do not.

Such enhanced radiation weapons are more familiarly called *neutron bombs*, and urban legend has it that they kill people but spare buildings. Unfortunately for buildings, this is not true. Neutron bombs are just as damaging to buildings, but are better able to seek out and kill any humans hiding inside.

Although neutron bombs are still fusion bombs, they are considered tactical weapons, rather than strategic weapons,[53] because they can be built with smaller kilotonnages (tens to hundreds of kilotons) than conventional fusion bombs (thousands of kilotons) and are, therefore, more the size of very large fission bombs. As a result, individual neutron bombs conceivably could be used tactically to take out portions of large cities, as opposed to very large fusion bombs (e.g., 50,000 KT; 50 MT) that would completely destroy an area the size of the state of Rhode Island.

DIVER DOWN

The native Bikinians have still not returned to Bikini Atoll because of concerns about residual radioactivity that remains in the wild foodstuffs. In 1997, however, Bikini Atoll was opened to brief visits and tourism. With environmental radiation levels just slightly above the normal background, the Bikini Islands are now quite safe to visit and have become a popular destination for both scuba divers and fly fishermen. Bikini offers divers the opportunity to explore hundreds of different wrecks of historic American warships that were sunk in the atoll lagoon as part of the nuclear testing program. The lagoon waters are shallow and warm, with magnificent underwater visibility. For fly fishermen, Bikini offers an abundance of game fish, which have thrived and multiplied, unthreatened by any human presence.

CHAPTER 9

AFTER THE DUST SETTLES: MEASURING THE CANCER RISK OF RADIATION

When you can measure what you are speaking about
and express it in numbers you know something
about it, but when you cannot . . . your knowledge
is of a meager and unsatisfactory kind.

—*Lord Kelvin*

The average human has one breast and one testicle.

—*Des MacHale*

THE COLOR PURPLE

Purple spots. That's what the Japanese atomic bomb survivors feared most. In the years following the atomic bombings of Japan, the survivors began to realize that they were not completely out of the woods as far as health risks were concerned. Even those who never had any symptoms of radiation sickness were still prone to higher than normal rates of death. And the prelude to their death often took the form of small purplish blotches (petechiae) on their skin. Just as a sneeze was once feared to be a precursor to the plague by sixth-century Europeans, purple skin spots were thought to be harbingers of leukemia by the postwar Japanese. The difference was that the Europeans were wrong about the sneeze.

Dr. Fumio Shigeto was exposed to the bomb's radiation while waiting for a streetcar on that fateful August morning of 1945. He had just arrived in Hiroshima the week before, having recently been appointed the new vice-director of the Red Cross Hospital. Suffering only minor injuries from the blast, he made his way to work, where his subordinate, Dr. Terufumi Sasaki, was already busy treating bomb victims.

It was Dr. Shigeto who first discovered that all the hospital's x-ray film had been exposed by the bombing. In his younger days, he had spent some time studying radiology. He immediately grasped the implication of the exposed film, and was one of the first doctors at the scene to understand that many of the bomb victims were suffering from radiation sickness. Still, this would not be the only health effect of radiation that he was first to recognize.

Dr. Shigeto continued to work at the hospital for many years. Since the patients came from the immediate vicinity, over time he saw many patients who had survived the bombing. In 1948, he noticed before anyone else that leukemia rates were elevated among the bomb survivors.[1] Even the ones who had never previously exhibited any signs of radiation sickness sometimes succumbed to leukemia and died.

Initially, Dr. Shigeto's reports of elevated leukemia rates among bomb survivors were met with skepticism from the Japanese Ministry of Health and Welfare, as well as American public health officials. The Americans were particularly suspicious of reports from Japanese doctors about new health problems related to radiation. It was thought that the Japanese had a vested interest in exaggerating these problems, because they could then push for bomb victim reparations from the US government. In the postwar years, the bar was high for new claims about radiation health effects, especially if they came from the Japanese. Anticipating the doubters, Shigeto had religiously kept his own statistics on leukemia incidence among his patients, and his data suggested that leukemia was, in fact, on the rise.[2] With time, as the leukemia rates reached unprecedented levels, it finally became well accepted that radiation can produce leukemia.

By 1953, the elevated leukemia rates in the atomic bomb survivors had peaked, and began to subside the following year. Nevertheless, starting in 1955, increased rates of many other types of cancer were becoming apparent. The rates would keep rising until finally, in 1982, these cancers too would start to abate. In the end, a wide variety of cancers seemed to be elevated among atomic bomb survivors, and radiation came to be seen as an unselective carcinogen, with an ability to induce cancer in most all body tissues. This distinguished radiation from the various chemical carcinogens, each of which tended to produce cancer in a limited spectrum of tissues. Radiation did not discriminate. Virtually all tissues seemed to be at some level of risk for cancer.

When a 19-year-old Marshall Islander, Lekoj Anjain, who had been an infant on Rongelap Atoll in 1954 when the Bravo hydrogen bomb test occurred, developed an acute case of leukemia in 1972, it was widely acknowledged that his radiation exposure was the likely cause.[3] He was flown to Brookhaven National Laboratory in New York, where the leukemia diagnosis was confirmed, and was then transferred to the National Institutes of Health in Maryland, where he was treated. Unfortunately, his disease was incurable at the time, and he died at the end of the year.

Contracting leukemia at age 19 is not unheard of, but it is rare. (The median age of leukemia occurrence is 65.) His was the first Marshall Islander death attributable to radiation, and cause of death was leukemia, consistent with the Japanese experience. This supported the contention that leukemias are the first cancers to appear in a population exposed to high radiation doses.

THE MOST IMPORTANT PEOPLE LIVING

By the time the atomic bomb survivors first started coming down with cancer, mankind's experience with high-dose radiation exposures was vast: uranium miners, the first x-ray and radium patients, early radiologists, radium dial painters, the two atomic bombings in Japan, and Marshall Island fallout. In all of these encounters with radiation, the legacy was a lot of suffering for many people. Yes, it was quite evident that high doses of ionizing radiation could produce cancer. But exactly how dangerous was it? And what about exposures to lower doses of radiation that everyone is subjected to? Were they something to worry about as well?

The different populations of radiation victims held the answers to these questions, but many scientists, many years of study, and a tremendous amount of money would be required to gather the necessary data and wring out useful information about cancer risk. Nonetheless, the epidemiologists—scientists who study the patterns of disease within populations—saw a great scientific opportunity to answer fundamental questions that people had wondered about since radiation was first discovered, and to answer those questions in a measureable way.

Of all the radiation-exposed populations, the atomic bomb survivors held the most promise for highly precise and reliable findings. There are a number of scientific reasons for this that are beyond our scope here, but we can elaborate on the two main reasons: (1) Hiroshima and Nagasaki had very large numbers of people who represented all ages and both sexes, and were exposed simultaneously to a wide range of radiation doses over their whole bodies; and (2) individual doses for bomb survivors could potentially be determined very accurately based on an individual's exact location at the time the bomb detonated (with some corrections made for building structures and ground contours).[4] Just as most New Yorkers know exactly where they were standing when the World Trade Center was attacked on September 11, 2001, most atomic bomb survivors knew their precise location when the atomic bombs were detonated, and that information revealed their radiation dose.

These two factors—large numbers of people who had reliable individual dose estimates—translated into very good statistics. And good statistics meant that the epidemiological studies would have the power to answer scientific questions without having to worry that the study's findings were driven merely by chance.[5] When you're trying to determine the association between any particular exposure and any particular health outcome, statistical power is the name of the game. The atomic bomb population had more statistical power than any radiation population study either before or since, and it is unlikely ever to be surpassed in either the quantity or quality of data.

All that was needed to capitalize on this unique opportunity to better understand radiation's risk to human health was to track the exposed people over their entire lives; then it could be determined exactly what health effects cropped up among these people and, further, the rates at which these health effects appeared could be measured. No small feat. If the rates proved to be elevated in a radiation-exposed population relative to an unexposed control population, radiation could be inferred to be the likely cause. Should the study be successful, the product would be a complete catalog of all of radiation's health effects. What's more, there would also be a reliable estimate of the risk per unit radiation dose (i.e., the risk rate) for each health outcome.

This was an extraordinary chance to determine how the most feared agent on Earth affects health, and the more astute scientists recognized it as exactly that. As radiation scientist Dr. Robert H. Holmes said of the Japanese atomic bomb survivors to a journalist, "These are the most important people living."[6] And so, after some fitful starts, much controversy, and intense battles over who would pay for it, on November 26, 1946, President Harry Truman directed the National Academy of Sciences to begin a long-term study of the effects of atomic bomb irradiation on humans. This directive lead to the establishment of the Atomic Bomb Casualty Commission (ABCC), funded by the United States. The cornerstone of this research effort was the massive Life Span Study (LSS), which has been following the medical outcomes of 120,000 bomb survivor and control subjects up to the present day (i.e., over 65 years).[7] The LSS is considered to be the definitive epidemiological study on the effects of radiation on human health.[8]

☢

Radiation may have been a new hazard for mankind, but interest in health risks was nothing new. People have always been preoccupied with measuring their risk of death from various activities and exposures. In fact, the first known calculation of life expectancy by profession was compiled by Ulpian (170–228 AD), a Roman jurist. His calculations allowed for the determination of the risk of death from various occupations, and thus let people gauge how dangerous those occupations were. (Gladiator was a bad one!)

Then, in the seventeenth century, John Gaunt (1620–1674), a London shopkeeper with an avocation of dabbling in public health statistics, decided to use readily available public mortality data to calculate the risk of death from various health ailments. In Gaunt's own words: "Whereas many persons live in great fear and apprehension of some of the more formidable and notorious diseases, I shall set down how many died of each; that the representative numbers, being compared with the total [mortality], those persons may better understand the hazard that they are in."[9]

What Gaunt ended up producing was the first known *life table*, which is simply a tabulation of the probabilities of surviving to each year of age. In addition to satisfying people's curiosity about how long they might expect to live, it was soon realized that a life table provided a rational means to price life insurance. So John Gaunt is often credited with being the originator of the modern life insurance industry. It seemed that public health statistics held much more valuable information than people had previously realized.

After a couple of centuries passed, the "bean counters" working in public health turned things up a notch, and shifted what had been merely descriptive public health statistics into the new branch of science called epidemiology. London physician John Snow (1813–1858) was their patron saint. Snow was a physician and public health advocate who believed in the value of counting and measuring. Consequently, in 1854, when a cholera epidemic swept through London and killed a large proportion of the population, it was not surprising that Snow was out counting. He was counting deaths. By counting cholera victims, calculating death rates by neighborhood, and precisely mapping the locations of victims' residences, he was able to show that those dying at the highest rate were likely obtaining their drinking water from a residential underground water supply system that had a public pump on Broad Street.[10] He had thus identified a contaminated drinking water source as the cause of the epidemic. In an apocryphal anecdote to an otherwise accurate account, Snow was reputed to have singlehandedly ended the epidemic by absconding with the pump's handle, thus cutting off that source of disease-contaminated water to the community and forcing the residents to use other, cleaner water sources.

Following in the tradition of Snow, radiation epidemiologists would count and map disease incidence in relation to the two atomic bombings' ground zeroes; by analogy, they could show that ground zero was the epicenter for radiation sickness and various cancers, just as the Broad Street pump had been the epicenter for cholera. The radiation scientists, however, would even go further than Snow. He never measured an individual's cholera risk in terms of the amount of water drunk (i.e., the *risk per unit dose*). That was a mistake that the radiation epidemiologists would not repeat.

LET'S DO LUNCH

On December 6, 1946, less than two weeks after President Truman had authorized the National Academy of Sciences to begin research, a team of American scientists was already in Hiroshima and ready to begin work. But first they attended a formal lunch that was hosted by the assistant mayor of the city, Hisao Yamamoto, and held in a primitive building near ground zero. The building had been hastily constructed to replace the totally ruined city hall.[11] The assistant mayor apologized for the shabby accommodations, but he was nonetheless able to treat the Americans to a lavish three-course meal.

Dr. Masao Tsuzuki joined the lunch; he was Japan's leading expert on the health effects of radiation and the future attending physician for the fallout-exposed fishermen from the *Lucky Dragon No. 5*. Local physician Dr. Ikuzo Matasubayashi was also present. Matasubayashi reported to the Americans that he and a statistician Masato Kano were already gathering data on the survivors of the Hiroshima bombing. Their goal was to identify all survivors and their exact locations at the time of the bomb's detonation, and to also include a brief medical description of each person's injuries. During the 15 months since the bombing, with the help of volunteers, they had already registered tens of thousands of people (out of an estimated 200,000 survivors) in their study, with a file for each victim. Although they did not have the facilities to catalog the data properly, they had secured a large room with many tables and stacked the files on tables in sequential piles that corresponded to the distances the survivor had been from ground zero; this was reminiscent of Snow's mapping the distances of cholera victims from the water pumps. They knew that any new ailments that might show up in this population over time should be traceable to the bomb's epicenter (i.e., ground zero) if radiation were the cause. Thus, the Japanese had already broken ground on what would become one of the greatest epidemiological cohort studies ever conducted.[12]

A *cohort study* is the strongest and most reliable study design that epidemiologists have at their disposal. The reasons for its superiority are quite involved, and beyond the scope of our discussion. Having said that, we can still get a sense as to why a cohort study is superior to other study designs by comparing it to a somewhat weaker type of study—the case-control study—that is often used as a second-best alternative when a cohort study is not possible. But before doing this, we must first understand exactly what a cohort study is and how it works.

To best appreciate the strategy of a cohort study, it is useful to consider the origin of the word cohort. A cohort was the basic fighting unit of a Roman legion, starting with precisely 480 recruits, averaging about 20 years old.[13] The soldiers in the cohort were all enlisted into service at the same time and usually were not replaced as they died off (unless battle casualties were very heavy), so the fighting strength of a cohort tended to diminish slowly with time.[14] The soldiers in the cohort spent their entire term of enlistment together (typically 20 years), with few reassignments in or out. They lived, ate, slept, fought, and died together. And most importantly, they aged together. So the average age of the soldiers in any particular cohort was anywhere from 20 to 40, depending upon the year the cohort had originally been formed.

Over time, the weak or sickly, and the poor fighters, tended to die off first, leaving only the stronger soldiers to fulfill their enlistment terms and retire into civilian life. Any beneficial exposures or experiences that tended to increase a soldier's fitness (e.g., a diet rich in protein or strengthening exercises) would tend to increase that soldier's lifespan. In contrast, anything that adversely affected a soldier's fitness to fight (e.g., drunkenness or asthma) tended to be associated with a shortened lifespan. It can be seen that by recording the different exposures and behaviors of the individual soldiers in the cohort, and then waiting to see whose names appeared on the sick call and casualty lists over time, it should be possible to identify those exposures or behaviors that put soldiers at risk for illness and premature death. For example, drunkards might be expected to enter battle with a hangover, thus decreasing their chances of surviving to fight another day. This is why any study that follows a defined group of people over a long time to determine how their exposures affect their health outcomes is said to be a cohort study. In essence, such a study emulates tracking casualty statistics of a Roman cohort.

The term "cohort" is now more broadly considered to define any collection of people that is tracked as a single defined group. With a little reflection, we can appreciate that a single individual can simultaneously be a member of multiple cohorts. For example, he or she might be 1 of 400 babies born in a certain hospital in 1955, 1 of 50,000 children vaccinated for polio in some particular state in 1962, and 1 of 200 graduates of the class of 1973 for some specific high school. By studying each of these different cohorts, we might find a higher risk of sudden infant death for children delivered by caesarian, an increased risk of contracting meningitis for people who got a polio vaccine, or a higher risk of a fatal car crash for those who binge drank in high school.[15] The study of such cohorts can give us very reliable health risk information if we ask the pertinent questions and have the time and patience to wait for the data to roll in.

The LSS of the atomic bomb survivors is a classic cohort study. These survivors were recruited into the study at the time of the bombing, their exposures (in this case, their various radiation doses) were individually recorded, and their disease and death outcomes were tracked over time as the whole group (i.e., cohort) aged. If radiation doses were associated with different diseases and death, it would be readily apparent in such a study, given enough years of follow-up.

Now let's compare a cohort study to a case-control study, in order to illustrate why the former is considered superior.[16] A case-control study compares people with a certain type of disease (cases) to those without that disease (controls), and then asks the individuals in both groups what types of things they had been exposed to in their younger years. The expectation is that if exposure to a certain agent is associated with getting the disease in question, the cases will report more exposures than the controls will. Unfortunately, case-control studies often rely on the accuracy of the mental recall of the study subjects. The problem is that cases are sometimes more likely to recall an exposure than the controls, simply because cases are often seeking an explanation for their disease and have already considered various exposures that may have contributed to it. In other words, they are primed for the question. Controls, in contrast, don't ponder their earlier exposures because they have no reason to do so, and therefore, they tend to underreport their exposures. For example, you might find it difficult to recall whether your childhood neighbors owned a dog unless you still have the leg scar where it bit you. Injury and illness tend to sharpen the memory, and this is often the doorway through which bias enters a case-control study.

Case-control studies can suffer from other biases that are too complicated to be considered here.[17] Let's just say that the well-known propensity of case-control studies to contain hidden biases makes them suspect until their findings are replicated multiple times in different study populations. Only then will case-control study evidence be considered reliable.

Cohort studies, on the other hand, have much less potential for bias because the exposures are ascertained before the disease appears, so recall of exposures is not involved. Because cohort studies are less prone to biases, they are considered the most accurate of all population studies. Cohort studies are the gold standard. This is true to the extent that even multiple case-control studies are often considered merely suggestive until they have been validated in a subsequent cohort study.

We can thus see that all studies are not equal. They differ in the strength of their scientific design. Cohort studies are much more heavily weighted than case-control studies when it comes to scientific evidence.[18] Isaac Newton once famously remarked about judging scientific evidence: "It is the weight [of the experiments], not numbers of experiments, that is to be regarded." As in so many cases, Newton was exactly right about this too.

When the ABCC (Atomic Bomb Casualty Commission) team members saw the piles of folders, they were astonished at all their Japanese counterparts had already achieved. They later noted in their report that the data already accumulated "appear as accurate and reliable as could be expected in light of the circumstances, particularly the limited facilities."[19] Furthermore, the Japanese seemed to appreciate the obstacles to completing the work in a timely fashion, and welcomed the involvement of the Americans, even providing them with English translations of all their findings thus far. So the LSS hit the ground running, thanks largely to the up-front work of a few prescient Japanese physicians and scientists, and a small army of volunteers.

Unfortunately, the honeymoon between the Japanese and American scientists would be short-lived, and petty bickering about data control and publication authorship would soon follow. As so often is the case in scientific research, professional pride, infighting, and jealousy soon marred the progress and hampered the work. The details of the multiple feuds between the American and Japanese scientists have been well documented elsewhere and are not worth relaying here.[20] Nonetheless, in the end, both groups came to acknowledge that they needed the other for the LSS to succeed, and they put aside their differences enough to get the work done. This less-than-happy marriage nevertheless produced many children, in terms of scientific findings, and continues to provide new insight to this day on how radiation affects health.

In 1975, the ABCC was reorganized into the Radiation Effects Research Foundation (RERF), which now administers the ongoing LSS. The RERF, a joint venture of the Japanese and American governments, is headquartered in Hiroshima and remains the heart of radiation epidemiology to this day.[21]

FINDINGS OF THE LIFE SPAN STUDY

The LSS is now over 65 years old, as are its surviving study subjects. So the study has truly lived up to its name, following many subjects from nearly cradle to grave. Much has been learned. Let's look now at who these people are and what has happened to them.

The goal of the study was to include as many survivors of the Hiroshima and Nagasaki bombings as possible. An exhaustive search was conducted and 61,984 people were identified in Hiroshima, and 31,757 in Nagasaki. It is hard to say how many people were missed, but this total of nearly 94,000 probably represents about half of the total survivors who had been within the geographical study area at the time of the bombings.

Over 97% of these people received doses less than 1,000 mSv, so relatively few had actually experienced radiation sickness, which typically requires whole-body dose levels above 1,000 mSv. In fact, 94% had received doses of less than 500 mSv, and 80% had received doses less than 100 mSv. (For comparison, annual natural background doses in the United States average about 3 mSv.)

As this cohort aged, its members succumbed to various ailments, as all of us do when we get older. Few of these natural ailments could be easily traced to the victims' radiation exposures, with the notable exception of cancer. As of 2013, 10,929 people in the study had come down with cancer. Comparisons with cancer rates from control populations (i.e., people similar in other respects but lacking atomic bomb radiation exposure) suggested that only 10,346 cancer cases would have been expected (by 2013), in a group comprised of unexposed 94,000 Hiroshima and Nagasaki citizens. So it is assumed that the excess cancers, 623 cases, were due to the radiation exposure. Of the 623 cases, 94 were leukemia and 527 were other types of cancer. Therefore, the excess cancers in this bomb survivor study group are about 6% more than would have been expected for a reference population with the same age and gender distribution.[22]

There was also a clear dose response for radiation-induced cancer risk. That is, those who had been closer to ground zero, and therefore had higher radiation doses, suffered proportionately more cancers than those who were farther away. This convincing dose response relationship strongly supported the notion that the radiation was causing the additional cancer cases.

Scientists often rely on dose response data as one of several criteria to establish causal associations. The reasoning is simple. If one thing causes something else to then happen, increasing the first thing (dose) should also increase the consequence (response). For example, if eating excess food causes obesity, someone who eats large amounts of food should be more obese (all else being equal) than someone who eats less. Establishing a dose response is often considered essential evidence to support a claim of a cause and effect relationship. In other words, a study that demonstrates a dose response does not, alone, prove that the one thing causes the other, but failure to show a dose response is often considered evidence against a causal connection.

Dose response data gives us something else. It allows us to establish a rate for the response. For example, we may find that every cupcake eaten translates into one pound of weight gain, or we may find that a single cupcake results in two pounds of weight gain. A finding of "two pounds per cupcake" is twice the rate of "one pound per cupcake," suggesting that cupcakes are twice as threatening to a person's waistline.

AT THE END OF THE DAY

After more than half of a century of counting, what does the LSS have to show for itself? In terms of cancer, plenty! We now know the average cancer risk for a unit dose of ionizing radiation (i.e., the rate). And we know this very accurately, at least for higher doses. That is, we now know the percentage of increase in the lifetime risk of cancer per mSv of whole-body radiation dose.[23] And what is that value? It is **0.005% per mSv.**[24] And what can we do with this number? We can use the number to convert any known whole-body dose into a cancer risk estimate.[25] For example, the risk of cancer to a patient produced by a whole-body spiral CT diagnostic radiology scan,[26] which delivers a whole-body dose of about 20 mSv, is determined as follows:

20 mSv × 0.005% = 0.1% increased
lifetime risk of contracting cancer

Now comes the hard part, which is to interpret that risk metric. What exactly does it mean to us?

The risk percentage can be interpreted in many different ways, but most people find one of the following two interpretations to be the most comprehensible:

1. A risk of 0.1% is the same as saying that there is a 1 in 1,000 chance of getting a cancer from this radiation dose. To put it another way, if 1,000 patients received a spiral CT scan, only one of those patients would be expected to get a cancer from having had the scan. This interpretation raises the question: "Is 1 in 1,000 too much risk?"

2. We can alternatively compare 0.1% risk with the baseline risk of getting a cancer. The unfortunate truth that most people often fail to appreciate is that cancer death is quite

> common. Although the exact number is debatable, a figure of 25% (1 in 4) is a reasonable estimate of an average American's risk of dying from cancer. (It's much higher than 25% for a heavy smoker.) So the 0.1% risk from radiation needs to be added to that baseline risk we all share. This means that if your overall risk of death by cancer started at 25% before the spiral CT scan, then after the scan it's 25.1% because of the added radiation exposure. This interpretation raises the question: "Is moving your risk level from 25.0% to 25.1% too dangerous?"

Now, having a handle on the level of risk, you need to ask yourself a question: "Is this cancer risk acceptable *to me*?" And the answer probably depends on the benefit of the CT scan to you. If you've been in severe internal pain in your abdomen and a whole-body spiral CT scan (sometimes called a "cat" scan) represents the most informative procedure for diagnosing the source of that pain, you may feel that the risk is well worth the benefit. But if you have no symptoms of anything, and the procedure is being done just to screen for possible disease, you may not consider it worth the risk. But only you can decide that.

In any event, the ability to convert a radiation dose to a risk estimate levels the playing field. Armed with an accurate estimate of risk for a whole variety of exposure scenarios, we now have the tools we need to make decisions regarding which radiation exposures are acceptable and which are not. We can also compare dose levels from different types of carcinogens for the same level of risk. (For example, we can ask questions like: "How does the cancer risk of radiation compare with the cancer risk of cigarette smoking?") Most importantly, we now have the means to make an informed decision about controlling our radiation risks, and we can dispense with naive approaches, like the wife or daughter standard for radium ingestion mentioned earlier. We owe all this to the "most important people alive"—the atomic bomb survivors—for allowing us to learn from their tragic experience. Now it's up to us to be worthy stewards of this knowledge, and put it to good use for the betterment of public health.

As you ponder these risks, you may ask: "How relevant is this risk estimate, driven largely by the atomic bomb survivor data, to me and my situation? I haven't been exposed to such high doses. Are the estimated rates of cancer based on high dose bomb survivors even relevant to low doses that I'm exposed to?" Excellent questions! Let's explore this issue.

The overall higher doses of the bomb survivors could certainly make a difference to the accuracy of low dose risk estimates. Although it is often pointed out that most of the atomic bomb survivors received nowhere near the doses required to produce radiation sickness (i.e., greater than 1,000 mSv), the fact is that they still got a lot more radiation than you'll receive from something like a dental x-ray. When we talk about risk per unit dose, as we do above, we assume that there is a one-to-one relationship between dose and risk. That is, if you halve the dose you halve the risk, and if you double the dose then you double the risk. But is that true? Some scientists think that at very low doses this linear relationship doesn't hold true. They point out that many laboratory studies show cells can repair low levels of radiation damage, and it is only when damage levels get to the point where repair mechanisms can no longer cope that radiation risk then becomes directly proportional to dose. This is a fair enough criticism and, if it's true, our risk estimates inferred from the higher doses sustained by atomic bomb survivors may overestimate the risk from lower dose exposures, such as chest x-rays. But if the lower doses really aren't as dangerous because of repair processes, then the risk rates derived from atomic bomb survivors that are used to set radiation protection standards will be overprotective, not underprotective. We have certainly not underestimated the risk.[27]

The cancer risk rate of 0.005% per mSv, specified above, does not directly account for cellular repair and protection mechanisms that may be in play at low radiation doses.[28] For this reason, most reputable scientists believe the radiation risk estimates we use for purposes of radiation protections and risk benefit analysis, driven as they are by relatively high-dose atomic bomb survivor data, represent the worst case scenario. There is no credible scientific evidence to suggest that we have significantly understated the risk of cancer from radiation, nor is there any valid theoretical basis to suggest that we have done so. Since these risk estimates represent the worst case, when we set radiation dose limits to protect against the worst case, we are being highly conservative in terms of minimizing risk and, therefore, affording the highest level of protection. To put it simply, when radiation scientists aren't sure about something, they tend to recommend policy measures that overprotect the public, in order to compensate for that uncertainty.

This is not unlike the way civil engineers might approach the design of a suspension bridge to ensure the public is protected from its potential collapse. They would likely first calculate the strength of the steel cables needed to support the weight of the roadbed and vehicle traffic, based on fundamental static engineering principles and materials research, and then double that cable strength in building specifications, just in case the real cable that shows up at the construction site doesn't meet its theoretical strength specification. In effect, the project has been "over engineered" to provide a *margin of safety*. Radiation protection standards can be thought of in the same way. They have been over engineered by using worst-case atomic bomb survivor risk rates, coupled with added safety factors to account for uncertainties.

We'll further investigate cancer risk calculations and how they can be used to weigh risks and benefits in Part Three. But for now, let's just absorb the take-home message of this chapter. The message is that the LSS of the atomic bomb survivors is one of the strongest epidemiology studies ever conducted and has given us a very accurate estimate of cancer risk per unit of radiation dose. It is so strong that it is unlikely to be undermined by any other epidemiological study within our lifetimes.

To reiterate, it is important to remember that the LSS is so reliable because it is a large cohort study with many years of follow-up. In the epidemiology game, you want a cohort study—the gold standard—to provide the evidence. And the bigger the cohort study, the stronger the evidence. So the atomic bomb survivor cohort study, comprised of over 94,000 study subjects, recruited simultaneously and followed for over 65 years, is about the best evidence you can get and provides tremendous reassurance as to the reliability of the health risk estimates generated from its findings.

This also explains why epidemiologists pay little attention to the small case-control studies that occasionally appear in the scientific literature or popular press, claiming to contradict the findings of the LSS. For example, when the media reports on some newly published small case-control study that appears to show dental x-rays are producing brain tumors at an alarming rate, all it elicits from the epidemiologic community is a collective yawn. What authors of such reports often fail to mention is that it has been well established that people with brain tumors not only report more dental x-rays, but also report more exposure to hair driers, hair dyes, and electric razors.[29] The likely reason is that people with brain tumors are psychologically sensitized to any exposures related to their heads. So which is it, the dental x-rays, the hair dryers, the hair dye, or the electric razors that caused their brain tumors? In all likelihood, none of them. The best explanation for the apparent associations is simply that these case-control studies are tainted by recall or other biases. But at least these are known biases. In a well-designed, case-control study, there are ways of statistically adjusting for those biases that you can anticipate, but it is impossible to correct for biases that you don't even know exist. Alas, unknown biases are the Achilles heel of case-control studies.

So what do these contrarian case-control studies do to our established risk estimates for radiation-induced cancer? Nothing. The LSS is so strong that a small case-control study is never going to unseat its findings. We may improve the precision and accuracy of our current risk estimates as more information comes out of the LSS. We may even augment that evidence with data from some large ongoing case-control and cohort studies of medically exposed patients, and even larger studies of radiation workers with decades of exposure history.[30] But these studies will likely only improve the precision of our risk estimates at low doses, and allow us to confirm that the atomic bomb victim data is relevant to people who are exposed to radiation under different circumstances and at different dose rates. For the most part, we have what we have, in terms of cancer risk information, and what we have is pretty darn good.

In short, we know much more about the cancer risks of radiation than any other carcinogen.[31] So, if we, as reasonable and intelligent people concerned about our health and the health of others, can't come to a workable consensus about acceptable levels of cancer risk from the use of radiation, there is little hope that we will be able reach a consensus on any other cancer hazards for which very much less is known, and for which there is virtually no possibility that a huge cohort study will ever suddenly appear and provide the needed information. On the flip side, however, if we can agree on general principles by which cancer risks of radiation will be addressed, these same principles can likely be co-opted to comparable risk situations, where the data are scarcer, but the need no less compelling. Radiation is thus an excellent test case as to how best to handle environmental cancer threats in general.

With all this focus on radiation and cancer, one might presume that cancer was the main interest of the scientists who started the atomic bomb survivor studies. Remarkably, in the beginning, virtually no one was interested in measuring radiation-induced cancer risk. That was just a side activity. The original goal of the atomic bomb survivor studies was to measure the horrific genetic effects that certain people had predicted would occur among the progeny of the atomic bomb survivors. Some of these predictions of a population explosion of mutant children were the fantasy of science fiction writers with overactive imaginations, but others were based on the sound laboratory research of geneticists, radiation biologists, and other reputable scientists. Public interest and concern about genetic effects were extremely high, and cancer risks seemed to be of minor concern. Fortunately, the mutant progeny that had been predicted did not appear, although the survivor studies continue to search for them. In the next chapter we'll explore exactly why these mutants were expected, and why they never showed.

CHAPTER 10

BREEDING SEASON: GENETIC EFFECTS

When an atomic bomb is set off . . . and kills hundreds of thousands of people directly, enough mutations have been implanted in the survivors' [reproductive cells] to cause a comparable amount of genetic deaths . . . from now into the very distant future.

—Hermann J. Muller, 1946 Nobel Laureate in Physiology or Medicine

There is something fascinating about science. One gets such wholesale returns of conjecture out of such a trifling investment of fact.

—Mark Twain

LORD OF THE FLIES

Hermann Joseph Muller Jr. (1890–1967) was a man of humble origins. The grandson of European immigrants to New York City, he spent his childhood first in the neighborhood of Harlem in Manhattan and then in the Bronx. His father was a metal artisan, and his mother a housewife. They could offer him no more than overcrowded city public schools for his education, but Hermann made the most of those schools. He inherited from his father a love of the natural world, particularly all things biological. His aptitude for science combined with his strong work ethic won him a scholarship to Columbia University to study biology. At Columbia, his biological interests turned specifically to genetics and he eventually earned a PhD in the subject. He remained in academia, moving from student to faculty member, and climbed the academic ladder from one university to the next. Then, on December 10, 1946, at age 56, he stood in his formal jacket with tails to accept his Nobel Prize in Physiology or Medicine. Was his another story of living the American dream? Hardly.

The first half of the twentieth century had been the heyday for the radiation physicists. By 1946, there had been 21 Nobel Prizes in Physics awarded for discoveries related to radiation. Muller, a biologist, was being awarded a Nobel Prize in Physiology or Medicine "for the discovery of the production of mutations by means of x-ray irradiation." Radiation biology had arrived.

The experiments that won Muller his prize had actually been completed in 1927, well before World War II. Yet only 20 years later, after the war had ended and in the wake of the atomic bombings of Japan, had they caught the public's attention. The world had changed dramatically in those 20 years and so had Muller. He was not the man he was before. It was as though he were accepting his Nobel on behalf of a man who no longer existed.

The question that needed to be answered back in 1927 was whether radiation could produce inheritable mutations. An inheritable mutation is a permanent change to a gene that causes it to transmit an altered trait (i.e., a new trait, not found in the parents) to the offspring.[1] Muller showed that radiation could and did produce inheritable mutations, at least in his biological model, the common fruit fly (*Drosophila melanogaster*). But why study fruit flies?

Before fruit flies became the model of choice for genetic research, Gregor Mendel (1822–1884) had done groundbreaking hereditary experiments with ordinary garden peas. Mendel, by all accounts, was a washout at most everything he did. He had failed at an ambition to be a parish priest because he was too squeamish to visit sick parishioners. He spent two years at the University of Vienna studying natural history, but failed his examinations to get a teaching certificate. He even acknowledged that he had joined his order of Augustinian monks, in Brünn, Moravia (now Brno, Czech Republic), in 1843, largely to escape the pressures of working for a living.[2]

The abbot of the monastery had assigned Mendel gardening duties, likely because of his prior experience with natural history, which suited Mendel just fine. In his spare time, he started performing hybrid crosses between pea plants that displayed opposite traits (e.g., smooth vs. wrinkled seeds; green vs. yellow seeds; tall vs. short height) and was bemused to find that offspring of the crosses tended to have the trait of either one or the other parent plant, but not an intermediate trait.[3] For example, a hybrid cross between a tall plant (six-foot height) and a short plant (one-foot height) produced offspring that were either six-feet tall or one foot, but never three feet. This phenomenon may have been news to Mendel, but botanists of the time were well aware of it, although they couldn't provide an explanation.

But Mendel, like John Snow, appreciated the value of counting and measuring things. So he kept meticulous records of his pea hybridization experiments. In the winter, when there was no gardening to be done, he amused himself by studying his pea records. At some point he noticed that no matter what traits he crossed, there were always constant ratios of the two traits among the offspring, and the exact ratio depended upon whether it was a first generation or second generation cross. The constant ratios followed very consistent patterns, regardless of the trait being examined. Mendel reported his findings at the 1865 meeting of the Brünn Society for the Study of Natural History, but no one at the meeting was interested in Mendel's ratios. His report was published in the meeting's proceedings and forgotten.

Then in 1900, three botanists who were working independently with different species of plants and different traits, rediscovered the ratios. To their disappointment, their subsequent searches of the literature revealed that Mendel had scooped them 35 years earlier.[4] Nevertheless, they had demonstrated that Mendel's ratios were not restricted to just peas. Still, were they restricted just to plants?

Mendel's ratios were ultimately codified, along with other concepts of inheritance that he revealed in his writings, into the grander *Mendelian principles of inheritance*, a set of rules that defines how hereditary traits are transmitted from one generation to the next. The ramifications of the Mendelian principles to the science of genetics were completely revolutionary. Description of those principles is well beyond the scope of this book. Nevertheless, we can say that they have certain mechanistic implications as to how heredity works and suggest that inheritable traits are transmitted from parent to offspring in the form of discrete units of information that we now know as *genes*. Mendel's work implied that some versions of genes, known as *variants*,[5] were associated with dominant traits (e.g., smooth seed, yellow seed, and tall plant), while other genetic variants were associated with recessive traits (e.g., wrinkled seeds, green seeds, and short plant). Whenever a dominant gene was present it suppressed expression of the recessive gene; therefore, recessive genes could only be expressed in the absence of dominant genes. Mendel's fundamental conclusion—that the male and female parents could each contribute different versions of the same gene to their offspring, and that some variants dominated over others—was tremendously insightful. His peas had given him the awareness that genes control heredity; but, at that point, genes existed solely as an abstract concept that enabled scientists to predict the distribution of traits among offspring, nothing more.

Mendel was definitely on to something, but to progress further, a better biological model than pea plants, or any other type of plant for that matter, would be needed. Enter the common fruit fly.

Thomas Hunt Morgan (1866–1945) was a prominent American biologist on the faculty of Columbia University, which was considered by many to have the premier biology department in the United States at the time. He was among the first to recognize that fruit flies are an ideal model for performing heredity research in animals.[6]

In many respects Morgan was the antithesis of his future protégé, Muller. They had very different personalities, perhaps owing to their notably different backgrounds. While Muller was self-conscious about his first generation immigrant status, Morgan had a pedigree that was none too shabby. His father's side of the family was originally of Welsh descent. They had settled in Kentucky, opened a store and trading route, and soon became one of the wealthiest families in the state. Not only that, virtually all the men in the family had been war heroes in either the Revolutionary War or the Civil War (fighting on the Confederate side). His uncle, John Hunt Morgan, in particular, was nearly worshipped as a fallen hero of the Confederacy in his hometown of Lexington. Morgan's mother also had blue bloodlines. She came from the aristocratic Howard family of Maryland. The Howards had long been involved in Maryland politics, with Morgan's grandfather once serving as governor. And his great grandfather was Francis Scott Key, the famed lyricist of the *Star Spangled Banner*. None of this pomp phased Morgan. He never talked much about his family heritage, preferring instead to discuss biology, his true passion.

Morgan's famous "Fly Room" was a laboratory where he conducted his genetic research on fruit flies.[7] He had previously experimented with dozens of different types of animal species as models for inheritance studies, but each had its drawbacks. Ultimately, he settled on the fruit fly as the most practical genetics research tool, for reasons that will soon become apparent.

Morgan was a questioner by nature, but his doctoral training had made him an even greater skeptic. In 1886, he was among the first graduate students enrolled at the then ten-year-old Johns Hopkins University in Baltimore, a new type of graduate university founded exclusively upon a research premise, which was a novel educational concept at the time. There were no formal lecture classes, only laboratories. As a learning exercise, before starting any original research on their own, students were required to select some important published research in their field and repeat the experiments themselves to either verify or refute the findings.[8] All scientific dogma was questioned, and nothing accepted without replication. Given this mindset, it is not surprising that Morgan was skeptical that Mendel's findings in pea plants were relevant to animals. He decided to repeat Mendel's inheritance experiments, but using his fruit flies instead of pea plants as the model organism. He didn't expect to find much.

Morgan soon proved himself wrong about Mendel's work. Using his fruit fly model, he discovered, to his own astonishment, that the traits of fruit flies transmit from generation to generation in a pattern that exactly corresponds to Mendel's principles. That is, the pattern of inheritance that Mendel saw for tall versus short pea plants was exactly the same as Morgan saw for long versus short wings. And all of the second-generation hybridizations and back-crosses with the flies produced offspring in the same ratios as found for Mendel's peas. Morgan then analyzed other traits in the flies and they behaved similarly. At this point, Morgan, too, became an unabashed disciple of what had become known as *Mendelianism*.

Morgan did more, however, than just show that Mendel's work in plants could be repeated in an animal model. He took Mendelianism to the next level. He was able to convincingly demonstrate that the pattern of gene transmission from one generation to the next corresponded to the pattern of chromosome transmission, causing him to make the conclusion that genes were not just a scientific conceptualization; they were actual physical entities that resided on chromosomes, the threadlike structures found in the nucleus of cells. This insight proved to be absolutely correct, and would earn him the Nobel Prize in Physiology or Medicine in 1933. Genes were real things that could potentially be isolated, manipulated, and studied. They were also things that could be damaged.

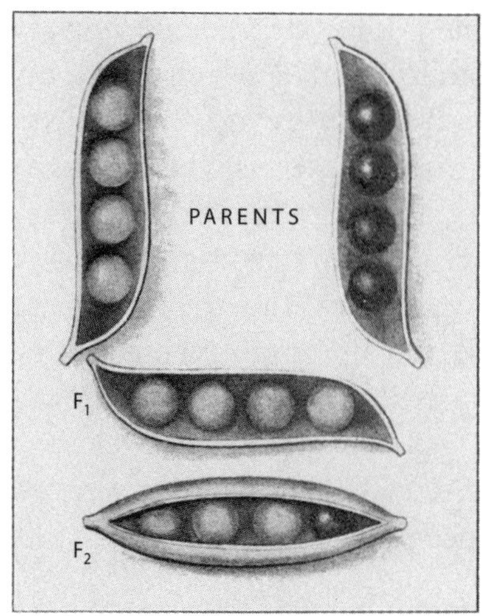

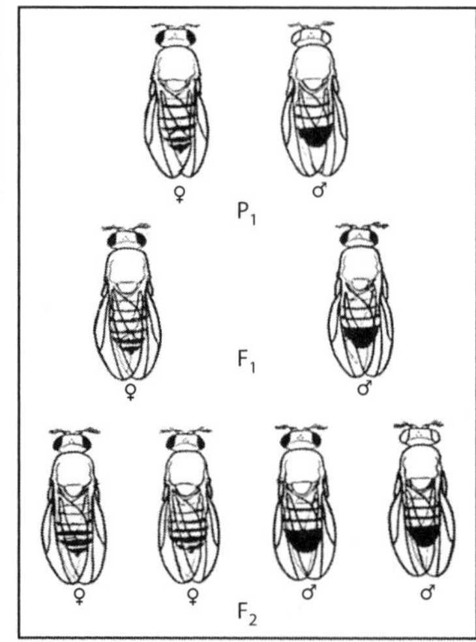

FIGURE 10.1. IDENTICAL GENETIC TRANSMISSION IN PEAS AND FRUIT FLIES. Thomas Hunt Morgan found the exact same genetic inheritance patterns in fruit flies that Gregor Mendel had previously reported in peas, suggesting that the fundamental mechanisms of genetics were universal for all species, whether plant or animal. When Mendel crossed yellow peas with green peas, the first generation (F1) of peas were all yellow, but the second generation (F2) were 3/4 yellow and 1/4 green. Morgan found the same ratios for hybrid crosses of fruit flies with different eye colors. Crossing red-eyed flies with white-eyed flies resulted in all red-eyed progeny in the first generation, but 3/4 red-eyed and 1/4 white-eyed progeny in the second generation. These findings ultimately resulted in the fruit fly becoming the standard animal model for studying the genetic effects of radiation.

After completing his undergraduate studies, Muller stayed on at Columbia to obtain his PhD, and Morgan became his doctoral mentor. The two men even ended up coauthoring a book entitled *The Mechanism of Mendelian Heredity*.[9] It was considered by many to be the seminal textbook on Mendelian genetics, articulating Mendel's principles better than even Mendel had. Morgan even sold Muller on the idea that fruit flies were the best tool for genetic research. Nevertheless, the two men would never become friends. Muller felt that his mentor, Morgan, did not give him enough recognition for the brilliance of his ideas. Morgan, who was data driven, thought ideas were cheap to come by and preferred to heap his praise on those students who had actually toiled to produce the data needed to test hypotheses. Ironically, Morgan, for all his aristocratic roots, identified more closely with the workers than the intelligentsia.[10]

Tension between Morgan and Muller persisted, and eventually the two men went their separate ways. Still, both remained devoted to fruit fly research for the rest of their lives. What was it about fruit flies that made them so special?

If you want a pet, buy a dog. If you want meat, raise pigs. If you want to do genetic research, however, get yourself some fruit flies. For fruit flies, their virtues are in their numbers. They have very short generation times (10 days), produce large numbers of offspring per generation (about 500), and are very easy to care for and manipulate. Give them a piece of banana to eat and a small jar for a home, and they do quite nicely. It is also easy to distinguish the males from the females (by a dot on their abdomen). Best of all, unlike dogs and pigs, they don't bite.

Fruit flies enabled genetic research that could not be directly performed in higher animals or humans. Consider this: humans have generation times that average about 20 years, so the number of generations of humans since 15 AD (i.e., 2,000 years ago) has been about 100. In contrast, an equal number of generations of fruit flies can be recapitulated in less than three years. And the offspring from a single pair, magnified over those 100 generations, would produce many times more descendants than there are people on Earth. In short, if a lot of numbers in a brief period of time are required, fruit flies can deliver.[11]

This is particularly important if your interest is in inheritable mutations. These altered traits are sometimes benign or even advantageous, but usually are harmful to the offspring. Mutations occur naturally and randomly in all populations of plants and animals, but they are very rare. It took two years for Morgan to find his first mutant, a white-eyed fly, among the hundreds of thousands of flies that he reared in his lab.[12] (The eyes are normally red.) So, if you want to find mutants, you have to look very hard. This is why fruit flies are useful for studying mutations. Since there are so many flies available to examine, if you have the patience to sit and look at each one, you will be able to find even the rarest of mutations, provided that your butt can hold out. This is reminiscent of Rutherford's atomic "bounce back" experiment, where the recoil of alpha particles off of gold foil was so rare that it took many hours of watchful waiting, looking through a microscope in a dark room, to be able to measure them. Nonetheless, mutations, like alpha particle recoil, were considered so important to understanding the science that it was worth the wait.[13]

But Muller was frustrated with the very low frequencies of mutations in the flies.[14] The rarity of mutations was slowing the pace of genetics research to a crawl. And it was sheer drudgery inspecting flies, one by one, looking for . . . what? Mutants could take any form, and screeners never knew exactly what body alterations to look for. If progress were to be made, the way they screened for mutants needed to change.

Then Muller had a brilliant idea. He knew (as did all other geneticists) that most mutations of genes are typically masked because chromosomes come in pairs (one coming from the father and one from the mother). Unless a gene undergoes a mutation on both chromosomes, the presence of a normal gene usually masks the mutant trait, because most mutations are recessive; hence, the mutation cannot be detected in the offspring. The exception to this is the X chromosome, the female sex chromosome. In females, the X chromosome is present as a pair, XX, the same as for the nonsex chromosomes. In males, however, it is accompanied by a male Y chromosome, forming a dissimilar XY pair. The Y is much smaller than the X chromosome, and only carries genes related to maleness, so there are no copies of X chromosome genes on the Y chromosome to mask X chromosome mutations.[15] This makes the X chromosome an ideal target to screen for mutations. Even the re-

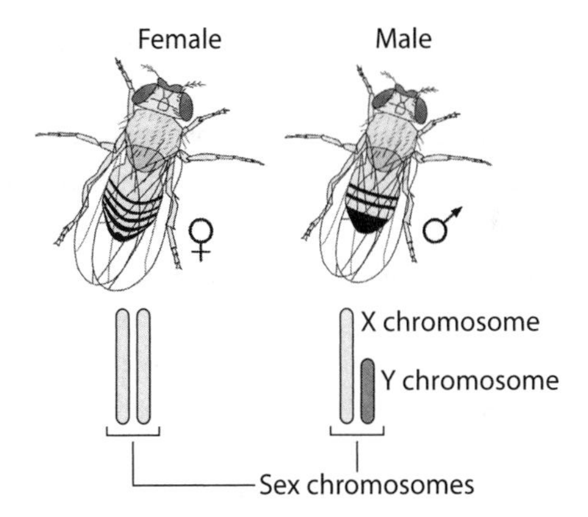

FIGURE 10.2. MALE FRUIT FLIES HAVE A SINGLE X CHROMOSOME. Similar to many other animal species, female fruit flies have two X chromosomes while males have only one. The key to Hermann Muller's success in measuring radiation-induced mutations in fruit flies was his insight that this dissimilarity in the number of X chromosomes between genders meant that females had two copies of those genes located on the X chromosome. Muller was able to experimentally exploit this genetic gender difference by specifically studying X-linked lethal mutations of male flies. The increased sensitivity of his male fruit fly mutation assay allowed him to measure radiation mutagenesis down to very low doses, which had tremendous implications in terms of protection against radiation-induced cancer in humans.

cessive mutations on X should be revealed in male offspring because males only have one X chromosome, not two.

The second part of Muller's insight had to do with the type of mutant most likely to be produced. He reasoned that, since the X chromosome is relatively large, it likely carries many essential genes. Mutating an essential gene on an X chromosome should usually result in death of the offspring (i.e., an X-linked lethal mutation), but the female offspring would not die because they have a second X with a normal gene to mask the mutation. Males, in contrast, would get no such reprieve. For males, a mutation in an essential gene on their lone X chromosome would spell certain demise.

Putting this all together, it followed that in the broods of mated flies, where the mother passed down an X-linked lethal mutation, there should be missing males. That is, the ratio of males to females would not be the typical 1:1 sex ratio, since the males with the mutated X would not be viable. In those cases, the brood hatchlings would be predominantly females. With this realization, Muller had a screening test for X-linked lethal mutations that didn't involve inspecting every fly with a jeweler's magnifying glass, one at a time. In this new screening test he could detect mutants simply by counting the females and males in a brood hatch and calculating the sex ratio. Broods with lower than 1:1 ratios (males to females) indicated an X-linked lethal mutation. Ironically, these postulated lethal mutations were not revealed by the presence of any mutants in the culture, but rather by their absence. The progeny of X-linked lethal mutations were male ghosts.

Muller's sex-ratio assay amounted to a revolution in the way mutation screening was performed and quickened the pace of the work by orders of magnitude. By unmasking the X-linked lethal mutations, there were now mutation frequencies high enough to allow different chemical and physical agents to be tested to see how they modified the background mutation rates. This fast-paced approach had the potential to provide insight into the underlying mechanisms involved, which was always the fundamental goal of the research. Unfortunately, it would be six more years before Muller decided to test whether radiation could modify mutation frequencies. This is because one of the first sex-ratio screening experiments he performed suggested that radiation couldn't be involved.

The first agent that Muller tested was temperature. He held flies at different temperatures and showed that the mutation rate modestly increased with temperature. The shape of the dose response curve revealed that the effect of temperature on mutation rate was not a straight line (i.e., linear), but rather was bent (i.e., curvilinear) in a way that suggested the relationship between temperature and mutation rates was exponential.[16] Since it was well established that the rates of chemical reactions varied exponentially with temperature, Muller reasoned that the underlying mechanism of mutation production in cells must, therefore, be chemically based, not physically based. So he proceeded to test various chemicals for their effects on mutation rates. He ignored radiation simply because it was a physical agent. What he did not realize at the time was that radiation interacts with the biochemistry of the cell.

Nonetheless, Muller eventually decided that it was a mistake to dismiss radiation without testing it. This was largely because it had previously been reported that radiation can damage chromosomes.[17] Could something that damages chromosomes not have some effect on mutation rates? It didn't seem likely, so Muller got a friend to set up an x-ray machine in his lab, and he used his sex-ratio screening test to measure mutation frequencies. What he found was astounding.

He started out with relatively high radiation doses but found that the radiation was causing sterility, so he wasn't getting any offspring at all. Since there can't be mutants if there are no offspring, he lowered the dose to just below the threshold for sterility in fruit flies.[18] What he found was that this radiation dose increased the mutation rate to about 150 times the normal mutation rate!

Then he went further. He halved the dose and found half the number of mutants. When he halved the dose again, he again found half the mutants. In short, because Muller had so many flies, there was no radiation dose so low that he couldn't detect some mutants, even when the frequency of mutations was extremely small. The sheer magnitude of the numbers of flies that he could irradiate and inspect allowed him to find mutations, regardless of how low the frequency. All he had to do was screen more flies.

The ability to detect mutations no matter how infrequently they occurred permitted him to test low doses of radiation that only rarely produced mutations. In the end, Muller concluded that there was some probability of producing inheritable mutations in fruit flies at all radiation doses, no matter how low, and that the frequency of mutations was directly proportional to dose (i.e., there was a linear dose response).[19] Accordingly, there was no tolerance dose below which no mutations occurred! The armor of the concept of a tolerance dose for all health effects, which the radiation protection professionals relied on to provide absolute protection, had gotten its first chink. There would be more chinks to follow, but inheritable mutation production in fruit flies was the first. A paradigm shift had occurred in radiation biology, and everyone felt it.

DEAR COMRADE STALIN

Soon after Muller completed his radiation experiments, the Great Depression hit, and affected him particularly hard. He had struggled financially all of his life, and he had become greatly disillusioned with capitalism . . . and the United States. By this time, Muller was on the faculty of the University of Texas. Some of Muller's students at Texas had introduced him to communism, and he helped them publish an underground student newspaper, *The Spark*, that promoted the virtues of communism to the wider university community. Always the enthusiast for social experimentation, Muller saw the burgeoning Soviet Union, and its great experiment with communism, as the best hope for the world.

In 1933, Muller was invited to live and work in the Soviet Union by his Russian geneticist friend, Nikolai Ivanovich Vavilov (1887–1943). This was quite an honor, because Vavilov was among the most famous and respected scientists in the Soviet Union. So Muller enthusiastically immigrated to Moscow, where he thought he would find utopia. He took a position as senior geneticist at the Institute of Genetics of the USSR Academy of Sciences. By 1936, he felt confident enough in his position to write a letter directly to Soviet leader Joseph Stalin (1878–1953), to unfold his ideas for a biological revolution to augment the social revolution of communism. This biological revolution he termed "positive eugenics." The plan called for the dedicated communist proletariat to forgo their own reproduction in favor of artificial insemination with sperm stock that was provided by the state and certified to have superior "genetic equipment."

Muller told Stalin:

> True we have today, rooted in the traditions from the bourgeois society of our past, the idea that our child must be derived from our own reproductive cells. . . . But with the gradual growth and

> understanding of the great social responsibilities and duties of reproduction, and of the separability [*sic*] of reproduction from the sexual act, these feelings will more and more come to be replaced by others equally strong and effective in furthering a high type of family.[20]

He then went on to promise Stalin that such a program would convey rapid benefits:

> After 20 years, there should already be noteworthy results accruing for the benefit of the nation. And if at that time capitalism still exists beyond our borders, this vital wealth of youthful cadres, already strong through social and environmental means, but then supplemented even by means of genetics, could not fail to be of considerable advantage to our side.

Muller included with his letter a copy of his recent book, *Out of the Night: A Biologist's View of the Future*, which spelled out the details of his positive eugenics plan.[21] Stalin read it, but was not pleased because the book was filled with Mendelian ideas. It seems that an anti-Mendelian geneticist, Trofim D. Lysenko (1898–1976), had Stalin's ear and had convinced him that Mendelianism had anticommunist undertones.[22] That Mendel himself was a Catholic monk was also seen as an affront to atheistic communism. Even putting these issues aside, the major problem with the book for the Soviets was that it argued against the possibility of inheriting characteristics acquired during life (e.g., exercising might make you stronger, but you wouldn't have stronger offspring).[23] The Soviets had drastically reformed their agricultural agenda and instituted programs to "teach" plants how to better acclimate to the Soviet Union's harsh winters so they could produce hardier seeds for planting future crops. If acquired characteristics could not be inherited, it meant that the Soviet agricultural reforms were doomed to failure. Time would prove this to be so, but in the meantime the Soviets were sensitive to any criticism of their agricultural policies, even if made obliquely, as in Muller's book. Because of this, Muller's proposal for positive eugenics was dead in the water unless it could somehow be made compatible with genetic transmission of acquired characteristics, which it obviously could not. Muller's letter and book had opened up a political can of worms that he was unable to close back up. His life in his new homeland would now become very precarious.

Stalin condemned Muller's book as being unfit for communist citizens and moved to suppress it. Fearing for his life, Muller left the Soviet Union in a hurry, traveled through different European countries for a couple of years, transporting 250 strains of fruit flies along with him, and ultimately returned penniless to the United States in 1939. He was finally offered an untenured position at Amherst College, where a heavy teaching load limited his research activities. Ultimately, he would find a long-term academic home at Indiana University, where he resumed his research in earnest.

Communism had not created the utopia that Muller had foreseen. He found that communist social policy had polluted the science of genetics, largely through the efforts of Lysenko. While Lysenko and his fake genetics thrived under communism, Vavilov, Muller's legitimate geneticist friend who had originally invited Muller to the Soviet Union, did not have the option of escaping to the United States when the crackdown on Mendelianism arrived. Instead, Vavilov became a marked man because of his "capitalist vision" of genetics and because of his close friendship with Muller. In 1940, while on a field trip to the Ukraine, Vavilov found that his research assistants had all vanished and were suddenly replaced by security police with a warrant for his arrest. Vavilov was accused of being a spy for the British and sentenced to death. Protests from scientific colleagues got his sentence reduced to 20 years of forced prison labor. He died in prison of malnutrition in 1943.[24] Muller was devastated. He kept a picture of his friend Vavilov on his desk for the rest of his life.[25]

Communism had politicized Mendelianism and perverted the science of genetics. Soon Muller would learn that genetics was equally capable of perverting social policy, and for that he would have to share some of the blame.

While Muller had been in the Soviet Union, interest in genetics had risen in the United States. Although the public was largely ignorant of the Mendelian principles of inheritance, most educated people understood that genes were some type of biological component that allowed certain traits to be transmitted from one generation to the next. If a woman had a large nose, for example, just like her father's, it was presumed that some gene was likely involved, and that the gene had been transmitted from father to daughter. Although the public was still fuzzy about exactly what genes were made of (as were the scientists), since the 1920s it had become well understood that the genes resided in the chromosomes of the father's sperm and the mother's egg.

People also had some appreciation that there could be altered (mutated) forms of genes associated with certain distorted traits and these could be passed down from generation to generation. What little they did know about the inheritance of mutated genes and the effects on the progeny came largely from the ideas of Charles Darwin (1809–1882), the famous naturalist. Darwin's theory on the evolution of species was based on the more likely survival of individuals carrying variations associated with favorable traits, and the likely demise of individuals carrying variations for unfavorable traits. Through this preferential selection of individuals carrying advantageous traits, Darwin argued that a species becomes more and more highly adapted to its environment with each new generation.

Darwin became notorious in some quarters because his proposed biological mechanism for the creation of new species contradicted the literal account of creation in the Bible, and was, therefore, seen as being an affront to God's role as creator of all things. Darwin published his theory in 1859, and it was hotly debated by the scientific and religious elites for some time. However, it didn't really become socially controversial in the United States until questions arose about whether it was appropriate for Darwin's ideas to be taught in public schools. These controversies ultimately resulted in highly publicized court dramas that were well covered by the newspapers in the 1920s, the most famous being the 1925 "Monkey Trial" of science teacher John Scopes (1900–1970), in Tennessee.[26] With events like that, by the time Muller returned to America in 1939, hardly anyone had not heard about evolution, even if they didn't have a very clear idea of what genes actually were. The public was also aware that most mutations were unfavorable, typically resulting in deformities, and it was these kinds of mutations that people feared in their own offspring.

DARWIN'S KIN

Further contributing to public consciousness of genetic defects was the growth in eugenic ideas, similar to the ones that Muller had promoted to the Soviets. Eugenics wasn't original to Muller, he simply used his scientific credentials to promote it. The eugenics movement was actually the brainchild of Darwin's cousin, Francis Galton (1822–1911), who was interested in the implications of Darwin's theory on the human condition. He concluded that the human species was no longer benefiting from the survival of the fittest because social practices allowed even the most physically unfit individuals to survive and reproduce, which permitted disadvantageous genes to be continually maintained within the human gene pool,[27] rather than being selectively eliminated. This allegedly contributed to the genetic deterioration of the human species.

As we've already seen, Muller was a disciple of eugenics. Not content to simply advance the basic science of genetics, he always sought to promote the practical implications of his work. The practical implications of Roentgen's discovery of x-rays were readily apparent to everyone, scientist and nonscientist alike. The human relevance of discoveries in fruit fly genetics, however, was more obscure, and Muller saw eugenics as a vehicle through which his work might have much broader implications for mankind.

Unfortunately, Muller was as naive about eugenics as about communism. He promoted the concept of *positive eugenics*, where men with supposedly good genes would be encouraged to donate their sperm, and socially conscious women with purportedly good genetics would volunteer to be receptacles of the donations. He was convinced that once people realized that they were, in effect, the repositories of genetic information for future generations, they would choose reproductive behaviors that benefited, rather than harmed, their future descendants. But it was *negative eugenics* that was the more straightforward approach to achieving the same results. Muller acknowledged that negative eugenics would work equally well, yet he somehow thought this nefarious form of eugenics would never be employed in a "civilized" society. However, while positive eugenics required people's voluntary and ongoing cooperation, negative eugenics could be irreversibly imposed upon people in the form of forced sterilizations, castrations, and even murder of the genetically undesirable. In retrospect, it's hard to understand how Muller and other prominent scientists who promoted eugenics were so naive to think that negative eugenics would never be used.

The reality was that the eugenic movement eventually sought to improve the human gene pool by actively preventing those with apparently bad genes from reproducing. In practice, eugenics fanatics promoted mandatory sterilization of anyone they judged genetically unfit.[28] The Nazis used eugenic ideas as a rationalization not only for their purges against Jews and "Gypsies"—racial and ethnic groups they judged to be weak—but also for their persecution of Aryan Germans they considered genetically defective, such as people with mental disabilities. (Muller himself would have been seen as genetically unfit by the Nazis, since his mother was Jewish.) The Nazis took their eugenic policies well beyond forced sterilization,[29] and used it as justification for all types of atrocities, including mass executions of the "unfit."[30]

Nazi atrocities made eugenics a dirty word, and the alleged science underlying it has been largely debunked as severely flawed,[31] so few people talk about eugenics anymore. Nevertheless, the legacy of eugenics is that the common people who had witnessed its heyday in the 1930s and 1940s were now well aware of the notion that it was possible for gene pools to be contaminated with various mutations that contributed to deformities and disease. It was bad enough to think that these mutations could be introduced to the gene pool through random genetic errors, and more terrifying still to think that radiation could actually produce an added mutation burden that would be passed around forever, producing diseases and deformity in anyone unlucky enough to be randomly dealt a mutant joker from the deck of all available genes. Radiation's mutagenic effects were thus seen as adding more jokers to the deck, and thereby increasing everyone's probability of drawing a hand with a joker. It was believed that radiation, because of its alleged ability to pollute the human gene pool, put everyone at increased risk of producing defective offspring, even those who were never directly exposed to the radiation themselves.

This was the mindset of the public in the years right after the atomic bombing of Japan. Everyone was expecting a breed of monsters to be released on society when the bomb victims reproduced, and Muller's public statements were not allaying any of their fears.

Winning the Nobel Prize in 1946 retrieved Muller from his relative obscurity and put him directly in the public eye. He leveraged his new fame. Putting his disreputable ventures into communism and eugenics behind him, he started telling the enamored public about the social implications of his work.[32]

In March of 1949, Muller addressed the US Conference of Mayors in Washington, DC, and told them he expected more deaths in the aftermath of the atomic bombing in Japan than had occurred at the time of the bombings. He said this would be due to debilitating hereditary defects that would occur in subsequent generations, and that they could affect whole populations for thousands of years. In short, the atomic bombing of Japan had released a genetic calamity on the world. He warned: "With the increasing use of atomic energy, even for peacetime purposes, the problem will become very important of insuring that the human germ plasm [i.e., gene pool] . . . is effectively protected from this additional and potent source of permanent contamination [i.e., radiation]."[33] Such a warning from a Nobel Prize winner gave people pause. But, alas, as he had been about communism and eugenics, Muller would be wrong about radiation too. His dire predictions made to the mayors, and sensationalized by every major news organizations, would never happen.

Although he acknowledged that "many of the [mutation rates] are only very roughly known even for *Drosophila* (fruit flies), and we are admittedly extrapolating too far in applying this to man," Muller nevertheless came up with some scary estimates of radiation-induced mutation rates for humans based on his fruit fly data. He combined his latest fruit fly findings, which suggested the existence of a phenomenon he called *partial dominance* even for recessive genes, with some equations of theoretical geneticist C. H. Danforth; these suggested a slow rate of elimination of deleterious mutations from a population, and Muller concluded that every newly arising mutation ultimately resulted in the death of at least one descendant. Since he also estimated an extremely high mutation rate of "one newly arisen mutant gene in ten germ cells," the implications of excess mutations in the population caused by radiation exposure were enormous.

Despite the fact that Muller acknowledged he was working from a "state of ignorance" about mutation rates in humans, he nonetheless felt there was a need to "remain on the safe side." Other scientists were concerned that Muller's extrapolation to humans went well beyond what could be gleaned from the fly data. The chairman of the Radiological Committee of the British Medical Council cautioned Muller: "It is a pity to apply the results of experiments on insects directly to man without any qualification. . . . In the absence of proof it would greatly strengthen the argument if a suitable proviso were inserted before the question of applicability to man was discussed." But Muller refused to attach any caveats to his assertions regarding the allegedly high risk to humans of inheritable mutations. Rather, he retorted, "The burden of proof is on the other side, with those who wish to show that man is *not* like a fruit fly in these matters."[34]

In the absence of any data other than for fruit flies, it was impossible to challenge Muller's assertions. The blathering back and forth about whether fly data could be equated with higher organisms amounted to no more than idle talk in the absence of data. Only data would settle the issue. Luckily data soon appeared, and it came from the Chicago Health Division of the Manhattan Project.

OF MICE AND MEN

In 1941, as a consequence of Muller's 1927 findings with fruit flies and the concerns it raised, the Chicago Health Division of the Manhattan Project received a new secret assignment:[35] to determine how much radiation a man could absorb daily, without increasing his chances of having abnormal children.[36] The scientists then began an inheritable mutation project with mice to duplicate Muller's radiation studies on fruit flies. The project, headed by Donald R. Charles, was located at the University of Rochester in New York State. Starting with over 5,000 male and 20,000 female mice, the researchers bred hundreds of thousands of offspring, examining each for 170 different traits that might be susceptible to mutation.[37] When some deformity was noticed that might be a mutation, the affected mouse was further bred to determine whether that defect could be inherited by its offspring. If the defect was inheritable, it was counted as a *bona fide* inheritable mutation.

What they found was not reassuring. The lowest dose tested was 130 mSv.[38] The results showed that offspring of male mice that had been irradiated to 130 mSv and bred with unirradiated females had mutation rates approximately three times the natural background mutation rate. At the highest dose tested (2,380 mSv) there were nearly nine times as many radiation-induced mutations as background mutations. It was not possible to push doses much higher than this, because higher doses produced sterility. Ninefold was high, but it wasn't nearly the 150-fold increase in mutations that Muller had seen at doses that nearly produced sterility in fruit flies.

The interpretation of these findings was that there was a practical limit to the level that mutations could be increased by radiation, and the ceiling for the mutation level was determined by the threshold dose for inducing sterility. Once sterility was induced, there would be no progeny, either normal or mutated.

These findings in mice came out just as the Life Span Study (LSS) was planning the design for its study of inheritable mutations in atomic bomb survivors. Muller was asked to consult on the project. He and James V. Neel (1915–2000), the lead epidemiologist on the inheritable mutation project of the LSS, sat down and crunched the numbers with regard to what they might expect to find among atomic bomb survivors if mutation rates in humans were similar to those in mice.[39] They estimated that there would be about 12,000 children born in Japan for which either one of both parents had been exposed to bomb radiation. And they estimated that the average dose for the exposed parent would be 3,000 mSv. (The real average dose turned out to be less than 300 mSv.) Using these assumptions, coupled with the mouse mutation rates, they estimated that between 36 and 72 children would be born with an abnormality due to the bombing. These would be in addition to the background inheritable mutations expected among a group of 12,000 children, which was about 120 (1%).

This meant that they could expect to find only 192 children out of 12,000 children (1.6%), or less, with inheritable abnormalities in the bombed cities (Hiroshima and Nagasaki), versus 120 (1.0%) in the unbombed control city (Kure). Based on these numbers, Muller became skeptical that any epidemiological study, no matter how well designed, could detect such a small difference (1.6% versus 1.0%). He told the other scientists that the planned epidemiological study represented "a dangerous situation from the standpoint of the general and scientific public, who would be prone to assume that if no effect could be demonstrated, there must not have been one."[40] Since the human study would most likely find nothing, Muller suggested that a parallel study in animals be conducted, perhaps in monkeys, where all scientific parameters could be more tightly controlled. Such a study would have the higher sensitivity needed to detect this anticipated level of difference.

An enormous monkey study was quickly ruled infeasible for a variety of practical reasons, so mice, once again, became the only option. It was decided that another mouse study, even larger than the Rochester study, with irradiation conditions that more closely mimicked human exposure conditions could settle the question. With this, the megamouse project was conceived.[41] It would ultimately be conducted at Oak Ridge National Laboratories in Tennessee, by Liane B. Russell and William L. Russell and would involve 7,000,000 mice (nearly the current human population of New York City); this was a mouse study that would be almost 100 times larger than the human LSS.

It took 10 years to complete the megamouse project, and the LSS is still ongoing. True to prediction, the LSS never found a significant increase in inheritable mutations. The mouse study, however, did show a significant increase, but it was only twofold, as opposed to the ninefold increase seen in the Rochester mouse study. Why the difference?

For the most part, the Oak Ridge findings agreed with the earlier Rochester work, but a few very key additional facts emerged that greatly lowered the anxiety level. First, researchers found that by lowering the dose rate to which the mice were exposed from 900 mSv per minute (the Rochester dose rate) to 8 mSv per minute (i.e., much closer to the dose rates that radiation workers would actually experience), the mutation rates dropped dramatically.[42] Second, if the male mice were not immediately mated with the females following irradiation, the mutation rate also dropped precipitously. The longer the time between irradiation and mating, the lower the mutation rate. Some reduction in mutation rate could be seen if mating was delayed even a few days, but the rate continued to drop even up to a two-month period after irradiation. The likely explanation was that newly produced sperm replenishes the radiation-damaged sperm during this delay period, so that nearly all the damaged sperm had been replaced two months after irradiation. Also, postirradiation cellular repair processes probably mitigated the mutation production when dose rates were lower and conception times delayed.

Of the two modifying factors—dose rate and postirradiation conception interval—the lower dose rate was the more important factor in mitigating inheritable mutation risk in a radiation worker setting. Based on the new mouse data collected at the lower dose rate, it was determined that it would take a dose as high as 1,000 mSv to double the natural background rate of inheritable mutations, rather than just the 80 mSv suggested by the Rochester high dose-rate experiments. Also, although the time between radiation exposure and conception could not be controlled among humans, the reality is that very few of a male worker's offspring would be conceived immediately following a relatively high dose exposure to his genitals, whereas all of the mice in the Rochester study were conceived immediately following irradiation. This, too, was a factor that would significantly suppress the actual inheritable mutation rates in humans in real-life exposure situations. Thus, these new findings in mice did much to relieve the tremendous anxiety caused by the earlier fruit fly studies.

In the end, the only useful data for setting radiation protection standards for inheritable mutations came from the positive mouse studies and the one negative human study. Even though negative, the LSS's findings suggested that human sensitivity to radiation was not significantly higher than the sensitivity of mice, and that dose limits set to protect against inheritable mutations could be set based on the mouse data.

Ironically, identification of the exact dose limits for protection against inheritable mutations became a moot point. This is because human sensitivity to radiation-induced inheritable mutation turned out to be much lower than human sensitivity to radiation-induced cancer. So the dose-limiting factor for human protection standards came to be the cancer health effect and not the inheritable mutation effect. In other words, the dose limits set to protect against radiation-induced cancer more than adequately protects against the inheritable mutation hazard. In protecting against cancer, you protect against both.

Some people couldn't accept that it wasn't possible to find significant numbers of mutants among the atomic bomb survivors. After all, there were birth defects among the children of pregnant women exposed to bomb radiation (primarily small head sizes and mental deficiencies). Weren't they mutants?

No, they weren't. It is important to understand that not all children with birth defects represent new inheritable mutants. In fact, very few do. And this is particularly true for children who are born with defects because they were exposed to high doses of radiation while in the womb. When doses get high enough to cause cell death (~1,000 mSv), they can cause radiation sickness in adults due to the killing of cells in critical tissues, but they can also cause birth defects by killing off cells in critical tissues that are growing and developing within embryos. This can be thought of as a type of radiation sickness of embryos. Because there are many different types of developing tissues in embryos, each of which becomes relatively sensitive at different times during gestation, various types of birth defects can appear following irradiation of pregnant women. The major determinant of exactly which type of birth defect will arise is the specific day of gestation when the irradiation occurred. But the children born with these defects do not ordinarily have any genetic defects that would affect their fertility, reproduction rate, or mutation rate, because they are not genetic mutants (i.e., their deformity is not inheritable). This distinction is often misunderstood.

The bottom line is, once again, that high radiation doses kill dividing cells wherever they might be found, whether in the bone marrow of adults (producing anemia), the intestinal lining (producing GI breakdown), the developing embryo (producing birth defects), or the testicle (producing low sperm counts or sterility).

These health effects, in and of themselves, are not evidence of gene mutations; rather, mutations happen through a different process that typically does not involve cell death because dead cells cannot pass on their genes.

Testicular irradiation and effects on sperm production have been an unfortunate recurring theme in this book, with no mention at all about female ovaries and the effect of radiation on their eggs. This is because less was originally known about radiation effects on eggs, since it was much easier to get postirradiation sperm counts from men than egg counts from women. Nevertheless, we now know a good deal about how radiation affects the ovaries and eggs, and it seems that women get the worst of it.[43]

The reason has to do with the way a woman's eggs are produced. Sperm are produced continuously throughout a man's life by the spermatogonia cells in the testicles. Eggs, in contrast, are produced only during embryonic development, so that a woman has all the eggs for her entire lifetime on the day she is born. With time these eggs either ovulate individually during the menstrual cycle and find their way through the fallopian tubes to the uterus, or they randomly die off as they age (the fate of most eggs) until they are all gone at around age 55. Since production of the female hormone estrogen in the ovaries is dependent upon the presence of viable eggs, the complete loss of eggs means the loss of estrogen, and menopause ensues.

By killing off eggs, radiation hastens the day when the eggs will be completely gone, and thus produces premature menopause. In contrast, production of the male hormone testosterone in the testicles is not dependent upon the presence of viable sperm, so even complete and permanent sterilization of men with radiation does not eliminate testosterone production. Thus, radiation sterilization of women is the equivalent of female castration (i.e., removal of the ovaries) since both eggs and hormones are gone; but sterilization of men with radiation is not the equivalent of male castration (i.e., removal of the testicles), because the male hormone levels and libido remain normal, and primary and secondary sex characteristics are retained.

Science can sometimes be a self-corrupting process. In an attempt to give relevance to research that outsiders often perceive as arcane, scientists at times resort to public rhetoric. Repeated enough, scientists can begin to believe their own rhetoric, and that rhetoric then starts driving the direction of the science, rather than the science driving the rhetoric.

Muller had wallowed in his share of rhetoric, but he rose to his very best when he stuck with the science. That was his training and where his genius lay. He had recognized the reasons why fruit fly mutants were so hard to measure, and he had the brilliance to circumvent the problem with his X-linked lethal mutation assay. Muller later realized the newly identified viruses that could infect bacteria (*bacteriophages*) would provide a revolutionary new tool for taking genetics to its next frontier; this was the means to discover the stuff of genes. Such viruses behave like autonomous genes that move around independently through the environment, unencumbered by any cellular milieu, and then suddenly infect bacterial cells, hijacking those cells' internal machinery for viral gene reproduction. Like genes, viruses can also be mutated. Muller recognized that the supremacy of the fruit fly in genetic research was coming to an end, and he ventured to predict the new direction that experimental genetics would take:

> [Viruses provide] an utterly new angle [from which] to attack the gene problem. They are filterable, to some extent isolatable, can be handled in test tubes, and their properties, as shown by their effect on the bacteria, can be studied after treatment. It would be very rash to call these [viruses] genes, and yet at present we must confess that there is no distinction known between genes and them. Hence, we cannot categorically deny that perhaps we may [soon] be able to grind genes in a mortar and cook them in a beaker after all. Must we geneticists become . . . chemists . . . simultaneous[ly] with being zoologists and botanists? Let's hope so.[44]

Of all Muller's predictions, this was the most prescient. In this prophecy, at least, he would prove to be absolutely correct.

Muller also provided another major scientific contribution to radiation protection in that he made people rethink the concept of a tolerance dose—the longstanding idea that for every health effect of radiation there existed some dose below which there was absolutely no risk at all. We now know that this concept is applicable only to radiation effects where the health consequence is a result of killing cells. When doses aren't high enough to kill cells (i.e., greater than ~1,000 mSv), there can be no such health effect. In contrast, for inheritable mutations (and cancer), where cell death does not drive the health effect, the tolerance dose concept is meaningless. This is because, for the lower-dose health effects, risk is determined by the probability of damaging a specific biological target within a viable cell, a matter that we will deal with presently.

Muller remained committed to teaching the next generation of geneticists up until his retirement. He taught an annual course called Mutation and the Gene, which was very popular among biology students at Indiana University. Muller remembered one lanky young man in particular as being the brightest student he ever had, but also recalled him as having a nervous disposition and some difficulty with orally expressing his ideas.[45] Remarkably, in less than five years, that inarticulate student would coauthor a paper that would rock the entire world of biology, and give the work of Mendel, Morgan, and even Muller himself, an entirely new perspective. The young man was James D. Watson, and his revolutionary discovery, in 1953, pushed fruit fly genetics off center stage and provided new insight into fundamental genetic questions that Muller could only dream about. It would also help reveal the still unknown mechanism by which radiation produces its biological effects.[46]

CHAPTER 11
CRYSTAL CLEAR: THE TARGET FOR RADIATION DAMAGE

At once I felt something was not right. I could not pinpoint the mistake, however, until I looked at the illustrations for several minutes. Then I realized that the phosphate groups in [Linus Pauling's] model were not ionized. . . . Pauling's nucleic acid in a sense was not an acid at all . . . a giant had forgotten elementary college chemistry!

—*James D. Watson*

A scientist will never show any kindness for a theory which he did not start himself.

—*Mark Twain*

THE RED BADGE OF COURAGE

It is a common misconception that James Watson and Francis Crick discovered DNA. They did not. What they discovered was the structure of DNA, and the structure they identified revolutionized the way we now think about genetics. But if you've never heard of Watson, Crick, or even DNA, never mind. We're getting way ahead of ourselves. Let's put DNA aside for the moment. Our story of the cellular target for radiation damage doesn't start with DNA. Rather, it starts with pus.

Pus is that thick, yellowish stuff that oozes from infected wounds and soaks bandages. It is composed largely of white blood cells called leukocytes (from the Greek *leukos*, meaning white), which are in the process of attacking microscopic foreign invaders, be they bacteria, viruses, or yeast. In the zeal of battle, leukocytes swarm through the blood stream to the site of infection, reproducing wildly to spawn their own replacements, often bloating the area with their sheer numbers and leaking out of every orifice of the wound.

It was 1869, just four years after Mendel had published his seminal work on the genetics of peas, when Johannes Friedrich Miescher (1844–1895) began to study the chemical composition of human leukocytes.[1] Miescher had just graduated from Basel University Medical School in Switzerland, but was unenthused about becoming a practicing physician. His uncle Wilhelm His was a prominent physiologist on the medical school's faculty, who had made major contributions toward understanding the embryonic development of the nervous system. His suggested to Miescher that he might pursue a career in biological research as an alternative to practicing medicine, and further expressed to his young nephew his conviction that the key to understanding biology lay in determining the chemical composition of cells. Miescher admired his uncle's wisdom and resolved to devote himself to investigating the chemistry of cells. He moved to the nearby city of Tübingen, Germany, a center for scientific research at the time, and started working in the laboratory of Adolf Strecker (1822–1871), an organic chemist internationally recognized for being the first person to artificially synthesize an amino acid (alanine). Amino acids are the molecular links (monomers) that join with each other to form long-chained molecules (polymers) known as proteins. In their various forms, proteins are the major structural and regulatory molecules of all cells and tissues. Thus, Strecker had synthesized what was essentially a building block of life. Miescher had chosen wisely in deciding to work with Strecker and learned much chemistry from him.

After a while, however, Miescher became bored with the drudgery of amino acid syntheses. He began to believe that proteins themselves, rather than their amino acid building blocks, hold the key to life's secrets. So he left Strecker and started working with Felix Hoppe-Seyler (1825–1895), a pioneer in the new field of biochemistry. Hoppe-Seyler was among the first to directly study proteins, and had actually coined the term from the Greek *proteios,* meaning "the primary or main thing." Thus, by its very name, protein claimed preeminence among biological molecules.

Miescher reasoned that the most basic chemical constituents required for life would most easily be isolated from the most basic of cell types. Since leukocytes are among the smallest and simplest of all known human cells, he decided to work with them. Leukocytes tend to congregate in lymph nodes, those small olive-shaped immune glands found in the groin, armpits, and neck. Miescher recruited volunteers and tried to extract leukocytes from their lymph nodes with syringes, but he could never get quite enough for his needs. Then he had a brainstorm. Pus from wounds was laden with leukocytes. He decided to collect fresh bandage waste from hospitals and scrape off the pus to get his leukocytes.

With leukocytes now in hand, Miescher first developed methods to wash them free of debris. Then he designed techniques to separate their protein from their lipids (i.e., fat), the other main chemical constituent of cells. During these procedures Miescher identified yet another abundant substance that was neither protein nor lipid. He was further able to show that the substance originated exclusively from the nucleus of the cell, so he named it *nuclein.*

Miescher did an elemental analysis of nuclein and found that it was comprised of carbon, hydrogen, oxygen, nitrogen, and phosphorus; in effect, these were the same elements found in protein. Nevertheless, the proportions of the elements varied tremendously from protein. Unlike protein, which contained only traces of phosphorus, nuclein contained very large amounts. In contrast, nuclein completely lacked sulfur, which is a significant constituent of protein. Miescher's conclusion was that nuclein and protein were completely distinct biological molecules, with totally different structures, and presumably unique functions, whatever those functions might be.

Sadly, Miescher could make no further progress with his studies of nuclein. The tools of a biochemist at that time were only capable of studying small molecules containing but a few dozen atoms, not large molecules with many atoms. Nuclein, as a long-chained polymer, was a giant molecule with millions of atoms. Lacking any techniques to further probe its structure, there would be no further insight into its biological role. Unfortunately, Miescher's published findings on nuclein joined Mendel's genetic work with peas among the archives of intriguing but neglected science, waiting for a future generation of scientists to revisit with more powerful analytical tools.

AN OCTOPUS'S GARDEN

William ("Willie") Lawrence Bragg (1890–1971) was a handsome child, with a very kind and considerate disposition. He was the apple of his mother's eye, and the pride and joy of his father. Although he tended to be shy and humble, his intellectual talents were evident to anyone who shared his company. From his elders, Willie's genius earned him special attention and access to engaging adult conversation, which further stimulated his young mind. Among his peers, his genius simply made him the target of bullies, and exacerbated the loneliness that plagued his youth.

Willie's first scientific love was biology, and in this he was bucking a family tradition. His father, William Henry Bragg (1862–1942), was the most prominent physicist in Australia. An Englishman trained at the Cavendish Laboratory by J. J. Thomson, Bragg was offered, at a very young age, a prestigious professorship at the University of Adelaide in Australia; he had obtained this offer largely through the influence of his mentor, Thomson. The elder Bragg arrived in Adelaide early in 1886, just when laboratory training was beginning to be seen as essential to the education of scientists; so university administrators were more than happy to provide their new young professor with a state-of-the-art science laboratory. The laboratory contained the latest scientific equipment, and copious windows admitted the sunlight he needed to perform optical experiments. It also included four dozen high-quality lenses, one dozen prisms, and an assortment of mirrors. The laboratory was an ideal educational facility for teaching Newtonian optical physics, a specialty of Bragg's at the time.[2] The notable reforms that Bragg subsequently made to physics education at Adelaide won him a measure of fame throughout the scientific community of Australia, as well as a good deal of prestige within high society.

Then there was Willie's grandfather on his maternal side, Charles Todd (1826–1910). Grandpa Todd was an astronomer, meteorologist, and electrical engineer of even greater fame within Australia than Willie's father. He was a contemporary of Edison's, and like Edison, was deeply involved with introducing electricity to his country.

So, both Willie's father and grandfather were leading figures among the scientific intellectuals of Australia, and both foresaw great things for Willie in the field of physics. Together they would have a profound influence on Willie's scientific career, an influence that would shepherd him in a direction that seemed, at least at the time, far afield from his beloved biology. Of this period, Willie would later recall that his father's "bedtime stories were always the same—about the properties of atoms; we started with hydrogen and ran through a good part of the periodic table."[3] For the moment, however, Willie was indifferent toward physics and occupied his private moments exploring the natural world near his home in Australia, a country that possessed some of the most unique wildlife in the world, much of it still undiscovered.

Willie's interest in nature, coupled with his friendless existence, led him to take long, solitary walks on the beach, and to eventually collect seashells as a hobby. In fact, over a few short years, this young boy had amassed one of the finest collections of South Australian seashells in existence and had become well versed about all aspects of South Australian sea life.[4] It therefore might not be too surprising that the first scientific discovery of Willie's life would be a new species.

This happened one day as Willie was combing the beach for shells and found the bone of a cuttlefish. Cuttlefishes (*Sepia*) are really not fish at all. Rather, they are a specific class of mollusks (*Cephalopoda*) closely related to octopuses. They also are the premier camouflage artists of the ocean. Somehow they have the ability to manipulate their skin texture and modify the reflection of light to the point that they can assume the appearance of virtually any object they encounter in their environment. By blending in, they escape torment from predators.

A cuttlefish's skeleton consists of a single, porous, and brittle bone, roughly the shape and size of a shoehorn. Most people have seen cuttlefish bones, but not on the beach. They are routinely hung in birdcages by bird fanciers, to perform the double duty of providing an object for gnawing that sharpens the pets' beaks, while simultaneously providing a dietary source of the essential element calcium, the major constituent of bone.

When Willie first saw the cuttlefish bone, he immediately recognized what it was, but was unable to identify the exact species. It was unlike any cuttlefish bone Willie had seen before, so he showed the bone to a local expert, who confirmed Willie's expectation. It was from an unknown species, and Willie, as its discoverer, was entitled to give this new species its scientific name. Thus, the cuttlefish became *Sepia braggi*—Bragg's cuttlefish.

Willie was thrilled. He would have liked to continue exploring the wilds of Australia for the rest of his life, making more natural discoveries in a country that still had a largely pristine environment . . . but this was not to be. Willie later said, "biology rather than physics might have been my trend had there not been such a strong family tradition [in physics]."[5] For better or worse, Willie, like his namesake cuttlefish, would need to blend into his environment if he were to survive as a scientist. And Willie's scientific environment was physics, not biology.

SIMPLICITY RULES

Despite its still unknown function, nuclein occasionally attracted the curiosity of various chemists over the first few decades following its discovery. These chemists each contributed their own small piece to the puzzle of nuclein's chemical composition. By the late 1920s, when biologists started to show an interest, chemists had formulated a succinct chemical description, if not an exact structure. The nuclein molecule appeared to be a very long chain (i.e., a polymer) of phosphate-containing monomers, which the chemists called *nucleotides* to reflect their nuclear origin. Each nucleotide was comprised of just three components: (1) a nucleobase (a single- or double-ringed structure that resembled a honeycomb cell); (2) an unusual circular sugar structure (specifically, deoxyribose—a five-carbon sugar that was missing one of its oxygen atoms); and (3) a phosphate group (a phosphorus atom bound to oxygen atoms). What's more, nucleotides used their phosphate groups like hands grabbing the sugar of their neighboring nucleotide to form the chain, similar to humans forming a conga dance line, with all the nucleotides in the chain facing in the same direction. Curiously, only four different chemical versions of the nucleotides were present in this huge biological polymer. They were called cytosine and thymine (the single-ringed nucleobase forms collectively known as the *pyrimidines*), plus guanine and adenine (the double-ringed nucleobase forms collectively known as *purines*).

Chemists also had determined that the phosphate groups were electrically charged. That is, the phosphates were devoid of any hydrogen ions (positively charged) to neutralize their negatively charged oxygen atoms. This meant that the nucleotide chain was highly negatively charged along its entire length. Such molecules that shed their positive hydrogens to become negatively charged are called acids. Consequently, nuclein could reasonably be described as a nuclear acid polymer. To reflect this new chemical understanding, nuclein was given a new name, more fitting to its known chemistry. It became *deoxyribose nucleic acid* (now termed deoxyribonucleic acid), meaning an acid from the cell's nucleus that is derived from the sugar, deoxyribose. That's quite a mouthful, so it's usually just called by its acronym, *DNA*.

This polymer chemistry stuff might not sound all that simple. Admittedly, it is a little complicated, particularly when trying to visualize DNA's chemical structure in three dimensions. Nevertheless, because DNA is fundamentally just a very long molecular chain with only four different types of links, it was considered much too simple to be able to encode complex genetic information. How could genes write the very story of life, if they only had a four-letter alphabet to work with?

The easiest way to appreciate why DNA seemed too simple to be the stuff of genes is to consider the alleged superiority of protein as a potential vehicle for passing on genetic information to future generations. DNA's linear polymer of nucleotides is somewhat analogous to the structure of proteins, except that proteins are linear polymers of amino acids rather than polymers of nucleotides. DNA has just four very similar nucleotides that differ from each other in only subtle ways. In contrast, proteins have 20 versions of amino acids to work with when forming their polymer chains. And those 20 amino acids are very different from one another in terms of their structure, charge, solubility, and so forth. So, by linking vastly different amino acids together in innumerably different orders, you can come up with highly diverse protein structures as different as a rhino's horn is from the lens of a human eye. Not so for DNA. The structure of DNA seemed to always be the same, both between cells and between species. How can you encode the vast genetic complexity of life on Earth with such a simple molecule that has an unchanging form?

At first blush, it seems justified to suppose that simple molecules do simple things, and complex molecules do complicated things. It was, in fact, this intuition that led scientists to speculate that DNA must amount to some type of simple nuclear scaffold or skeleton to which the complex genes, allegedly consisting of proteins, are attached. This presumption about the informational deficit of simple things was a rather naive contention even for the time. Nevertheless, this groundless supposition about the functional limits of simple structures was taken for granted by many biologists. In a short while, however, their belief that simplicity was DNA's greatest weakness would prove to be absolutely wrong.

A TOUGH BREAK

Even before Willie discovered his cuttlefish, he had already made scientific history, although he would have much preferred if the honor had befallen someone else. Willie has the distinction of being the first person in Australia to receive a diagnostic x-ray. His story is not unlike that of Toulson Cunning, the hapless Montreal gunshot victim, who had his wounded leg imaged by the local physics professor with a Crookes tube just days after Roentgen had discovered x-rays.

In Willie's case, however, his assailant was his younger brother, a tricycle was the weapon, and the professor with the x-rays was his father. Six-year-old Willie was riding his tricycle when his younger brother Bob decided it would be good sport to take a running jump onto the back, and so he did, thereby toppling the tricycle and crushing the bones in Willie's left elbow. The family doctor ruled the break beyond repair, suggesting immobilization of the arm in its least obstructive position and allowing it to set stiff. Willie's father, the great Professor Bragg, and his mother's brother "Uncle Charlie" Todd, himself a physician, decided the stiff-arm plan was unacceptable. Instead, they pooled their professional talents to try and save Willie's arm.[6]

It was early 1896, and Professor Bragg had just recently learned of x-rays, which had been discovered by Roentgen a few months earlier. In fact, Bragg had been using a Crookes tube to reproduce Roentgen's results, just as so many of his fellow physicists had already done months before in less scientifically remote parts of the world. The professor had tested the device on himself and produced some quality x-ray photographs of the bones in his hand, so he knew that x-rays from Crookes tubes could reveal bone structure.[7] As for Uncle Charlie, he provided expertise in ether anesthesia. The two brothers-in-law collaborated to design a therapy plan that combined anesthesia, physical manipulations, and x-rays. The idea was to anesthetize Willie every few days and, while the boy was unconscious, manipulate the bones in his elbow and flex the arm in all directions. This was intended to coax the bones into aligning properly and to prevent the elbow joint from fusing into a stiff position.

Progress was assessed with regular x-rays of the elbow. Although Willie dreaded the arm manipulations, the x-rays equally frightened him once he saw the sparking Crookes tube with its strange glow. Nevertheless, after his fearless little brother, Bob, volunteered to have his own elbow imaged as a demonstration of safety, Willie suppressed his cowardice and submitted to the x-rays.

Ultimately, success was achieved. Willie's elbow function was preserved, although his left arm did end up slightly shorter than his right. Still, the whole experience had been traumatic for him. Even though he was very young at the time, Willie was able to faithfully recount the details of the incident many decades later. It had made a big impression on him, and it was his introduction to the power of x-rays to reveal hidden structure. It was also his first inkling into how physics and biology might cooperate to reveal both form and function.

SEX IS OVERRATED

That genes resided in the nucleus, and specifically on chromosomes, had been generally accepted ever since 1915, when Morgan had linked gene transmission to chromosome transmission in fruit flies. It was also well known that genes could be passed from one generation to the next during reproduction. But, in 1928, bacteriologist Frederick Griffith (1877–1941) discovered something about gene transmission that shocked geneticists. He showed that when he mixed a heat-killed strain with a live strain of bacteria, the progeny of the live bacteria would assume some of the genetic traits of the dead bacteria. It was as though autonomous free-floating genes within the detritus of the dead culture were being taken up by the living bacteria and altering their genetic makeup. Not only did it amount to gene transmission without sex; it was gene transmission without life. These genes were strange things indeed! What were these naked genes, viable even apart from their cells? Presumably they were some type of protein particles, but the question remained open.

THE SCIENTIFIC OUTBACK

Professor Bragg's foray into x-rays with the Crookes tube, fortuitous as is was for poor Willie's arm, had actually interrupted the physicist's ongoing work with radio waves, his main scientific interest at that time. In fact, in August 1895, just four months before x-rays would be discovered and less than a year before Willie would shatter his elbow, Bragg was working in his laboratory, preparing a demonstration of radio waves for his physics course. Unannounced, the 24-year-old New Zealander Ernest Rutherford paid him a visit.

Rutherford was en route to England to begin studies at Cambridge.[8] He had been a runner-up for a Cambridge scholarship offered to students in the British provinces, as New Zealand was at the time. But the original nominee was unable to commit to the terms of the award, so it was offered to Rutherford at the last minute. Rutherford had been working in the potato patch of his family's farm when news of his scholarship arrived. He is said to have thrown his shovel into the air proclaiming, "That's the last potato I'll ever dig!"[9] Within days, he set sail for England, but used the opportunity of his ship's stopping in Adelaide, Australia, to look up the famous Professor Bragg and say hello.[10]

Bragg was actually tinkering with his Hertz wave oscillator at the moment Rutherford showed up. Coincidently, Rutherford had brought along his own toy. He showed Bragg an electromagnetism detector that he had built himself. It was the device that he would ultimately employ in Cambridge to ring bells at long distances without wires, just as Édouard Branly had first demonstrated four years earlier. The two men talked radio waves all afternoon. They found a kindred spirit in each other and became lifelong friends. It is unlikely that Rutherford encountered six-year-old Willie during his brief sojourn in Adelaide. If he had, the loud and gregarious New Zealander would never have imagined that this quiet and shy little boy would eventually become his scientific successor at the Cavendish Laboratory.

Enjoyable as it was for Bragg, Rutherford's visit had underscored the virtual impossibility of staying in the forefront of science while working in the scientific backwater that Australia was at the time. In fact, with no central electricity in his laboratory, Bragg had to electrically power his experiments with his own small generator. And while Bragg was just beginning to take his first x-ray photographs with his Crookes tube, Edison was already demonstrating his newly invented fluoroscope at the National Electric Exposition in New York, showing rapt crowds the real-time moving x-ray images of internal bones, rather than just still x-ray photographs.[11] The sad truth was that Bragg's research had thus far amounted to little more than reproducing the findings of Newton, Marconi, and Roentgen, with light, radio waves, and x-rays, respectively. And he had only belatedly become aware of the discovery of radioactivity by Becquerel.[12] Nevertheless, Bragg's relative isolation had forced him, through necessity, to become a master instrument maker. He and his technical assistant, Arthur Rogers, both had very high standards for design and construction of their homemade scientific instruments, a skill that would later serve Bragg extremely well.

Rutherford sympathized with the predicament of his friend back in Australia, and for many years exchanged regular correspondence with him about the latest scientific developments and shared experimental reagents whenever he could. And Bragg achieved some progress despite the obstacles. Following up on an observation by Marie Curie that alpha particles have a limited range that is dependent upon their energy, Bragg discovered that the rate of ionizations produced along the alpha particle's path reaches a maximum just before the particle exhausts all of its energy and comes to a complete stop. This peak level of ionizations, releasing an energy burst at the very end of a particle track, is now called the *Bragg peak*. The Bragg peak has been used over the years in a number of applications, most recently in radiation therapy. This type of therapy strategy uses a proton accelerator, of similar design to that used at the Cavendish to first split the atom, to treat tumors. Since the range of the protons in tissue is energy dependent and well defined, the beam can be positioned and the particle energy adjusted so that the protons stop within the tumor itself, rather than in the normal tissue beyond the tumor. This means that the cell damage caused by the Bragg peak is predominantly restricted to the tumor, which should result in better tumor-cell killing. Grubbe, Kelly, Kaplan, and other early researchers of cancer radiation therapy, could only have dreamed of the day when tumor-destroying radiation doses could be delivered so precisely as to demolish internal tumors without also decimating the surrounding normal tissues. In fact, because of the efficiency of the Bragg peak in killing tumor cells while sparing normal tissue, proton accelerators are currently being installed in many radiation therapy clinics around the world.

Notwithstanding his discovery of the Bragg peak and other research progress he made in Australia, Bragg began to find his remote location oppressive to his work. It typically took six months for news of any scientific discovery to make its way to Australia. That meant he was perpetually six months behind the scientists in less remote locations. Bragg knew that if he wanted to do cutting-edge physics, he would need to return to England. What's more, he believed that higher educational opportunities for his sons, Willie and Bob, were lacking in Australia. He wanted them to go to Cambridge, where he had studied. So he resolved that he would try to find himself a suitable faculty position at a university somewhere in England and move his family there.

LEAVING A TRACE

In the spirit of Miescher, who believed the fundamental chemistry of life would most readily be revealed by studying the simplest body cells, the leukocytes, many geneticists came to believe that the fundamental chemistry of genes could best be learned from studying life's simplest of organisms, the viruses. Recall that even Muller, in recognition of the limitations of his own fruit fly research, had to concede that viruses provided "an utterly new angle [from which] to attack the gene problem." Therefore, many geneticists became virologists, and it wasn't long before two of them, working together, were able to conclusively resolve the issue of whether genes were made of protein or DNA.

Viruses are so simple that they do not even carry the molecular machinery needed for their own reproduction. A virus must instead inject its genes into a host cell, and fool that cell into using its reproductive machinery to reproduce the viral genes as though they were its own. But does a virus inject protein or DNA into a cell? Radioactivity would provide the answer.

If you grow viruses in media containing radioactive elements, those radioisotopes get incorporated into the molecular constituents of the viruses. Just as ingested radium found its way into the spines of the radium dial painters because it chemically resembled the calcium component of human bone, radioactive phosphorus fed to viruses will find its way into the phosphate groups that comprise the "backbone" of the virus' DNA. Likewise, radioactive sulfur will find its way into the sulfur component of proteins. Since DNA contains no sulfur, and proteins contain only small amounts of phosphorus, such radiolabeling of viruses with radioisotopes of phosphates or sulfur will specifically result in radiolabeling the DNA or the protein, respectively. This is what virologists Alfred Hershey and Martha Chase did to resolve the question of which of the two biological polymers, DNA or protein, represented the stuff of genes. They radiolabeled viruses with radioactive phosphate and sulfur and then looked to see which radioisotope the viruses were injecting into cells when they reproduced. It turned out that the cells contained radioactive phosphorus only, suggesting that the viruses were only injecting their DNA and not their protein. This conclusively proved that genes are made of DNA.[13]

HOME AGAIN

In 1909, the chairman of the physics department at the University of Leeds, England, decided to retire. Rutherford, who had moved on from his first faculty position at McGill University in Montreal, Canada, was now physics professor at the nearby University of Manchester. He prevailed on the Leeds selection committee to offer the vacant position to Bragg. It was the opportunity Bragg had been waiting for. He accepted the position straightaway, and booked passage to England.

Upon arriving, Bragg immediately went to visit his friend Rutherford in Manchester, where they relived their first encounter of so many years earlier. This time, Rutherford played host and showed Bragg his latest scientific marvels, including a pump that he had designed to collect radon gas emanating from a radium solution. (This radon pump's design was similar to the one that Rutherford would later provide to Dr. Howard Atwood Kelly in Baltimore, so that Kelly could collect radon for his brachytherapy cancer treatments.[14])

Willie enrolled at Cambridge University, fulfilling his father's wish, and began studying physics at the Cavendish Laboratory. He was academically well prepared for Cambridge. In fact, he had been taking college courses at the University of Adelaide since the age of 16. When Willie arrived at the Cavendish, his father's old mentor, J. J. Thomson, was still in charge, and would remain in charge until Rutherford succeeded him a decade later.

If Willie believed he had left behind the ignorant bullies that haunted his childhood in Australia, he was sadly mistaken. At Cambridge he encountered the same prejudice against provincials that Rutherford had experienced years before. Willie was treated as a second-rate student by his fellow classmates, who harbored the widespread misconception that Australia was a country best described as a mixture of aboriginal wasteland and British penal colony. But Willie was as academically gifted as Rutherford, and just as Rutherford had soon outshone his detractors, Willie's tormenters were presently struggling to keep up with him. This, in turn, provided another reason for them to dislike him. Also, like Rutherford before him, Willie earned himself a prestigious scholarship. He did this by achieving a top score on a competitive mathematics exam that he took while confined to bed, battling pneumonia. And then in 1912, at the young age of 22 and in his very first

FIGURE 11.1. WILLIAM ("WILLIE") LAWRENCE BRAGG. Willie Bragg realized that x-rays bounce off crystals just as light bounces off mirrors. His discovery of this phenomenon and his description of its associated mathematics (Bragg's law) won him a Nobel Prize in Physics (1915) when he was only 25 years old. He remains the youngest scientist ever to win a Nobel Prize. (*Source:* Photograph courtesy of the AIP Emilio Serge Visual Archives, Weber Collection, E. Scott Barr Collection)

year of research work at the Cavendish, Willie made a momentous discovery that further distanced himself from his peers. He discovered a property of x-rays that no one else had ever appreciated before. They bounce!

The x-ray question that Willie solved had been puzzling physicists for some time. It had to do with the mysterious effect that crystals have on a narrow x-ray beam passing through them. A crystal is a special type of solid in which all the individual molecules are lined up in perfect rows and columns, in three dimensions. Crystals further have the remarkable ability to self-assemble, such that if you have a pure solution of some particular chemical dissolved in water, and then you allow the water to evaporate away, the molecules left behind will tend to organize themselves in regular rows and columns as they begin to solidify. The forces that drive crystal formation are governed by the laws of thermodynamics (think entropy, enthalpy, and such), but we need not get into all that. Simply know that crystals are all around us. Table salt and sugar granules are good examples of small crystals, but crystals can also grow to large sizes as in rock candy and diamonds. The bottom line is simple. If a chemical solution evaporates into a crystal, you can be assured of at least two things: (1) It is essentially pure of molecular contaminants, because contaminants would disrupt the crystallization process; and (2) the molecules that have joined to form the crystal have a very regular structure, because if they didn't, they wouldn't be able to line up uniformly.

When a very narrow x-ray beam (as narrow as a laser pointer beam) is aimed at a crystal, most of the x-rays pass right through it, as x-rays tend to do, forming a spot on a photographic film placed directly behind the crystal in the line of the beam. Nothing unusual there. But some scientists had noticed that they also found multiple small spots of exposure in regular patterns, outside the line of the beam. Since it had already been established that x-rays could not be bent or focused with conventional optical lenses, what exactly was going on in crystals? No one knew.

Willie had been pondering this problem for a while when suddenly he realized something unique about the atoms in crystals, which made them different from other solids. In a crystal, because of its regular molecular structure, all of the component atoms are aligned along natural sheets, or geometric planes, like the surface of a table. For example, in common table salt (sodium chloride), there would be serial planes of sodium atoms alternating with planes of chloride atoms. In sugar crystals (comprised of only carbon, oxygen, and hydrogen), its three types of atoms would also fall along their own planes. All right, you say, but so what?

Willie might have had the same reaction, but he had just completed his natural science exam, a large part of which was devoted to light and optics. In preparation for the exam he reviewed his lecture notes from his optical physics class; in particular, he went over some lectures by Professor C.T.R. Wilson, where Wilson described white light as behaving like a mixture of electromagnetic pulses comprising a very limited range of wavelengths.[15] Just as these pulses of light could bounce off mirrored surfaces, like tennis balls off a court, Willie hypothesized that the shorter wavelength pulses of x-rays should bounce, or reflect, off the mirrorlike planes formed by the sheets of atoms within crystals. Also, he proposed that the deflection spots produced on the photographic film when x-rays passed through crystals were "due to the reflection of x-ray pulses by sheets of atoms in the crystals."[16]

Willie even thought he could prove his hypothesis. Certain types of crystals tend to fracture along their atomic sheets, leaving surface planes of atoms with mirrorlike surfaces. Mica, a silicate mineral responsible for the sparkling flecks seen in many types of rocks, is one such crystal. Willie showed that x-rays bounced off mica sheets onto photographic film just like light bounces off mirrors. Furthermore, as he varied the angle of incidence, the angle of bounce varied according to the same physical laws as the reflection as light. He took his freshly developed photographic evidence, showing the mica-reflected x-ray spots, straight to J. J. Thomson. Thomson was elated. When Willie's father learned of his son's achievement he proudly wrote to Rutherford, "My boy has been getting beautiful x-ray reflections off mica sheets, just as simple as reflections of light in a mirror!"[17] Hearing the news, the cautious Rutherford set his two latest protégés Harry Moseley and Charles Galton Darwin (grandson of the famous naturalist, and godson of eugenicist Francis Galton) to work on repeating and validating Willie's findings, which they promptly accomplished, thus corroborating Willie's claims. Rutherford, too, then reveled in the excitement. Just as Rutherford had once recognized that the angle at which alpha particles bounced off gold foil had profound implications regarding the size of gold's atomic nucleus, Willie recognized that the angle at which x-rays bounce off a crystal had profound implications regarding the distances between the atom sheets within that crystal. Both were momentous discoveries.

To fully appreciate the significance of Willie's insight to science, it is important to understand how his theory of x-rays bouncing off crystals could be practically used. Imagine you're at home at night in a dark room with a window to the outside, and you have a laser pointer. Let the laser's light beam simulate your narrow x-ray beam, and let the window simulate your crystal. Let's also assume that the window is doubled paned; that is, it has two panes of glass trapping an air space in between, as is commonly done to provide heat insulation. You now shine your laser pointer out the window at a 45-degree angle aiming at a tree outside. You will see a spot on the tree where the laser light has passed through the glass, but if you now turn your attention to the wall perpendicular to the window you should also see two spots. One of these is formed from the reflection of the beam off the inner windowpane, and the other from the reflection off the outer windowpane. If the widow were triple paned, you would see three spots. This is what Willie realized was happening when x-rays passed through crystals. They were being reflected off the multilayered natural "window panes" formed by the planes of atoms within the crystal.

Let's further suppose that it is important to know the actual distance between the two panes of the window, but there is no way to get a ruler into the window for a direct measurement. How would you do it? Willie realized that the distance between the two panes is related to the distance between the two spots on the wall and their angle from the window surface. Using his prodigious mathematical skills, he was able to generate a simple formula that allowed him to determine the distance between atomic planes in any

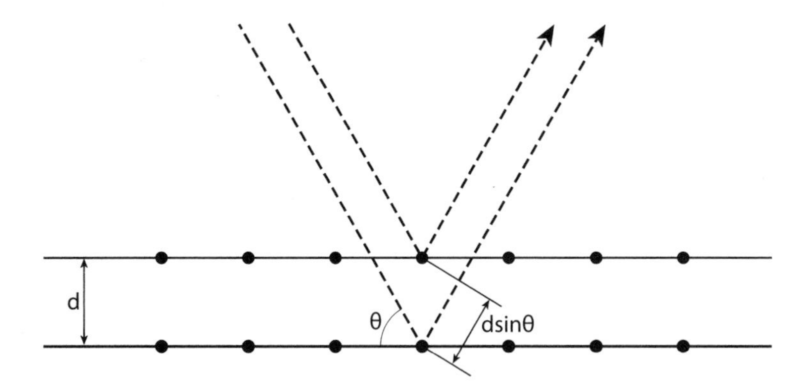

FIGURE 11.2. BRAGG'S LAW. When an incident x-ray (*dashed arrow*) enters a crystal, it can bounce off any one of the atomic layers. By measuring the angle of the bounce, the distances between the atomic layers (*d*) can be calculated. Bragg's law allows atomic distances to be determined very precisely for any molecule that can be crystalized, and provided James Watson and Francis Crick with the clue they needed to discover the structure of DNA.

crystal based on the distances between the reflected x-ray spots on the photographic film:

$$n \lambda = 2 d \sin \theta$$

where d is the distance between the sheets of atoms, θ is the angle of reflection, and λ is the wavelength of the incident x-ray. (The n is a term denoting some whole number corresponding to a wave interference concept known as the *order of reflection*, which is beyond the scope of our interests here.) We need not understand this equation, but it is important to appreciate its implications.

This simple mathematical relationship, now known as *Bragg's law*, basically says that reflected x-rays behave a lot like reflected light, and that the degree of reflection is dependent upon the wavelength of the radiation. Should this surprise us? After all, light and x-ray radiation are both electromagnetic waves, distinguished solely by their wavelengths. Nevertheless, it was an eye opener at the time, and it had extremely important practical implications. Bragg's law would ultimately provide a powerful tool for understanding the three-dimensional structure of any molecule that can be crystallized. In effect, the physicists had come to the rescue of the biochemists by providing a method to study molecules in three-dimensions rather than just two, even when those molecules are very large, as biological molecules tend to be.

Willie took his angle of reflection idea to his father, who was experienced in both optics and instrument-making, and together they built a device to measure the angle of x-ray reflection that emulated the original design of Newton's instrument for measuring light refraction. But instead of measuring light refraction through prisms, as Newton had done, the Braggs' instrument measured x-ray reflection off crystals. Then the two Braggs started irradiating different crystals and deducing their atomic structures from the angles of x-ray reflection. They spent the next two years determining one crystal structure after another, from lowly table salt to diamonds.[18] In doing so, they contributed mightily to the future understanding of molecular structure and bond lengths between atoms.

Willie first revealed his discovery of the x-ray reflection properties of crystals and their similarity to the reflection of visible light by mirrors at a meeting of the Cambridge Philosophical Society. Shortly thereafter his presentation was published in their *Proceedings*. It is notable that Willie began his explanation on the behavior of x-rays in crystals by drawing parallels directly from a textbook with which his audience was quite familiar; this was *An Introduction to the Theory of Optics*, by Arthur Schuster. The book itself drew heavily from the original optical work of Newton. Thus, the study of radiation had come full circle, and returned to its original roots in optics.

The physical principle of producing multiple reflections from objects that have their internal structures organized along multiple stacked planes is now called *Bragg reflection*, in honor of Willie. Unfortunately, it is often mistakenly attributed to his father, who was already famous for discovering the Bragg peak phenomenon of particulate radiation. It was a perpetual problem that the father and son's scientific contributions were often conflated, although each man made frequent attempts to set the record straight. The

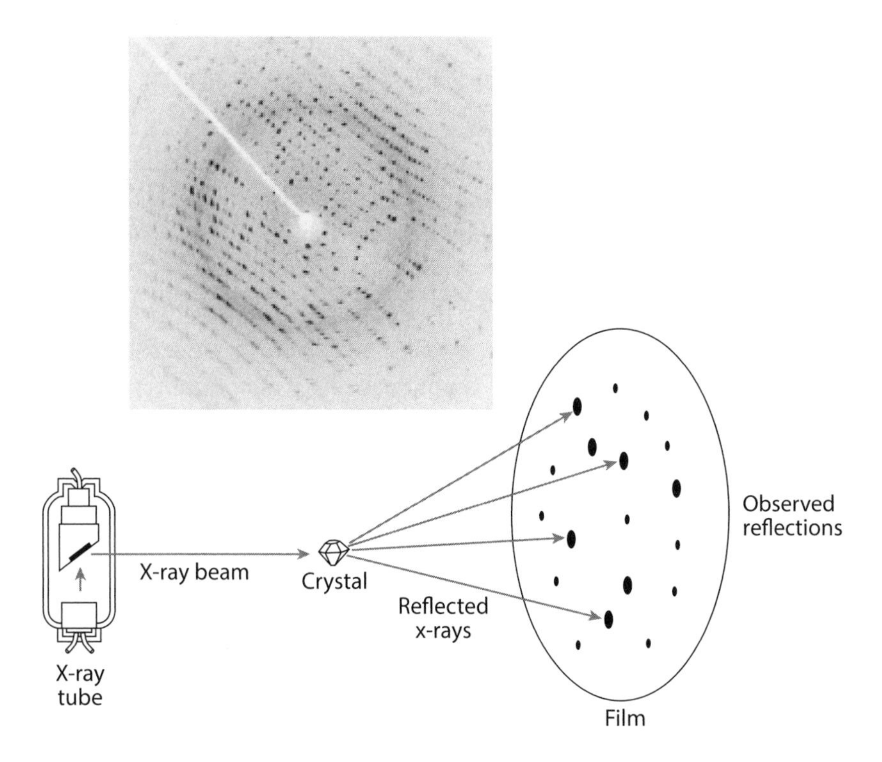

FIGURE 11.3. X-RAY CRYSTALLOGRAPHY. The technique of x-ray crystallography entails directing a narrow beam of x-rays through a crystal. The x-rays then reflect at different angles off various planes of atoms within the crystal's molecular structure (*bottom*). The reflected x-rays produce spots on photographic film. Analysis of the spatial distribution of the spots on the film provides the angle of reflection values, which are needed to calculate, using Bragg's law, the distances between atomic planes in the crystal. An actual film from a simple protein purified and crystalized from chicken egg whites (lysozyme) is shown (*top*), but large and complex molecules can also be deduced using the same approach. (*Source*: Image of crystallography technique provided by courtesy of Dr. Gert Vreind)

fact that they shared the same first name didn't help. Some even thought they were the same person. Willie would soon start going by his middle name, Lawrence, rather than his first, to help lessen the confusion, but it was of little help. Even today, the two men are often confused in the scientific literature.[19]

BULL'S EYE

While geneticists had been working to identify the chemical substance of genes, radiation biologists had been busy confirming that radiation produced its biological effects by damaging genes. The evidence was overwhelming and consistent. The scientists were able to show that, in order to kill a cell with radiation, you must irradiate its nucleus, the home of its genes. And if you irradiated a whole cell, you could rescue it from death by transplanting a nonirradiated nucleus from another cell into the irradiated cell. It seemed that the only determinant of whether there would be a biological effect was whether the radiation reached the nucleus. Furthermore, it could be readily seen under the microscope that radiation was able to break chromosomes in the nucleus, and the magnitude of the biological effects was proportional to the amount of chromosome breakage. Likewise, chemicals that protected the chromosomes from breakage mitigated the biological effects as well. In short, there was a direct connection between the gene damage and the biological effects. Given that the substance of genes had now been determined to be DNA, the logical implication was obvious: DNA is the target for radiation's biological effects.[20]

The term *target*, when used in a scientific sense, does not imply that radiation is actually aiming to damage DNA. Rather, it means that DNA is the target that must be hit (damaged) in order for radiation to produce a biological effect. Other biological molecules, including proteins, are damaged by radiation just as readily. In fact, the vast amount of molecular damage in cells occurs outside of their DNA, but only the damage to DNA is important to producing biological consequences. In retrospect, the reason seems clear. All the other molecules of the cell are replaceable, and the DNA—the cellular director that is giving all the orders—governs their replacement. But if a cell's director is damaged, replacement is impossible, control is lost and the cell's functions go awry, sometimes resulting in cell death and sometimes in mutation. If the mutations happen to occur in sex (germ) cells, the consequence can be inheritable mutations. If instead, those mutations happen to occur in nonsex (somatic) cells, the consequence can be cancer. The cell type and the radiation dose determine the specific type of health effect, but damage to DNA is always the mechanism.[21]

THE CUP RUNNETH OVER

Other than brother Bob, to whom he was devoted despite the unfortunate elbow incident, Willie had no close friends. Always shunned by his peers, Willie suffered profound loneliness. But all that changed when he met Rudolph "Cecil" Hopkinson. Cecil was an electrical engineering student at Cambridge, and his father was a leader in the development of England's electric power industry, similar to what Willie's grandfather Todd had done on a much smaller scale back in Australia. Despite his aristocratic family ties, Cecil had none of the social pretentiousness of the other Cambridge students, and he and Willie hit it off immediately. Interestingly, it was an attraction of opposites, since Cecil was bold and loved adventure spiced with danger, while Willie was relatively meek. Cecil's personality was actually more like Willie's brother, Bob. However, unlike Bob, Cecil was somehow able to get Willie out of his shell. Willie would later say: "What [Cecil] gave me was like water in a thirsty land. He dragged me into adventures that bolstered up the self-confidence in which I was so sadly deficient."[22] Cecil introduced Willie to skiing, hunting, sailing, and other vigorous outdoor activities. With the daring duo of Bob and Cecil in his corner, Willie was now able to enjoy pleasures of companionship of which he had been so cruelly deprived as a boy. He was lonely no more. And then war broke out.

STRETCHING EXERCISES

Among the biologists, DNA had become the center of attention. As both the substance of genes and the target for radiation damage, no one any longer doubted its fundamental importance to understanding how life works. Nevertheless, its physical structure was still unknown, as was an explanation of how it replicated itself, as genes were known to do.

The next breakthrough came from another graduate of Cambridge University, William Astbury (1898–1961). Astbury was working as a textile chemist at the University of Leeds, specializing in the determination of the structures of fibrous textiles, including wool and animal hair. Wool and hair—both comprised of long fibers made of proteins—would not form crystals, and so were not amenable to x-ray crystallography in the traditional sense. Still, Asbury found that by stretching the fibers out, they would line themselves up in a more or less ordered structure that roughly approximated the highly ordered structures found in crystals. Astbury was familiar with x-ray crystallography techniques because Willie's father had been his doctoral advisor. While studying under Bragg, he had done some x-ray crystallography work on tartaric acid, one of the main acids found in wine. Since the stretched fibers represented a simulated crystal structure, he decided to give the x-ray crystallography technique a try. He did not get the same high resolution images that resulted from real crystals; nevertheless, his x-ray photographs of the fibers were good enough to rule out some hypothetical structures as inconsistent with the x-ray data, thus shortening the list of potential structures that needed to be considered.

The one thing he couldn't rule out from the x-ray data was a coiled structure. In fact, by altering the force applied to the fibers, Astbury could alter the distribution of the spots on the photographic film. The movement of the spots was consistent with the notion that the added force was stretching out the coils and, thereby, separating the atoms from each other.

When Astbury published his findings with these proteins, others scientists were impressed, and some noted that DNA was also a fibrous material that could be stretched into alignment. So Astbury obtained a purified DNA sample from a colleague and repeated what he had done with the wool. Surprisingly, he got comparable results. Again the data suggested a regular structure that could be altered by stretching, but the data was not good enough to determine atomic distances. Nevertheless, others were inspired to pick up the ball and try x-ray crystallography on more highly purified samples of DNA.[23]

THE WELL RUNS DRY

When World War I broke out, Cecil, Bob, and Willie, all joined the British army and went to the front. Cecil and Bob were given combat duties, but because of his scientific background, Willie was assigned to work on a technical project, the goal of which was to determine the exact location of enemy cannon by triangulation of the sound waves they generated when fired.[24] This involved scrambling around the front lines and placing microphones at various locations to collect the sound wave data. So all three men did their service on the frontlines of battle, fully in harm's way.

Under horrific battle conditions and mounting casualties, Willie conscientiously performed his duties, although his melancholy was steadily on the rise. Then came a devastating blow. News arrived that his beloved brother, Bob, had died in the Battle of Gallipoli, in Turkey. He had his leg blown off by enemy cannon fire and died shortly thereafter. He was buried at sea from a hospital ship. Willie was devastated.

Soon Willie received better news. He and his father had been jointly awarded the 1915 Nobel Prize in Physics for their work with x-ray crystallography. The announcement was a surprise to everyone because the rumor had been circulating that Edison and Tesla would share the Nobel for their multiple electrical inventions based on DC and AC currents, respectively. The relatively unknown Braggs, and the little x-ray reflection instrument they fabricated, had displaced two of the most famous inventors in the world. But Willie, in light of the ongoing war, now found such awards somewhat banal, a viewpoint shared by his fellow soldiers who were not impressed with their illustrious trench mate. They started contemptuously referring to Willie as "The Nobbler."[25] Once again, the only thing that Willie's intellectual skills had earned him from his peers was disdain. His superiors, however, did take notice. Willie, in a letter home, amusingly noted that now "generals humbly ask my opinion about things."[26]

As Christmas approached, the war seemed no closer to an end, but Christmas itself was peaceful enough for Willie and his fellow soldiers to enjoy a modest Christmas dinner that even included plum pudding sent from their families back home. Nevertheless, as they ate, there was a continuous rumble of guns in the distance, reminding them that not everyone was able to enjoy their pudding.[27]

The new year brought no relief from the fighting. The war dragged on and on. Then one day came the news that Willie had so dreaded. Cecil was dead. He had been at the Battle of Loos, France, when poison gas released by the British had blown back onto their own lines. He survived that debacle, but was severely wounded a few days later in combat outside of Loos. Transported back to England, he lingered in agony and died ten weeks later, on February 9, 1917, at his family's home.[28] He was 25 years old. Cecil's death was a tragic blow to Willie's psyche. Of Cecil, Willie remembered: "He was the warmest-hearted and most loyal friend it was possible to imagine."[29] To his mother, Willie wrote: "I am in such despair over it."[30] Willie would never have another friend so dear.

Willie himself survived the war. He returned to Cambridge University, but it too had suffered great losses. Of its graduates from the five years immediately preceding the war, half had been wounded and a quarter had been killed. Everyone at Cambridge after the war had someone they mourned. The grief was palpable. Rutherford had no sons, but was grieving for his young protégé Harry Moseley who, like Bob Bragg, had lost his life in Gallipoli. Rutherford maintained that the loss of the brilliant Moseley was a national tragedy. American physicist Robert Millikan would later say of Moseley: "Had the [war] no other result than the snuffing out of this young life, that alone would have made it one of the most hideous and irreparable crimes in history."[31] That Willie had survived was nothing short of a miracle. As a second lieutenant, he had certainly dodged his share of bullets. At one point during the war, the survival time for a second lieutenant on the front lines averaged just six weeks.[32] Willie had been at the front for over two years.

After the war, Willie became a professor at the University of Manchester, taking the position that Rutherford had vacated when he assumed directorship of the Cavendish upon J. J. Thomson's retirement in 1919.[33] Perhaps because of their shared grief over the loss of Cecil, Willie found a soul mate in Cecil's cousin, Alice Hopkinson. They courted, married, and had four children, two boys and two girls. The couple lived a tranquil existence of a college professor and his wife for nearly two decades. And then another death changed all that.

THE KING IS DEAD, LONG LIVE THE KING

It's said there is no such thing as minor surgery. True enough. Rutherford died in 1937 at the relatively young age of 66, while still a highly productive scientist. He died from complications of surgery for a strangulated hernia—a procedure that was considered routine even in 1937. It was a shame. He was a kind and generous man, and left a loving wife, as well as many devoted friends, colleagues, and former students.[34] Despite his good-natured jibes at theoretical physicists, he had good personal relationships with most of them. Einstein was an admirer, calling Rutherford the "second Newton."[35] Appropriately, Rutherford's ashes were buried in Westminster Abbey, very near the grave of Isaac Newton, the man whose study of radiation in the form of visible light started it all, more than 200 hundred years earlier.

Rutherford's unexpected death left the Cavendish's leadership position open. It was offered to Willie. He assumed directorship of the laboratory in 1938, again succeeding Rutherford as he had at Manchester. But he wasn't there long before Germany invaded Poland, and England was again at war. With another war in full swing, Willie was called back into service by the British army to resume his work of locating enemy cannon by sound wave triangulation. Meanwhile, his father, still spry in his late 70s, became famous for his visits to the bomb shelter beneath the Royal Institution during German air raids, conversing with the frightened inhabitants and trying to boost their morale.

The war was not going well for England. The allies were taking a beating while the United States remained on the sidelines, trying to maintain an isolationist policy. Then the Japanese attacked Pearl Harbor and America was all in, declaring war on both Japan and Germany on December 8 and 11, respectively, of 1941. The entry of the United States would eventually turn the tide against the Axis powers, but Willie's father would never see that. He suddenly became ill and died on March 10, 1942, three long years before the war's end. With his father's death, all of Willie's confidants, other than his wife, were now gone, but Willie had become accustomed to personal loss. He would persevere.

By the time the United States dropped its atomic bombs on Hiroshima and Nagasaki, thereby bringing World War II to a complete end, Rutherford and the elder Bragg had been dead for eight and three years, respectively. Rutherford had once refused to shake the hand of Nobel Prize winning chemist Fritz Haber (1869–1938),[36] because he had worked on the development of chemical weapons during World War I, even though his own British army had used such weapons against the Germans. We can only wonder how Rutherford, a great humanitarian, would have felt had he lived to see his beloved nuclear fission put to use as a weapon.

WORTH A THOUSAND WORDS

World War II disrupted research on the structure of DNA, and it wasn't until the early 1950s that work resumed in earnest. One might expect that the Cavendish, with its expertise in x-ray crystallography, would have taken the lead, but two factors inhibited DNA research. First, the biologists at the Cavendish had already committed themselves to determining protein structures through x-ray crystallographic techniques. They were on the verge of great breakthroughs in that area; however, their focus was on the structures of myoglobin and hemoglobin, the body's oxygen carriers, rather than on genes, since no proteins had yet been (or ever would be) shown to be genes. Solving these protein structures would ultimately earn the Cavendish more Nobel Prizes, but it certainly turned the laboratory's attention away from genetics. Also, x-ray crystallographic studies of DNA structure were already underway by Maurice Wilkins (1916–2004) and his colleague Rosalind Franklin (1920–1958) at nearby Kings College, in London. British social protocol at the time called for scientists to respect each other's research turf, so Cavendish scientists interested in DNA structure had to settle for persuading the Kings College scientists to reveal any data they might be willing to share.

It turned out that the only two Cavendish scientists with a passion for studying DNA's structure were newcomers to the laboratory. These were an American postdoctoral fellow James Watson (born 1928), with a background in genetics, and a Cavendish graduate student Francis Crick (1916–2004), who had migrated into the study of biochemical structures from a prewar dabbling in physics. Both were passionate about the importance of DNA structure to understanding genetics, and both were frustrated at the lack of attention to this at the Cavendish and the slowness of the Kings College scientists in producing DNA crystallography data.

It didn't help matters that Nobel Prize winner Linus Pauling was taking a strong interest in DNA. Pauling was one of the most brilliant protein structural chemists in the world and the discoverer of the *alpha helix*, a major structural feature of almost all proteins. In his work on the alpha helix, he had used a combined approach of model building and x-ray crystallography to correctly deduce the alpha helix's structure, and there was no reason to believe that he would not be successful using that same approach for DNA. The fact that Pauling had actually written to Wilkins, asking to see some of his x-ray photographs of DNA, had sent a shiver down Wilkins's spine. But Pauling was no more successful in getting a look at Wilkins's films than Watson and Crick had been.

The reason why neither model building nor x-crystallography alone could produce a definitive answer was because in model building, there were just too many possible conformations to consider, with no clear way to discern which conformations might be correct. For x-ray crystallography, the data for stretched fibers were just not good enough to calculate all the bond angles required to define the exact structure. But the x-ray photographs were good enough to eliminate some theoretical models as being contrary to the x-ray reflection data. In that way, the crystallography limited the number of models that needed to be considered, and thus aided deduction of the correct structure. This, essentially, was the scientific landscape that existed in 1951, the year Watson and Crick teamed up to discover the structure of DNA.

The molecular structure that Watson and Crick were most curious about was also a helix. The structure of a helix is analogous to a spiral staircase. For DNA, the helix structure would amount to the nucleobases (i.e., the "stairs") spiraling around themselves. Pauling had found that proteins formed helixes, in which amino acids were the spiraling stairs. There was also a prevalent theory, subscribed to by Watson, that helical structures were the most efficient structural conformation for all polymers.[37] Watson and Crick wanted to know whether the x-ray photographs of DNA were consistent with a helical conformation. Once they confirmed that, they could discard all other molecular conformations and focus on playing with the models of nucleotides, which the Cavendish machine shop had made for them from sheet metal, until a workable helical structure was found. How would they know when they had found the correct structure? They weren't quite sure, but they had the strong belief that "the truth, once found, would be simple as well as *pretty*."[38]

This delicate dance between the Wilkins laboratory in London, and Watson and Crick in Cambridge, went on for some time. Meanwhile, Watson and Crick considered data generated by Erwin Chargaff (1905–2002) that suggested the nucleobases in DNA were paired with each other; that is, adenine (A) paired with thymine (T), and guanine (G) paired with cytosine (C). They also considered the possibility that multiple polymer strands of DNA were twisted around each other, similar to the way a rope is formed. But how many strands were in this "rope"? If they were lucky it would be a small number, ideally no more than three.

Then their worst fear was realized. Pauling reported that he had discovered the structure of DNA! His proposed structure had three strands wrapped around themselves, with the nucleobases facing outward and the phosphate backbones on the interior. But Watson quickly realized that Pauling had it wrong. The phosphate groups were ionized and negatively charged. If they were on the inside, their negative charges from the different backbone strands would repel each other and force the strands apart. Also, Pauling's DNA structure did not account for the base pairing suggested by Chargaff's findings. Pauling had gotten close to the correct answer, but his proposed structure for DNA was, nevertheless, completely wrong. Watson and Crick still had time to find the correct answer before Pauling realized his mistake and went back to work on the problem.

Then Watson got the information he needed. In discussing the folly of Pauling's proposed structure with Wilkins at Kings College, Wilkins let his guard down and showed Watson one of Franklin's x-ray photographs of DNA. Watson was no crystallographer. As he would later quip, "I was even ignorant of Bragg's law, the most basic of all crystallographic ideas."[39] Nevertheless, he knew enough. Crick had described to him what an x-ray photograph of a helical structure of DNA should look like, and Franklin's photograph matched Crick's description. The bottom line was that Franklin's crystallography data suggested a helical structure with a diameter consistent with two, not three, intertwined strands—a double helix.

Putting everything together—a two-stranded helix, paired nucleobases, and the phosphate backbones on the outside—the unmistakable conclusion was that the two strands of nucleobases were somehow physically joined with each other interiorly in a conformation that would allow their phosphate studded backbones to spiral around each other in an endlessly repeating pattern reminiscent of a barber pole. After several days of playing with the sheet-metal nucleotide models back at the Cavendish, they settled on the only possible structural conformation that accounted for all the data. They called their final DNA structure the "B-form" of a double helix, as opposed to an alternative "A-form" that they had first considered but ultimately abandoned. They rejected the A-form because the DNA needed to be dehydrated for it to occur, and this wasn't consistent with the fact that the DNA in cells is surrounded by water molecules. The A-form was, therefore, unlikely to be the correct structure for DNA in its native state. But they also disliked the A-form because it wasn't "pretty."

Something beyond physical beauty, however, spoke to the validity of Watson's and Crick's DNA structure; this was its functional simplicity. As every geneticist well knew, genes are replicated during cellular reproduction. The relative complexity of proteins, the very feature that allegedly made them likely harbingers of complex genetic information, was also seen as an obstacle to their easy replication. In contrast, just as form is said to follow function, the nucleobase pairing that was critical to holding the two strands of Watson's and Crick's double helix together, also formed templates with all the information needed to replicate one strand etched within the other. In other words, the two strands were not identical to each other, but instead were complementary. Just as a handprint in the sand is complementary to the palm that produced it, one strand of the DNA was complementary to the other. And just as a palm reader uses the lines on a person's palm to read his fortune, the lines in the palm print should be equally informative in revealing that fortune.

In short, Watson's and Crick's double helical structure accounted for all the data, was physically beautiful, and functionally simple. How could it not be true? Higher quality x-ray photographs subsequently generated by various laboratories showed that the Watson-Crick structural model of DNA was absolutely correct, and that it was, in fact, very pretty.

Watson and Crick had kept most of their DNA activity out of the view of Willie, who thought the two had been working on protein structures. The duo misled Willie partly because they feared that the more people who knew what they were doing, the more likely information would leak out to Pauling or other competitors. They also knew that Willie disapproved of the Cavendish scientists infringing upon the research territory of Wilkins's group at Kings College. Nevertheless, Watson and Crick, unlike Pauling, did not have the clout to rush their discovery to press. They needed Willie to get the attention of the editors at *Nature* (one of the world's premier scientific journals) in order to publish their paper before Pauling. By virtue of his lofty scientific reputation, Pauling would certainly enjoy expedited publishing once he had realized his earlier error about the phosphate charges and corrected his mistake.

Willie found Watson and Crick's findings sound and agreed to write a cover letter highlighting the importance of the manuscript they were submitting to the editors of *Nature*. Actually, it was hardly a manuscript, as it was only one printed page in length and included no data. Nevertheless, that one simple page contained all the important information. Pauling would certainly read it and weep, having come so close himself. In what was likely one of the rare understatements of Watson's scientific career, the paper coyly concludes, "It has not escaped our notice that the specific pairing [of nucleotides] we have postulated immediately suggests a possible copying mechanism for the genetic material."[40]

Encouraged by Willie to write a popular book about finding the structure of DNA so that the public could share in the excitement of scientific discovery, Watson set to the task. But the product, slated to be published by Harvard University Press and entitled *Lucky Jim*, was panned by many, including both Crick and Wilkins. As Watson's colaureates for the Nobel Prize, they found the book to be both inaccurate and offensive to just about everyone other than its author.[41] Crick told Watson that the book showed "such a naïve and egotistical view of the subject as to be scarcely credible," and that his "view of history is that found in the lower class of women's magazines."[42] The book was particularly insulting to Rosiland Franklin, who had been treated shabbily by Watson while alive, and who had since died of cancer.[43] Harvard cancelled the publication. In response to the firestorm, Watson softened the prose, and the book was ultimately published as *The Double Helix* by Athenaeum Press, with a foreword written by Willie. Indeed, Watson himself believed that the book would never have been published at all had Willie not endorsed it by providing a foreword.[44]

Many scientists were appalled that Willie would lend his prestige to the book. Even Pauling, who had always had a good professional relationship with Willie, couldn't understand his tolerance of Watson's malicious grandstanding. Willie's wife, Alice, was livid that Watson had described her husband as a "curiosity out of the past" who was "completely in the dark about what the initials DNA stood for" and got "more pleasure . . . showing his ingenious motion-picture film of how soap bubbles bump each other."[45] She strongly argued against Willie providing a foreword for a book that showed him such disrespect. But Willie, accustomed to abuse from upstarts, took a philosophical stance. He said the book was just "a record of [Watson's] impressions rather than historical facts," and explained that "the issues were often more complex . . . than he [Watson] realized at the time." To other scientists who felt that they had been wronged by Watson, he consoled, "One must remember that his book is not a history, but an autobiographical contribution to a history which will someday be written."[46]

It's said that all publicity is good publicity. Perhaps this partly explains Willie's tolerance of Watson's unflattering book. Or maybe he was so delighted that the structure of DNA had finally been discovered that no book, even one burdened with gossip, could detract from his joy. Watson had his own explanation for Willie's support of his book: "The solution to the structure [of DNA] was bringing genuine happiness to Bragg. That the results came out of the Cavendish [and not elsewhere] was certainly a factor. More important was the unexpected marvelous nature of the answer, and the fact that the x-ray method he had developed 40 years before was at the heart of a profound insight into the nature of life itself."[47]

If publicity for the Cavendish was Willie's objective, then he certainly got it. The book became a best seller and one of the most popular science books of all time. It has been in continuous publication since 1968, with multiple editions; the most recent edition was 2012, and there is even a BBC television dramatization of it.[48]

It is no small irony that the x-ray—that mysterious penetrating radiation that Roentgen had discovered so many years before and used to reveal the bone structure of his wife's hand—was ultimately used as tool to reveal the structure of DNA, its own biological target. It is unclear whether we would even now know the structure of DNA had x-rays not first been discovered. Remarkably, in pursuing the goal of finding the chemical structure of genes, mankind has learned not just that DNA is the substance of genes, but also how radiation can produce cell death, mutations, and cancer—by damaging DNA. We've even learned how agents that damage DNA, such as radiation, can be exploited to cure cancer. It is an important scientific advancement, and it was achieved over a relatively short period of time. It is a truly remarkable story with many heroes, most of them unsung.

Willie died in 1971, at age 82. He had lived a long and highly productive life that contained its fair share of triumph as well as heartache. At 25 years of age he was, and remains, the youngest scientist to ever win a Nobel Prize. It was awarded for his discovery that x-rays can actually reflect off stacked atomic planes just as light reflects off stacked panes of glass, a phenomenon now known as Bragg reflection. However, it is likely that Willie's proudest discovery was his first—his boyhood identification of a new species of cuttlefish, *Sepia braggi*, which he recognized solely by the unique structure of its internal bony skeleton. Paradoxically, Willie probably never saw a live specimen of the species he discovered, since that particular species of cuttlefish inhabits the ocean depths. Nonetheless, knowing his cuttlefish only by its internal structure was probably enough gratification for Willie.

In 2013, scientists at the University of California, Santa Barbara, discovered one of the tricks that cuttlefish, octopi, and other cephalopods use as a camouflage tool to reflect light and blend in with their surrounding environment.[49] It seems they have cells in their skin that can fold their cellular membranes into pleats, forming stacked planes of membranes that the animal can adjust at will, to modify its body's reflection of light. This newly discovered camouflage organelle has been termed a *tunable Bragg reflector* in acknowledgement of the man who originally discovered this reflection phenomenon; that is, William *Lawrence* Bragg.

CODA TO PART TWO

This brings us to the end of Part Two. In this part, we looked at the various health effects produced by radiation and learned the types of illness that radiation will produce is determined largely by two factors: (1) the dose level of the radiation; and (2) the exact type of tissues that are exposed. High doses (>1,000 mSv) produce cell death in exposed tissues, and the nature of the health consequence is determined by loss of the normal function of those cell types that are killed. Most of these various radiation illnesses are exceedingly rare, because, thankfully, few people ever experience doses this high.

At lower doses (<1,000 mSv), cell death ceases to be a problem and cellular mutations become the chief concern. Mutations are worrisome, because they are the precursors to both cancer and inheritable mutations. In contrast to the rare radiation sicknesses, cancers and inheritable mutations have natural background levels within the population that are higher than most people realize. Radiation adds to these high background levels in relatively small increments that are directly proportional to dose. Radiation does not produce any new types of cancer or inheritable mutations that are unique to radiation.

Regardless of the specific health effect or dose level, the target for radiation's health effects is always the same—cellular DNA. At high doses, the damage to DNA is so extensive that cells die. At low doses, the DNA damage is less and is mostly mended by cellular DNA repair systems. Nevertheless, there is a finite probability that some of the damage will scramble encoded genetic information, thus producing a mutation. We cannot completely prevent this. We can only lower the risk that it will happen.

Whether or not a low dose of radiation will produce any noticeable health effects largely comes down to luck of the draw. If you don't want to draw a joker, then ask the dealer to draw you fewer cards. In the same way, if you don't want a radiation-induced cancer, lower your radiation exposure as much as possible. Unfortunately, it's not as simple as it sounds, because if you are too vigilant about keeping your exposure down you will deprive yourself of some of the benefits of radiation technology. What's a person to do? We'll address that question next.

PART THREE

WEIGHING THE RISKS AND BENEFITS OF RADIATION

CHAPTER 12
SILENT SPRING: RADON IN HOMES

Probably one of today's most serious public health issues.

—Vernon Houk, former assistant surgeon general of the US Public Health Service

Get your facts first, and then you can distort them as much as you please.

—Mark Twain

POLYESTER

On December 2, 1984, Stanley J. Watras, an engineer working on construction of the new Limerick nuclear power plant near Pottstown, Pennsylvania, arrived at work. The plant, just seven miles from his home in Boyertown, was scheduled to begin generating power within three weeks, and the construction crew had just installed radiation detectors at the plant doors—a standard safeguard to ensure that nuclear workers don't exit the plant with any radioactive contamination on their bodies. When Watras arrived that day, he set off the alarms on the detectors as he walked *into* the plant. Over the following two weeks he would set off the alarms every morning. Further investigation revealed that his clothes were contaminated with radioactivity that he had picked up at his home![1]

When radiation safety personnel from the plant visited Watras's home, they discovered what they didn't think possible. There was more radon gas in the Wastras split-level house than was found in a typical uranium mine . . . about 20 times as much! Surprised, the radiation safety technicians checked the radon levels in the neighboring houses. "Our house," Watras remarked in consternation, "had perhaps the highest contamination level in the world, but our next door neighbors had none."[2] How could this be?

The Watras house was located on the Reading Prong, a geological formation full of uranium deposits. It extends from the town of Reading, in southeastern Pennsylvania, through northern New Jersey, and up into southern New York. Radon, a radioactive gas produced through the uranium decay chain, leaks from the ground in this area and then mixes with aboveground air.

The fact that radon seeps from the ground over uranium-containing earth had been known since 1908, when geologist Carl Schiffner first mapped the geographic distribution of radon gas in the Saxon region of Germany (see chapter 5). This map subsequently led to the discovery that air in the Schneeberg mines contained high levels of radon. Nevertheless, geologists hadn't previously appreciated how spotty radon leakage could be. Some ground locations can have virtually no leakage, while a spot a few hundred feet away could have huge amounts of radon streaming out. As it turns out, subterranean radon behaves like subterranean water. Just as underground water gathers in pockets and travels great distances along crevices in the bedrock, often to emerge in discrete locations on the ground surface in the form of a natural spring of water, radon travels along ground faults to emerge as a "spring" of gas. The Watras home was built right on top of a spring of radon gas.

FIGURE 12.1. STANLEY WATRAS'S RADIOACTIVE HOME. In 1984, Stanley Watras activated radiation detector alarms each morning when he entered his work facility at a Pennsylvania nuclear power plant. His radioactive contamination was ultimately traced to his home, which happened to have one the highest radon concentration levels ever recorded. This spurred the US Environmental Protection Agency to survey other homes, and led to the discovery that many homes in America had hazardous levels of this radioactive gas. (*Source*: Copyright Bettmann/Corbis/AP Images; image used with licensed permission from AP Images)

The Watras family discovered their radon exposure only because Stanley Watras happened to be monitored for radioactivity as a nuclear worker, and because he happened to be wearing clothing made from polyester. Polyester tends to produce static electricity, and that static electricity attracted radon-contaminated dust in the air of his home onto his clothes. The Watras family case was the first time people realized that natural radon levels in homes could be higher than in mines.[3] How many other houses built on radon springs were out there and, more importantly, how much danger did they pose to their residents?

PROGNOSIS

The Watras family had moved into their house in January of that year, so they had been exposed to the radon for less than a full year. Nevertheless, doctors told them, based on US Environmental Protection Agency's (EPA) risk estimates, their brief exposure to radon made them seven times more likely to die of lung cancer within 10 years compared to a person without radon exposure. Their children, Michael, 6, Christopher, 3, and Cynthia, 1, might not make it to adulthood.[4]

The family moved out of the house immediately and tried to resume their normal lives. As Stanley Watras explained at the time, "[If we] keep worrying about it, we might not live long enough to see whether the doctors were right, because depression and psychological pressure would be too much for us to survive."[5]

It had been firmly established since 1944 that breathing a lot of radon carries a substantial lung cancer risk. The risk of radon seems to be exclusively lung cancer. No other illness has been attributed to radon exposure, which makes sense given that only lung tissue receives a significant radiation dose when breathing this radioactive gas. No one seriously contests these basic facts. But after that, things get a little murky. Why? Because another major cause of lung cancer poses a much bigger threat than radon, and this of course is cigarette smoking. Cigarette smoke produces a statistical haze through which all radon data must be viewed.

Expert panels have evaluated mountains of data on miners in an attempt to precisely determine the amount of lung cancer risk that could be attributed specifically to radon exposure, rather than to smoking.[6] In addition to the historical Schneeberg findings on miners that we've already discussed, scientists had multiple modern cohort studies from radon-containing mines in various European countries, as well as in China and North America, including the mines of the Colorado Plateau. (This Colorado mining area includes the Paradox Valley, where the Flannery brothers of Pittsburgh, and Kelly and Douglas of Baltimore, had mined their carnotite ore in order to purify radium; see chapter 6.) In addition, there were also case-control studies of lung cancer risk from radon in homes, but serious methodological issues surrounding their design rendered them unreliable and of limited value for quantifying cancer risk, so the mineworker cohort studies became the primary data resource for the risk assessment.[7]

Despite the good fortune of having multiple cohort studies to work with—an epidemiologist's equivalent of striking gold—the analytical efforts of the expert panels were hampered. This is because most of the miners were also smokers, and smoking produces far more lung cancer than radon. Teasing apart the relative contributions of smoking versus radon to the lung cancer incidence in miners has been the greatest challenge to radon epidemiologists. This is not only because smoking was so prevalent among miners and nonminers alike, but also because smoking makes the lung cancer risk from radon worse—much worse. Smokers, therefore, represent a subset of the population that is particularly sensitive to radon.

Identifying sensitive subpopulations who may require special protection is an important task of any risk assessment process. Standards that protect the average person are inadequate for highly sensitive individuals. For example, pollen counts that would not bother most people might trigger anaphylactic shock in someone highly allergic to pollen. Sensitive subpopulations have a major influence when it comes to setting exposure limits for the toxins and carcinogens in our environment, because the limits chosen by regulators must protect everyone, even the most sensitive. In practical terms, this means that exposure limits set to provide sufficient protection for sensitive subpopulations amount to overprotection for people with normal sensitivity. When setting limits for radon, the subpopulation needing the most protection is the smokers, and it's the nonsmokers who enjoy overprotection.

It's not clear why smokers are hypersensitive to radon. It could be because radon and the chemicals in cigarette smoke damage DNA in somewhat different ways. For example, the ionizing radiation from radon tends to break DNA, while the highly reactive chemicals in cigarette smoke (e.g., benzo[a]pyrene) tend to attach themselves to the DNA and form what chemists call *bulky adducts* (because they represent large chemical additions to the DNA structure). These different forms of DNA damage may impede each other's repair, such that DNA breaks block the bulky adduct repair processes, and bulky adducts block the repair processes of the breaks. Thus, having both types of DNA lesions at the same time might be much worse that having one type alone.

Alternatively, smokers may be hypersensitive to radon because they have damaged their lungs to the point that their bronchi—the treelike tubes of air passageways—are no longer able to efficiently remove air particles that are breathed in. Radon often arrives in the lung attached to dust and other types of particles in the air. A nonsmoker's lungs are highly efficient at expelling particles up from the lungs and into the throat, where they subsequently get swallowed along with saliva and exit the body in feces. Smokers, however, slowly lose the cells lining the bronchi that perform this function, so particles tend to stay trapped in the lungs. In the case of radon, it could mean that particles laden with radon may stay in the smoker's lungs longer, resulting in a higher radiation dose to a smoker's lungs than a nonsmoker's lungs, even when they're breathing the same amount of radon.

It's not clear which of these two possible mechanisms is correct. It might even be both, or perhaps neither. There is no strong evidence that can explain exactly why smokers are hypersensitive to radon. Nonetheless, they definitely are at higher risk of lung cancer caused by radon—six to eight times higher.

HABER'S RULE

In order to set limits to radiation dose, you need to be able to measure it. For most types of radiation exposure, measuring dose is relatively straightforward. Since radon is a gas, however, measuring radiation dose to the lung is not that simple. It can be done, but relies on the use of mathematical models along with some assumptions about lung physiology. This is analogous to the bone physiology assumptions used by Robley Evans to calculate the incorporation of radium into the bones of the unfortunate radium girls. Also, lung physiology is very complicated and many of the parameters required to model dose cannot be directly measured, only estimated. The combination of multiple physiological assumptions and multiple estimated parameters increases the uncertainties about modeled radon lung doses to the point that confidence in their accuracy is compromised. Without reliable dose estimates, risk assessment falls apart. Fortunately, another strategy is available.

The alternative approach is to make use of a simple mathematical relationship known as Haber's rule for gaseous hazards:

$$D = C \times E$$

where D is dose, C is concentration, and E is exposure time. Haber's rule tells us that dose is approximately proportional to both the concentration of the noxious agent in the air and the duration of exposure to that agent. More scholarly dosimetric models may give more precise dose estimates, but Haber's rule seems to be remarkably accurate in predicting health risks from gases, without any of the complicated statistical hoopla that surrounds more erudite dose modeling.[8]

Haber's rule tells us something else. As we've learned, dose drives biological damage, so dose is what we want to limit. How can we do this? Well, in situations where we have limited control over radon concentration (i.e., in mines), we can measure the concentration in the mine air and then limit how long the miners are allowed to work in that particular concentration. In situations where we have no control over exposure time (e.g., the duration of a family's occupancy of a radon-contaminated home), we can achieve the same end by just assuming maximal occupancy time in the house as a worst-case scenario, and then limit the radon concentrations in the house to ensure that the dose limit will not be exceeded.

Thus, controlling the concentration of radon in the air of homes is a useful means to limit he radiation dose to the inhabitants. This is the approach the EPA has taken to protect the public from the residential radon threat.

For all of the above reasons, radon is one of the few radiation hazards not typically measured in millisieverts (mSv). Rather, Haber's rule has been invoked to provide a radiation dose unit unique to radon. Since most of the data comes from mine workers, the experts have chosen to calculate lung cancer risk from radon using a unit called the *working level month* (*WLM*). Although it has a technical and precise definition, for our purposes it can just be defined as the radon dose that a typical miner might receive from working in a typical radon-contaminated mine for one month.[9] Simply stated, it is just the radon concentration of the mine, expressed in working levels (WL), multiplied by the exposure time measured in months (M), à la Haber's rule. To be perfectly accurate, it's a measure of exposure, not dose. (As we know, ionizing radiation dose is measured in mSv.) But WLMs should be proportional to the true dose, so it can be used as a surrogate for the actual dose.

Expressing radon dose in WLMs also has another practical advantage. It allows us to depict other radon exposures, such as living in a radon-contaminated home, in terms of the equivalent mining work it would represent.

GETTING THE FACTS STRAIGHT: LNT

Because "the dose makes the poison" (remember Paracelsus), we need dose-response data to identify the dose where radon levels become a health concern. Since lung cancer is the health effect of interest, "doses" (i.e., WLMs) are typically plotted against lung cancer incidence rates, and fitted to a straight line. The fitted line must be drawn straight, rather than curved, because radon epidemiologists use the same assumption about the cancer risk from internal radon exposure as the atomic bomb epidemiologists use for estimating cancer risk from external exposure: *Cancer risk is directly proportional to radiation dose at all dose levels, no matter how low those doses are.* Epidemiologists call this the *linear no threshold* (*LNT*) model of cancer risk assessment. Adopting this linear, dose-response assumption for radon provides consistency across both internal and external radiation risk assessments. And consistency is always good policy for risk assessment.

The LNT model also has a mechanistic justification. Unlike the radiation sicknesses that we learned about earlier, where a certain threshold number of cells must be killed before health effects can be seen, scientists think that cancer induction behaves differently. Most scientists believe that there is a finite probability of contracting cancer from any exposure to a carcinogen because there is no dose so low that DNA damage does not occur, and DNA damage is the precursor to cancer. Thus, conceptually at least, even one interaction of radiation with a single cell's DNA could cause that cell to become a cancer cell. The probability of that happening is admittedly extremely low, but it is not nonexistent. Consequently, epidemiologists reject the concept of a threshold when dealing with carcinogens, and just assume risk is proportional to dose in a straight-line (i.e., linear) relationship, regardless of how low the dose is (i.e., they assume there is no threshold).

As we mentioned earlier when we discussed the cancer risk calculations generated from atomic bomb victim data, not all scientists favor linear models. Some strongly believe that the LNT approach to radiation protection has exaggerated the risk of cancer from low radiation doses, and their arguments have some validity.[10] In contrast, no credible scientists believe that the LNT risk assessment approach underestimates low-dose risk. Thus, the LNT model is generally considered the most conservative and, therefore, the most responsible way to determine cancer risk, precisely because it assumes the highest theoretical risks at the lower doses. Since such low-dose risk estimates are at the high end of reasonable predictions, dose limits based on LNT models afford the greatest degree of protection to the public. For this reason, the dogma among public health professionals is to use LNT as the default model for assessing cancer risk for radiation, and to resist any divergence from LNT modeling unless there is overwhelming evidence that it is inapplicable—a situation that seldom exists.[11]

CUTTING TO THE CHASE

We've taken the long route to get to the final radon risk estimates, but our incursion into the underlying risk assessment methods was important. Understanding the methods and approaches used in risk assessment does three things. It allows us to appreciate the strengths of the risk estimates, to recognize their weaknesses, and to gauge our level of confidence in their accuracy. Now that we know how radon risk estimates are generated, let's take a look at the final numbers.

The scientific experts' best estimate of the lung cancer risk per WLM is approximately 0.097% for smokers, and 0.017% for nonsmokers.[12] Stated simply, working for one month in a typical uranium mine increases the odds of getting lung cancer by about 1 in 1,000 for a smoking miner, and 1 in 6,000 for a nonsmoking miner.

Another way of expressing the risk of one month of work is as follows: For every 10,000 smoking miners that worked in a mine for one month, we would expect 10 of them to eventually contract lung cancer from their radon exposure. Alternatively, if 10,000 nonsmoking miners worked for one month, we would expect fewer than 2 of them (i.e., 1.7 miners) to get lung cancer. As you can see, the smokers have a much higher cancer risk from radon. Many people find this way of depicting risk to be intuitively comprehensible because it allows them to immediately see how frequently the bad outcome happens to a group of exposed versus unexposed people.[13]

MI CASA ES SU CASA

All right you say, so much for the mines. What about my home? Here's the EPA's logic on home radon risks. The EPA has decided that a home resident's radon dose should be no more than 2% of a typical mineworker's. That means the concentration of radon in home air should be at or below 0.02 WL (4 pCi/L; 150 Bq/m^3). To go much lower than 0.02 WL would be technically difficult (i.e., extremely expensive) and practically unwarranted, given that even outside air has an average radon concentration of 0.0025 WL.

What is the risk level to residents living their entire life in a home with radon concentrations at 0.02 WL? According to the EPA's calculations, if a person lived in that home every day of his life, occupied the house for up to 17 hours per day (70% of the time), and lived to be 75 years old, then that person's lifetime risk of lung cancer from the radon in his house would be 6.2% if he were a smoker, but only 0.73%, if he were a nonsmoker.[14] But exactly how bad are these levels of risk? Let's explore this question.

NUMBER NEEDED TO HARM

Instead of starting with 10,000 people and tracking their health outcomes as we did above for the miners, another way to portray risk is simply to ask the following question: "How many people would need to receive this particular dose level before one of them would be expected to have a harmful outcome (e.g., lung cancer)?" In other words, what number of people need to be exposed in order to harm just one? When risk estimates are framed around this question, the risk metric is called the *number needed to harm* (*NNH*).[15] It is an underutilized metric for risk characterization, but can be highly valuable because it gives a vivid sense of the magnitude of the risk and allows easy side-by-side comparisons of risk levels for various exposures. As an example of how NNH values work, let's use the approach to characterize the lung cancer risk in a home with radon levels at the EPA's exposure limit.

Restating the lung cancer risks from lifetime residence in a home at the EPA radon limit in terms of NNH units yields the values of 16 and 137 for smokers and nonsmokers, respectively. This means that out of 16 smoking lifelong residents of such radon-containing houses, one would be expected to develop lung cancer from the exposure, while it would take as many as 137 nonsmoking lifelong residents for one to become afflicted with lung cancer. As you can see, the lower the NNH, the greater the risk; so it's much more dangerous for a smoker to live in the radon-contaminated house than the nonsmoker.

It's important to remember that there's nothing magical about the NNH. It's just one of many ways to statistically depict risk. But studies have shown that characterization of risks as an NNH makes complicated risk scenarios comprehensible to people from a variety of backgrounds, and helps them develop a more accurate and realistic understanding of their personal level of risk.[16]

We'll see more uses for the NNH and its mirror-image twin, the *number needed to treat* (*NNT*), which is a metric of *benefit* that is sometimes used to assess the effectiveness of various medical treatments. We'll also learn how the NNH and the NNT can be considered together to weigh the risks and benefits of a particular type of radiation diagnostic procedure. We'll even learn how to calculate the NNH and NNT on our own when the risk assessors fail to provide them to us. But for now, let's just start thinking in terms of NNHs and NNTs as a way to clear our minds of the confusion that other risk metrics, such as *odds ratios* and *relative risks*, can often cause.[17] Confusion is not a good starting point for making valid risk decisions.

ALL TOO COMMON

If these radon risk levels still seem high to you, consider this. The chances of a nonsmoker contracting a fatal cancer of any type during her lifetime is approximately 25%, a number that is unfortunately high. The additional risk from living in the house with radon at the residential limit, according to the numbers quoted above, would increase that risk from 25% to 25.73%. But smokers, in contrast, already have an elevated baseline risk of developing a fatal cancer approaching 50% (depending upon how much they smoke), apart from any radon exposure they might have. For a smoker, the risk of developing a fatal cancer during a lifetime thus moves from about 50% to 56.2%.

Also, consider the fact that virtually no one matches the EPA's extreme exposure assumptions. Rather than living in a single home for an entire lifetime, typical Americans have more than 10 different residences during their lifetime. Since radon contamination of homes is rare, the chances that more than one of these residences would have high radon levels are fairly remote. Also, the prospect that a person would spend as much as 70% of his 75-year lifetime inside any single house is also remote. Thus, in all likelihood, no one in the United States living in a radon-contaminated house fits the EPA's worst-case description for residential radon dose (i.e., 70% of a 75-year lifetime occupying one house). Thus, the actual doses to residents living in houses at the EPA exposure limit of 0.02 WL are probably less than a tenth of the theoretical worst-case doses and, likewise, their lung cancer risk would also be less than a tenth of the risk levels calculated above.

WHO'S DOING THE DYING?

Based on the EPA's risk assessment numbers and their underlying assumptions, the agency has claimed that high levels of radon in homes theoretically kill as many as 21,000 Americans every year.[18] This amounts to 13% of the 160,000 annual lung cancer deaths in the United States, and 3.5% of the 585,000 total annual cancer deaths of all types.[19] But who are these "theoretically" dead people anyway?

Remember that smokers are six to eight times as likely to get radon-induced lung cancer, and consider that the prevalence of smokers in the US population is about 18%.[20] This large difference in sensitivity between smokers and nonsmokers, coupled with the fact that smokers are fairly common within the American population, allows us to make yet another prediction: Most victims of radon-induced lung cancers will be current smokers or former smokers. This fact, which expert panels have stressed time and again, greatly influences an individual's personal risk of getting lung cancer from radon.

Death from radon-induced lung cancer is actually quite rare among nonsmokers. Even if you accept the EPA's estimate that 21,000 annual lung cancer deaths in the United States are due to radon exposure in homes, about 19,000 of those dead would be smokers. Only 2,000 deaths would occur among nonsmokers, just 0.3% of the total annual cancer deaths in the United States. The 2,000 deaths per annum would put the rate of radon-induced lung cancer among nonsmokers at about one tenth of the annual deaths from influenza.[21] In short, you must be very unlucky indeed to die from radon-induced lung cancer if you're not a smoker.

THE EPA CONTROVERSY

When the EPA first issued its radon guidelines in the 1980s, the public was slow to comply. The same public that was in a panic over asbestos and formaldehyde contamination of their homes seemed unperturbed about radon. Social psychologists say that is likely because radon is a natural hazard, as opposed to a manmade hazard.[22] For whatever reason, people seem to be less afraid of natural hazards than manmade ones. Regardless of the explanation for the blasé attitude, the EPA warnings were not being heeded. So the EPA increased its hype with an advertising campaign that focused on the extreme situations and downplayed, and even concealed, that the lung cancer risk of radon was largely restricted to smokers.[23]

The EPA's well-publicized claim that radon is the second leading *known* cause of lung cancer, after smoking-related causes, was also misleading. Although the statement is strictly true, it implies that the leading cause of lung cancer among nonsmokers is radon exposure. This interpretation is false. Radon is estimated to account for 26% of lung cancers among nonsmokers. The other known causes are second-hand smoke (27%), occupational exposures to airborne carcinogens (4%), or breathing outdoor air pollution (2%). The largest proportion of lung cancers among nonsmokers is actually due to other causes that are currently *unknown* (40%).[24] In fact, cigarette smoke, even when breathed second hand, is taking its toll on the nonsmokers too. And the only reason that radon ranks second among the so-called known causes is because the bulk of the causes of lung cancer in nonsmokers remains unknown. So, lung cancer among nonsmokers is rare, and radon-induced lung cancer among nonsmokers is rarer still. See any pattern here? Smoking is extremely bad for your health. If you want to prevent lung cancer, get rid of smoking, and the radon problem will largely resolve itself.

Many of the scientists and risk assessors felt that the EPA had twisted the facts to advance its own agenda and were using scare tactics as public policy. This resulted in some pretty ugly fighting between the agency and the scientific community. The controversy spilled over into the public arena and led many people to the erroneous conclusion that the risks from home radon were not real; this was similar to the way climate controversies have led many to conclude global warming is a myth. Radon risks, like global warming, are not myths. They are real. But the level of the radon risk to homeowners was certainly overstated by the EPA, at least in the early years.[25]

Why the EPA became obsessed with eradicating residential radon exposure is not completely clear. Some have suggested there were political motivations. For example, it has been charged that home radon eradication was a convenient foil for the Reagan administration because it allowed the administration to appear aggressive in fighting environmental hazards without having to confront industrial polluters.[26] Some have cited personal interests of the politicians who pushed for strong antiradon regulations, including Senator Frank Lautenberg. His home in Montclair, New Jersey, was on a lot that included radon-emitting landfill, generously provided years earlier by US Radium's dial painting factory in nearby Orange, New Jersey. The landfill was a half-century-old "gift" to Montclair that kept on giving. It released radon to the overlying homes long after the company and all its workers were gone.[27] Regardless of the political motivations, it clearly wasn't the science that was driving the EPA's radon scare campaign. In fact, many of the scientists who had helped write the reports that the EPA was citing believed that the agency's national radon policy was misguided.[28]

The EPA's actions caused some scientists, including Philip H. Abelson, former science advisor to the American Association for the Advancement of Science and editor of the journal *Science*, to lose faith in the agency. In 1991, Abelson published a particularly vitriolic attack on the EPA's radon policy and the agency itself, in which he suggested that risk assessment studies for radon should be taken out of the EPA's hands and given to the National Institutes of Health. He rightly claimed that the EPA's radon policy caused "needless anxiety for millions of people," and that it was "inexcusable [that the EPA] did not differentiate between radon's effects on smokers and nonsmokers."[29] Abelson's contempt for the agency was palpable: "One of the weaknesses of the EPA is that it seems to be unable to learn."[30]

Thirty years later, the rhetoric has died down and the EPA is still responsible for protecting the public from radon risk, but it does seem to have learned its lesson. A visit to the EPA's public radon website today reveals that it is doing a better job at accurately characterizing the risk of radon, and it is more candid about the fact that radon is primarily a problem for smokers. The EPA has not, however, relaxed the 4 pCi/L (150 Bq/m^3) action level for radon in homes, which remains in line with the most recent report of the World Health Organization (WHO) that recommends home radon concentrations worldwide not exceed 8 pCi/L (300 Bq/m^3), and preferably be reduced to 2.7 pCi/L (100 Bq/m^3).[31] The WHO report, at least, acknowledges it is "important information to communicate [to the public that] the majority of radon-related lung cancer deaths occur in current and former smokers."[32]

ANOTHER KIND OF HARM

The Watras family members were probably the main victims of the EPA's gross miscommunication of risk. They were led to believe that their lives were in serious peril, and they carried that psychological stress with them even as they left their home for a safer environment. The truth is they suffered a moderate increase in cancer risk from their radon exposure, which was hardly a death sentence. If this had been a medical diagnosis, their erroneous prognosis would have been called a *false-positive* result (i.e., falsely determined to test positive for a fatal disease). Presently, we'll learn how diagnostic radiology is grappling with the problem of false-positive diagnoses, which often result in very high financial and psychological costs for both the individuals misdiagnosed and for the health care system in general. The environmental protection community has not been held accountable in a similar way for the consequences of their overly conservative risk predictions. But the societal costs are equally as large.

Whatever happened to the Watras family? Once the family moved out of the house, the EPA used it as a testing laboratory for various radon remediation technologies. After many months, the agency was able to lower the radon levels to 4 pCi/L, the recommended limit, and the Watras family moved back in. Stanley and his wife Diane are still living in the house (as of 2015). Their children are now grown. After 30 years, none of them has died of lung cancer, although the EPA risk estimates for the one year that they had lived in the house suggested that nearly all of them would have done so by now. How could the dire predictions for the Watrases have been so off the mark?

This is one of the trade-offs of using multiple, highly conservative assumptions in risk assessment. It may seem prudent to inflate the risk in order not to underestimate it. Nevertheless, by adopting high-end estimates for every uncertain risk parameter, the cascade of high-end risk assumptions can compound to the point where the final predicted risk levels become incredulous and may even defy common sense.[33] For example, according to the EPA's risk estimates, just one year of living in the Watras house (16 WL) would put a smoker's lifetime lung cancer risk from radon exposure at 56%,[34] in addition to his lung cancer risk from the smoking (15 to 50%, depending upon his smoking level), making his overall lung cancer risk as high as 100%. But as Naomi Harley, a professor in the Department of Environmental Medicine at New York University has pointed out, the lung cancer death rates for miners with the highest radon exposures ever recorded (nearly all of whom were smokers) were never greater than 50%.[35] Thus, estimates of lung cancer risk from radon in homes that exceed 50% must be considered suspect.

WHERE TO NEXT?

In this first chapter of Part Three, we began our journey into weighing radiation's risks and benefits by considering radon in homes. Why start with radon? We did this for three reasons: (1) radon was among the very first radiation hazards identified; (2) the exposed populations have been recognized—miners and residents of radon-contaminated homes—and their exposures have been measured; and (3) nearly a century of cohort studies defines the dose ranges where human health problems would be expected to occur. These are three legs of a four-legged table that professional risk assessors use to evaluate health risks and draw conclusions. The fourth leg is to accurately characterize the risk level to the people who have some interest in the risk assessor's findings; they are often referred to as the *stakeholders*, whether they are health professionals, government regulators, or the public at large. It is this fourth leg of the table that has proved to be the biggest challenge to professional risk assessors and risk managers. Even when the professionals get it right, their message is often terribly mangled by the time it reaches the public arena. That was the situation with radon.

Without a strong fourth leg, the risk assessment table wobbles, and all the legislation, regulation, and public safety initiatives that table supports are in jeopardy. There is also a danger of spawning exaggerated fear and even panic, which further exacerbates the problem.

Another reason to start with radon in homes as a first example of a risk-benefit analysis is because determining the benefit side of the equation is relatively straightforward. It's zero! There is absolutely no benefit from living is a radon-contaminated house. So for radon, the question is simple: How much risk is acceptable in the face of no benefit? In the next chapter, we'll look at diagnostic radiography, for which the benefits to health are potentially enormous, and see how that affects the risk-benefit balance.

CHAPTER 13

A TALE OF TWO CITIES: DIAGNOSTIC RADIOGRAPHY

Statistics can be made to prove anything . . . even the truth.

—*Noel Moynihan*

BOYS WILL BE BOYS

It had been a good long snowboard run, without incident, through the exhilarating mountain air of Vail, Colorado. With the dangers of the slopes behind him, 13-year-old Matthew Piscator headed for the momentary safety of the chairlift to begin the journey back to the mountaintop for what would be his second run of the first day of a Rocky Mountain winter vacation. As he approached the lift line, he decelerated rapidly, and then culminated his ride with the abrupt turning motion that snowboarders use to hit the brakes. This was a maneuver he had not yet quite mastered. As he made the quick turn, his board stopped on command, but his torso failed to notice that his feet had changed direction. Falling backward, he reached behind his body with his left arm to break his fall. But his fall was not the only thing that was broken. When he heard the cracking sound, Matthew immediately knew his vacation had ended, even before he became conscious of the pain.

If you're going to break a limb, Vail is a good place to do it. The medical personnel in this city see broken bones every day, and they know just how to handle them. Within minutes, the ski patrol had Matthew's arm immobilized and he was towed off to the resort's medical center. That bones were broken was obvious, since his arm now seemed to have a second elbow just above the wrist. Such an unnatural bend could occur only if both bones of the forearm (the radius and ulna) had been affected. An x-ray of the arm confirmed the double break, and the doctor was able to ascertain that the breaks were clean, and, therefore, could be easily set. Anesthetics were administered, the broken bones coaxed back into alignment, and a cast applied from wrist to elbow. With pain pills in hand, Matthew was sent on his way. But his vacation wasn't over after all. He was out on the snow again that very afternoon, but this time he was enjoying the mountain sites from the cargo bed of a dogsled with a professional musher working the brakes.

Matthew's experience with his snowboard was not unlike Willie Bragg's mishap with his tricycle over 100 years earlier. And the remedy for their broken arms had also changed little; this amounted to bone manipulation guided by x-ray images. Both boys went on to regain full function of their arms, outcomes unlikely before the advent of x-rays. Two boys doctored with x-rays, and two boys benefited. Odds don't get any better than that.

But what about the cancer risks to the boys from those x-rays? For Willie, we don't know anything about his radiation dose, except that it must have been much higher than Matthew's. Modern x-ray machines give only a tiny fraction of the dose that the old Crookes tubes delivered. Consequently, Willie's risk of cancer would also have been higher than Matthew's. Willie did die of cancer at the age of 82, but from cancer of the prostate, an outcome hard to explain by irradiation of his arm.

We can do a much better job at estimating Matthew's cancer risk from his x-ray. Modern x-rays of broken arms give an *effective dose* of about 0.001 mSv to the patient. This is the first time we've encountered the term "effective dose." What exactly does it mean?

As you recall, the cancer risk estimates for radiation exposures are based largely on atomic bomb survivor studies. Unlike atomic bomb victims who received radiation doses over their whole bodies, most diagnostic radiography procedures expose only a small portion of a person's body.[1] In Matthew's case, it was just his left arm that was exposed, so his risk is much less than if he had received the same dose over his whole body. How do we account for Matthew's lower risk due to his partial-body exposure? We simply ask: "What fraction of a person's whole body does one arm represent?" It turns out that an arm is about 5% of total body weight. Therefore, the risk associated with Matthew's arm dose of 0.020 mSv is effectively the same fatal cancer risk as if he received 0.001 mSv to his whole body (i.e., 5% of 0.020 mSv). Simply stated, that's what effective dose means.[2] It's a representation of the cancer risk from a partial body dose in terms of the whole-body dose that would produce the same risk level. The importance of knowing the effective dose is that it allows us to use our most reliable risk estimates—those from the atomic bomb survivor studies—to calculate the risk associated with partial body diagnostic radiography procedures. So, what is the bottom line for Matthew's risk of contracting cancer at some point in his life from his arm x-ray?

As we've already learned (chapter 9), atomic bomb survivor studies tell us that the risk of contracting a fatal cancer from a whole-body radiation exposure is about 0.005% per mSv. Therefore, Matthew's lifetime risk of contracting cancer from his arm x-ray is:

0.001 mSv (effective dose) × 0.005% per mSv = 0.000005%
(or odds of 1 in 20,000,000)

This suggests Matthew is more likely to win the megalottery than he is to get cancer from the x-ray exposure of his arm. As we already know, the number needed to harm (NNH) is simply the inverse of the odds, so the NNH for an x-rayed arm is 20,000,000 people. In words, this means that if 20,000,000 boys like Matthew broke their arms, just one of them would be expected to get a fatal cancer from his subsequent x-ray experience. How many people break their arms in the United States every year? About 1,300,000.

So how many of these patients would we expect to develop a fatal cancer when the NNH for an arm x-ray is 20,000,000? None of them! How many will benefit from having an arm x-ray? All of them! Now you have all the information you need to decide whether arm x-rays for broken bones are worth the cancer risk.

Of course, life is more than broken arms. How much of this arm story is translatable to other radiography procedures? As it turns out, at least for the risk side of the equation, a lot of it is. Effective doses have been calculated by dosimetrists—the professionals who estimate radiation doses—for virtually every standard diagnostic radiography procedure. These effective doses are available to the public in various publications and on the Internet, and are usually depicted in tables. In fact, some of these tables will also show the corresponding cancer risk in either percentages or odds. Unfortunately, few tables will show the NNH, but we now know how to calculate that for ourselves from the risk percentages. For most diagnostic procedures the NNH will be very high, suggesting very low risk. So for diagnostic radiography procedures, getting a handle on the risk is not a problem. Quantifying the benefits is another matter; you will find no comparable tables for this. Why not?

It turns out that the benefits of diagnostic radiography are more slippery than the risk, which is simply and objectively defined as the lifetime risk of getting cancer. Definitions of the benefits of radiography are more elusive. If only all the benefits of diagnostic radiography were defined as simply as "restored use of an arm"! The problems with quantifying benefits are threefold: (1) there is a huge range of benefits from diagnostic radiography, all the way from finding a curable disease before it becomes untreatable to documenting a medical condition for litigation purposes; (2) no two people see benefits in exactly the same way; and (3) the research on quantifying radiation's benefits has sorely lagged behind research into measuring its risk. Because of these problems, when it comes to weighing risk versus benefits of diagnostic radiation procedures, it's hard to find a numerical benefit scale that isn't broken. In most cases, the risks are highly defined and quantified, but the benefits are in the eye of the beholder. Nevertheless, one thing can be said with confidence: When it comes to using diagnostic radiography for finding the cause of clinical disease symptoms, the benefits almost always far outweigh the risks, largely because the risks are so very low. The only exception to this would be when a doctor orders an unwarranted diagnostic procedure. In that case, the patient receives a risk with no potential for benefit. Not good.

TAKING YOUR LUMPS

By the time Matthew returned to his hometown of Bethesda, Maryland, his arm was already on the mend. He was in nearly top form by the time spring baseball season rolled around, and he resumed his position as catcher, no worse for wear. Now it was his mother's turn for some x-rays.

Matthew's mother, Teresa, is very health conscious. She eats well, exercises regularly, and worries a lot. Mostly she worries about her health. Teresa is in her 50s, and a number of her friends and acquaintances have contracted cancers of various types in the last few years. She knows that cancer risk increases with age, so she's taking no chances. She's an advocate of cancer screening, particularly cervical and breast cancer, both major killers of women. Other than her age, she is not in any elevated risk group for either type of cancer, but she feels "one can never be too careful when it comes to cancer." She is lucky in that regard. Bethesda is the home of the National Institutes of Health campus, which lies in the heart of the greater Washington, DC/Baltimore metropolitan area, and she has access to some of the best health care systems and hospitals in the nation, all of which provide cancer screenings.

Teresa started having regular mammograms when she turned 40 and has already had one false-positive result—a shadow on an image that turned out to be nothing. Nevertheless, the positive mammogram, the follow-up procedures, and the consult with the breast surgeon, had taken their toll on her emotionally. Intellectually, she knows that the majority of positive mammograms turn out not to be cancer (i.e., false-positive results), but knowing that doesn't make going through the experience any easier. Despite it all, when she's due, she dutifully musters her courage, submits to yet another mammogram, and holds her breath until the results come in.

Teresa's understanding of the mammography's high false-positive rate is actually better than many doctors'. When 160 gynecologists were asked at a Continuing Medical Education meeting what the chances were of a 50-year-old woman actually having breast cancer if her mammogram came back positive, 60% of them said either 8 or 9 out of 10.[3] The truth is that the odds the woman actually has cancer are only 1 in 10. Before you conclude that gynecologists don't know much about mammograms, you should know that radiologists don't do much better, consistently overestimating their ability to identify cancer on a mammogram. Some radiologists, practicing for decades, fail to notice that most of their patients with a positive mammogram do not have breast cancer.[4]

To add to all her other concerns, Teresa now worries that the radiation from her mammograms has actually increased her chances of getting breast cancer, making it even more important that she keep up her screening regimen. As her age-related breast cancer risk increases with each birthday, so do the cumulative risks of her multiple mammograms. To beat it all, now she hears that mammograms haven't lived up to their expectations in terms of saving lives. Ugh!

Teresa's situation is indeed more complicated than Matthew's. But on the risk side of the equation, determining the risk of a breast x-ray is no more difficult than determining the risk of an arm x-ray. A quick Internet search tells us that the effective dose for a typical mammogram for a woman of average breast size is about 0.5 mSv.[5] As before, we simply multiply this effective dose by the cancer risk per mSv to get the overall cancer risk associated with the procedure:

0.5 mSv (effective dose) × 0.005% per mSv = 0.0025%
(or odds of 1 in 40,000)

Stated in terms of the NNH, for every 40,000 women having a mammogram, one would actually get cancer from the procedure. It follows that, if a woman has 10 annual mammograms in a row, then the NNH for the entire series of 10 would be 4,000 (i.e., the risk would increase 10-fold).

The first thing we notice is that a mammogram has more cancer risk than an arm x-ray. This is because mammograms require much higher image resolution than bone x-rays. Increasing resolution means more radiation and higher doses, and therefore, higher risk. There is also a sensitivity correction made in calculating the effective dose for breast irradiations because breast tissue has somewhat higher radiation sensitivity than other tissues.[6]

Another thing that's different for mammography compared to an x-ray for a broken arm is that, for the breast, the nature of the benefit and the risk is the same. Whereas the price of a restored broken arm is the risk of cancer, the benefit and risk for mammography are both breast cancer—either finding one or causing one, respectively. Since the benefit and risk are identical, one would think that the risk-benefit analysis simplifies to just the numbers. If the chance of finding a cancer is greater than the risk of getting a cancer, then the benefit outweighs the risk, and the matter should be closed. Fortunately, in the case of mammography, there are numerous studies that quantify the benefits of mammography. Let's just calculate the chances of benefiting from mammography and we should be done. Or so one would think.

Determining the value of mammography has, unfortunately, been quite a challenge and highly controversial, largely because the benefit depends on age. The older the woman, the greater the likelihood that a cancer will be found and, therefore, the greater the benefit. The increased likelihood of finding breast cancer with age is due to a combination of the fact that breast cancers are more common in older women and that it's easier to see cancerous growths on older women's mammograms because the higher breast density of younger women makes tumors harder to spot. But we need not get into these details. These are the kinds of issues that expert panels consider when trying to formulate recommendations as to what age women should begin having mammograms. This is also why the recommended age for beginning mammograms keeps creeping up. But for our purposes here, we just want to know the overall success rate of mammograms saving lives in the general population. What is the ultimate societal benefit? To put it even more simply, we want to know how successful mammograms are in doing what we want them to do. Do they prevent breast cancer deaths in women?

It is the answer to this question that has spawned the controversy, because the answer turns out to be, unfortunately, that mammograms have not met expectations. They don't save nearly the number of lives that everyone had hoped for. Let's review some recent numbers to illustrate this.

If 10,000 women, 50 years of age, begin receiving annual mammograms for 10 consecutive years (i.e., a total of 10 mammograms each), then 10 curable breast cancers would be found because of the mammography. If we frame the question as to how many women need to be screened in this way to save one life—that is, the number needed to treat (NNT)[7]—we can see the answer is simply the 10,000 screened, divided by 10 lives saved, or an NNT of 1,000. It is this number (1,000) that has disappointed so many people. It was anticipated that the NNT would be much lower than 1,000 (i.e., you wouldn't need to screen so many women to find a lethal breast cancer). In fact, given the reality that breast cancer is a leading killer of women, it seems counterintuitive that lethal cancers would be so hard to find when you are actively looking for them. After a 40-year history with mammography screening, and a lot of anecdotal testimonies from breast cancer survivors, it came as quite a shock that mammography wasn't doing any better than that at saving lives.[8]

It is important to note that this 1,000 number is not without controversy.[9] In particular, some say that mammography has been getting better with time, so that any long-term study of screening in the past will always underestimate current benefits. This is the "moving target" argument, in that it postulates that you can never get an accurate assessment of the value of any contemporary technology because, by the time you analyze the data, the state of the art for that technology has moved on to a different place. But even if this moving target criticism is valid, it probably will not have a huge effect on the NNT number. In any event, more and more breast cancer advocacy groups are coming to grips with the idea that an NNT of 1,000, even if it isn't exactly correct, isn't too far from the truth either. But for now, let's get back to our risk-benefit analysis of mammography radiation.

Even though mammography screening seems to be an inefficient process because 1,000 women need to be screened annually for 10 years in order to save one, at least you are saving that one life. Furthermore, if you need to screen 4,000 women to cause one breast cancer from the radiation (NNH), but only 1,000 women to prevent one breast cancer death (NNT), it would mean that the chances of finding cancers are better than the chances of causing it. So mammography wins the day, right? Yes, that would all be true if radiation-induced breast cancer were the only harm caused by screening. Unfortunately, for every one life saved by a decade of annual mammography screening, 613 of the 1,000 screened women (61%) will experience at least one false-positive reading over those 10 years, just like Teresa did.[10] To reiterate, a false positive is a mammogram that is read as positive for cancer when there actually is no cancer. Nancy Keating, a breast cancer physician at Harvard-affiliate Brigham and Women's Hospital in Boston, says that mammogram false positives are so common that "if you choose to participate [in a screening program], you should assume this is going to happen to you."[11]

False-positive mammograms set into motion a cascade of follow-up medical procedures that each takes its toll on the patients. In fact, out of those 10,000 women in the 10-year screening program, approximately 940 (nearly 10%), will end up having unnecessary biopsies and surgeries; some of them will also develop postsurgical complications, including infections and possibly death.

The bottom line on the radiation risk of mammography is this: If it were just the radiation dose alone that were driving the risk-benefit analysis of mammography, screening would get the thumbs up, albeit by a much smaller margin than what we found for x-rays of broken arms. It is not, however, the radiation risk alone that is turning up the heat on the mammography screening debate these days. Rather, it is the complete spectrum of harm caused by false-positive findings (added to the modest radiation risks) that is causing the medical community to rethink routine mammography screening. So, the issues with mammography are legion, and only one small part of the downside is the risk of causing cancer from the radiation. In fact, the radiation risk is usually considered so small that few critics of mammography even feel the need to mention it.

There is one caveat before you draw any premature conclusion about the usefulness of mammography. The above screening scenario does not apply if women have risk factors for breast cancer. Since women with risk factors—particularly a family history of breast cancer—are at higher risk going into the screening than the average woman, they would be expected to benefit more from mammography. And the greater their breast cancer risk factors, the more they should benefit.

Lest you think that mammography is the only radiation screening procedure facing this false-positive challenge, consider lung cancer. A study of radiographic screening for lung cancer found many early-stage lung cancers among smokers before they experienced any symptoms. This early detection should translate into better treatment responses and better cure rates. Great news, right? Here's the problem. The same numbers of early-stage lung cancers were also found among nonsmokers, a population with far fewer lung cancer deaths.[12] This suggests that most of the supposed lung cancers being found by the radiographic screening would never have developed into clinical disease. It seems that some cancers just don't progress while others do, and no one knows why. What we really need to do with cancer screening is to be able to distinguish the bad tumors from the nonthreatening ones (the majority). X-rays alone simply can't do that. They just reveal lumps (abnormal tissue growths) but can't predict which of those lumps will progress to clinical disease.

SUPERSIZE ME

If diagnostic radiography of breasts and lungs can find cancer in those organs, why not screen every organ in the body, and do it all at one time? That's the idea behind using spiral whole-body computed tomography scanning (spiral whole-body CT) to perform a complete body screen for diseases in patients with no apparent symptoms. These scanners are called spiral CT because they take a helical path around the body in order to produce a computer-generated x-ray image of internal organs.

This spiral whole-body CT is not the same as a traditional CT scan. A traditional CT scan (which may or may not use spiral technology) is used by doctors to look at cross-sectional "slices" of a person's body where disease is suspected to be located, based on symptoms. For a coughing patient, it might be the lungs. For a patient with a headache it might be the brain. The point is that the image is restricted to the region of the body of clinical interest, and the dose is delivered to a limited number of narrow cross sections through a particular organ rather than every organ (i.e., a small amount of total body tissue is exposed). The CT x-rays produce cross-sectional images, similar to how a meat slicer cuts cross sections of bologna. The slices are comparable to the images, and only the sliced parts of the bologna that get a radiation dose, not the whole sausage. Granted, bologna is a crude comparison to a human body, but actually it's not that far off the mark. Everyone has probably seen a cold-cut slice that has some curious imperfection in it. Is it an indication of a bad sausage, or just some normal variation in the texture of the meat? This is the challenge facing physicians reading CT scans.

Since the slices of a standard CT scan are relatively thin, the amount of tissue receiving the dose is a relatively small portion of the body. As such, the effective doses of a standard CT scan are significantly smaller than the local tissue dose. Spiral whole-body CT scans, in contrast, are quite different because the imaging device spirals around the body and slices the entire body into hundreds of two-dimensional images that are then stacked and reconstructed by computer into a three-dimensional image of all the internal organs. Because the whole body is imaged, and not just part of the body, the effective dose is equal to the actual body dose, typically about 20 mSv[13]—nearly 40 times the effective dose of a mammogram and 20,000 times that of an arm x-ray. And this means, of course, that spiral CT scans entail 40 times and

20,000 times the cancer risks of mammography and arm x-rays, respectively.

It is important to interrupt our story here, in order to point out that these whole-body diagnostic CT scanners used to screen for disease are completely unrelated to the full-body scanners used to screen travelers for weapons at airports. Although they both use x-rays, the doses from the airport scanners are negligible. In contrast to the dose of 20 mSv that a diagnostic whole-body CT scanner delivers to a patient, an airport scanner delivers only 0.000001 mSv (one millionth of a mSv) to the traveler. In fact, the x-ray dose received from the airport scanner is comparable to the dose from cosmic radiation that the traveler will receive from just 12 seconds of high-altitude flight, or just 1.8 minutes of background radiation while standing at ground level. Ironically, travelers accumulate a higher radiation dose while waiting in the long airport lines to be scanned than they do from the few seconds they spend in the scanner.[14]

As you might have already guessed, spiral CT images are no better at finding life-threatening cancers in the rest of the body than mammographic images are at finding lethal tumors in the breast. They also entail all of the same false-positive problems that mammography does, with one big exception. Mammography, at least, has 40 years of research behind it measuring its benefits. Although the benefits are not as big as we had hoped, they are there nonetheless, and lives are saved. The benefits of a spiral whole-body CT scan as a screening tool for cancer are entirely hypothetical. There is currently no evidence that these screens are saving anyone's lives.

You may ask: But what about all those commercials where people claim their lives were saved because a cancer was found during spiral whole-body CT screening? Weren't their lives actually saved? Maybe, maybe not. Perhaps their cancer was one that would never have progressed to the point that it gave them any clinical symptoms. Perhaps their cancer would have been cured even if it had

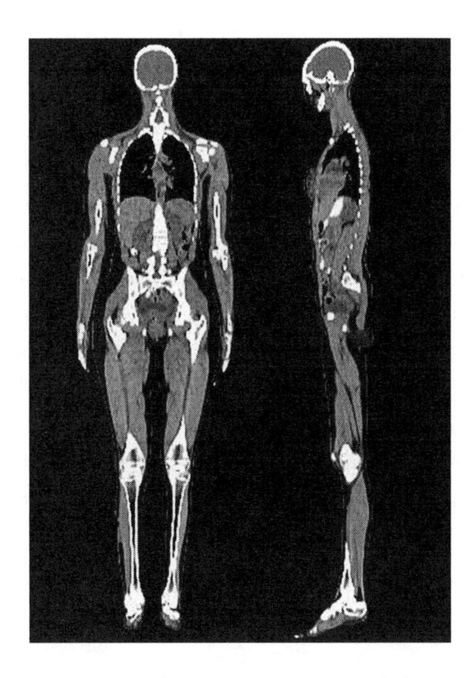

FIGURE 13.1. WHOLE-BODY COAXIAL TOMOGRAPHY (CT) SCAN. X-ray scans of the whole body are sometimes used to find and diagnose sites of disease. Physicians can use this technique to visualize slices of the body at any desired body depth by producing a series of two-dimensional (2D) images in either frontal (left) or side (right) view orientation. Serial 2D CT scan slices, like the ones shown, can also be combined, colorized, and reconstructed by computers to produce 3D images that physicians use to explore the anatomy of inner organs at any depth and from any angle. An example of such a reconstructed 3D image can be seen in video format at the following website: https://www.youtube.com/watch?v=sEnr6FJZOJM. (*Source*: The 2D images shown here are from the website http://radiologystudio.com/ and are used with permission generously provided by Dr. Xinhua Cao.)

been discovered later, when symptoms appeared. Or perhaps they aren't really cured, and their tumor comes back some time later. The point is that no individual can definitively know whether or not her life was saved by cancer screening. In fact, no one can point out the individuals that were or were not saved by screening. All we can say is what happens to patients on average, and from that we infer that there are benefits to individuals. The truth of the benefit is entirely in the numbers. That's why we need statistics. The truth about spiral whole-body CT scanning for cancer screening is that we currently lack the statistics required to demonstrate a benefit. At this point, the benefit is purely theoretical.

Therefore the benefit side of the risk-benefit equation is questionable. Now let's take a quick look at the box score for spiral whole-body CT scan risk:

20.0 mSv (effective dose) × 0.005% per mSv = 0.1%
(or odds of 1 in 1,000)

This means the NNH for each scan is 1,000. If a man has ten annual screening scans, his effective dose from the multiple procedures would, therefore be, 200 mSv (10 times 20 mSv), giving an NNH value of 100 (1,000 divided by 10). This dose and risk level is comparable to the doses received by some of the lower-dose victims in the atomic bomb survivor studies. In fact, most of the people in the Life Span Study received doses lower than 200 mSv.[15]

The comparison with atomic bomb victims is not meant to frighten, but rather to emphasize an underappreciated point. Although we think of atomic bomb survivors as high-dose victims, and many were, the vast majority of the participants in the long-term atomic bomb survivor study—the ones that are actually the major drivers of the cancer risk estimates we now use—did not receive doses anywhere near what would be required to produce radiation sickness. In fact, they had no radiation symptoms at the time of the bombing, and their major health consequence was increased risk of cancer. They are normal people, just like everyone else, concerned about whether they will be among the unlucky few who develop cancer because of a radiation exposure they had many years ago.

Let's now put all the harm (NNH) and benefit (NNT) pieces together. With an NNH for multiple spiral whole-body CT scans at about 100—the greatest cancer risk we've seen yet from a diagnostic radiographic procedure—and with an NNT that is currently undetermined but not likely to be any lower than mammography's NNT of 1,000, the value of using spiral whole-body CT scanning to screen for disease seems questionable. (Remember, for NNH values, higher numbers are better because it means you have to expose a lot more people before one is harmed. In contrast, for the NNT, lower numbers are better because it means you don't need to treat as many people in order to get one person who benefits.) Thus, the NNH and NNT numbers suggest that overall it is more likely to harm, rather than benefit, the screened population. That's not to say the procedure doesn't have value for patients that present with symptoms of illness. It's just that the "worried well" should beware.

WHAT ELSE?

There are many diagnostic radiography procedures out there and we can't review them all. Nevertheless, the three procedures highlighted here—arm x-ray, mammography screening, and spiral whole-body CT screening—like all other diagnostic radiography procedures, fall into one of two categories of use: (1) finding disease in patients that present with clinical symptoms of disease; or (2) screening for disease in people who have no symptoms.

In the first category, the benefits nearly always far outweigh the risks, as long as the chosen procedure is appropriate for the clinical condition. This is because the penalty for not finding the underlying cause of the disease can be severe—a nonfunctioning arm or death from cancer—and the risks of partial body radiography procedures are quite low.

In the second category, regarding the screening of the worried well, we should take pause. Do the real benefits meet the claims? How does the NNT compare to the NNH? And what are the added perils apart from the radiation? All these things need to be considered in consultation with a knowledgeable physician before making any decisions.

The issues are complicated, particularly if a person has a family history of disease. It could be argued that a positive family history is, in and of itself, a clinical finding that warrants radiographic screening. But everyone—patients as well as doctors—needs to understand that such screenings also have *false-negative* findings, in addition to the false-positive findings we've discussed. In the case of mammography, a false negative is a mammogram report claiming a clean bill of health for a woman who actually does have breast cancer.[16] Although the false-negative problem is much smaller than the false-positive problem, and gets much less attention, false negatives do occur. For every one woman saved by participating in a screening program, six women will die despite their participation in screening.[17] Women need to understand that participating in a screening program does not guarantee breast cancer survival.

WHAT ABOUT RADIATION THERAPY?

Even with all its nuances, weighing the risks of diagnostic radiography procedures is a cakewalk compared to weighing the risks and benefits of therapeutic radiation. Why is that? Because radiation therapy is primarily used to treat people who already have cancer, and that makes all the difference. Every patient and every cancer is different, and even the same cancer in the same patient can respond differently to radiation therapy over time. Also, radiation is often combined with both surgery and chemotherapy, because cancer is a formidable foe. This further complicates things.

Despite all these complexities in weighing the risks and benefits of radiation therapy, most radiation risk calculations for cancer patients are a moot point anyway. Many clinical and personal issues need to be weighed when considering whether or not to undergo radiation therapy for cancer. Usually, however, the risk of getting a secondary cancer from radiation treatment for the first one is not one of the primary concerns. This is because a cancer produced by radiation today will typically not appear for 10 to 30 years. It makes little sense to forgo treatment of a real cancer today based on a concern about a theoretical cancer that won't appear for at least 10 years anyway. The risk of dying now is greater than your risk of dying then. Also, for many older people, the latency period of the second cancer is longer than their natural life expectancy anyway. So they most likely will not live long enough to experience the second cancer. Secondary cancers are of more concern for younger cancer patients who have longer life expectancies.

On an optimistic note, we should welcome the day when secondary cancer risks come to be the major concern for radiation therapy patients. Why? Because it will mean that radiation therapy has become so successful at curing patients that we now have significant reason to worry about what will happen to them decades down the road. The good news is that we already seem to be headed in that direction. Currently, there are more than 12 million cancer survivors in the United States, three times as many as there were in 1971, and the number grows by about 2% each year.[18] As this population grows, so will the number of secondary cancer diagnoses, and the medical community will find itself a victim of its own cancer therapy successes. Let's hope this trend continues and the risk of secondary cancers from radiation therapy becomes the main radiation concern for the next edition of this book.

CHAPTER 14
SORRY, WRONG NUMBER: CELL PHONES

The absence of evidence is not evidence of absence.

—*Martin Rees*

Facts are stubborn, but statistics are more pliable.

—*Mark Twain*

TROUBLE ON SCOTTOW HILL

Scarborough is a small town of about 20,000 people on the southern coast of Maine. Now a resort area, it was once the site of many a skirmish with the local Abenaki Indians who objected to European settlers squatting on their land. The attraction for the settlers was excellent fishing and farmland, and ample stream flow to power sawmills, so they resisted moving on, even though they were under constant threat of assault.

What the town needed was a means to communicate danger to everyone in the event of an impending attack. Centuries before the advent of telecommunications, they didn't have many options, but they saw potential on Scottow Hill, one of the few elevated points of land. The settlers built a tower of wood on top of the hill, to be lit in case of an emergency. If any residents saw a bonfire on top of Scottow Hill, they knew to take refuge, for danger was imminent.

Today the Abenaki Indians are no longer a threat, and the bonfire tower is gone, now replaced by a cell phone tower that serves a similar purpose of ensuring vital communication among the residents. It also protects them from danger, but perhaps not quite as well. It is one of three cell phone towers in Scarborough, but service in town is still spotty. Resident Donald Day is concerned about the well-being of his family. He says the spotty coverage is a safety issue when traveling after dark on icy winter roads through the Scarborough Marsh. Day is a supporter of a proposal before the town council to erect another tower to improve cell phone coverage.

But other residents have their own safety concerns. Elisa Boxer is worried about the effect cell phone towers have on the health of residents. At a town council meeting in the summer of 2014, she gave a statement to the commissioners: "When a cell phone tower goes up in a neighborhood, it becomes a sick neighborhood. When it goes up near a school, it becomes a sick school and the teachers and students become part of a cancer cluster." Boxer went on to claim that radio waves from cell phone towers have been "scientifically linked" to cancer. Resident Suzanne Foley-Ferguson shared Boxer's fears. She said the cell towers needed to be "kept as far from any residential neighborhood as possible." Another resident, Karen Tanguay, chimed in that more studies were needed to truly figure out the health dangers from cell phone technology.

The outcome of the meeting was that the town council ducked the controversy and sent the cell tower proposal back to the ordinance committee. "I am glad it's going back to Ordinance," Town Councilor William Donovan said. "As much as it's been worked through at several levels, I think the public needs to know more about it." If Day wants to keep his family safe on the roads, he'll need to drive around the Scarborough Marsh on wintry nights, because there will be no cell phone coverage there any time soon.[1]

AN EPIDEMIC IN SEARCH OF A DISEASE

In 2011, the International Agency for Research on Cancer (IARC) in Lyon, France—a wing of the World Health Organization (WHO)—announced that its expert committee had ruled that radiofrequency electromagnetic fields (i.e., radio waves) should be classified among "possible carcinogens." The announcement was based largely on the findings of a very large cohort study and five smaller case-control studies,[2] and the world's news organizations took notice. The cell-phone opponents proclaimed victory, while the cell-phone fans screamed foul. The committee's ruling was actually a split decision between a majority of scientists who felt the data were strong enough to say it was "possible," and a minority who felt that the data weren't strong enough to say even that.

Although cell phone towers afford a looming monument to modern technology and a focal point for community anxiety, the highest doses of radio waves to humans actually come from use of the cell phones themselves, not the towers. This is because despite their weaker energy output, the cell phones are held next to the head. As we know, health risks are driven by dose. So if the radio waves were to cause cancer, the most likely place to find an association would be in the tissues receiving the highest doses. If you can't show that cell phones cause cancers of the head, it is not likely that you will be able to show that other cancers, in tissues that receive only a tiny fraction of the head's dose, are caused by radio waves. So most cell phone studies focus on cancers of the head, more specifically brain cancers, since the volume of the brain is quite large and absorbs most of the radio wave energy.

The cohort study was huge, including 420,095 Danish cell phone subscribers between 1982 and 1995.[3] This is much larger than the atomic bomb cohort study, which only follows 120,000 survivor and control subjects. The Danish study showed no association between brain cancers and cell phone use.

Cohort studies, as we've learned, are the gold standard of epidemiology and are seldom unseated by the less reliable case-control studies. Normally such a large cohort study with negative findings would have pushed the smaller case-control studies to the sidelines, but there were issues with the cohort study's design. The chief problem was that doses were never actually measured. Rather than using an exposure metric that can be directly related to dose, just having a cell phone subscription was taken as an indication of a person's exposure to cell phone radio waves. Perhaps the person owned a cell phone but never used it, or perhaps someone who was not a subscriber was using a cell phone owned by another person. The assumptions that the cell phone subscribers were the users and that their doses were proportional to the length of their cell phone subscription were seen to be weaknesses of the study, and this caused it to drop from gold to bronze status in the eyes of some committee members. This dosimetry problem opened the door for this extremely large negative cohort study to be shelved in favor of case-control studies that had better dose information.

Several small case-control studies suggested that there might be an association between gliomas, a specific type of brain cancer, and cell phone use, but they provided no evidence for an association with meningiomas (tumors of the brain's surface membrane), salivary gland tumors, leukemias, or lymphomas. Unfortunately, most of these case-control studies had their own issues, including the continual concern regarding all case-control studies; that is, there might be hidden biases in the study design. But one large case-control study was considered robust enough to be given serious attention. The INTERPHONE study, the largest cell phone case-control study to date, compared glioma patients (2,708 cases) to healthy individuals (2,972 controls). The study employed hours of cell phones use as a relative measure of radio wave dose to the brain.[4] The study found that for those in the top dose decile[5] (i.e., top 10%) of cell phone use (>1,640 total hours), there was a statistically significant increase of 40% in the risk of gliomas, compared to people who never used a cell phone. In addition, the elevated risk was found only in patients who had been exposed for at least seven years. Shorter exposures showed no increased risk, which was consistent with the idea that if an association were genuine there should be a lag, or latency period, before tumors would be expected to appear, as happens for all carcinogens. Worrisome findings.

But when the risk at the lower doses was examined the picture looked much fuzzier. Not only was there no elevated risk at any of the lower dose deciles, but for some deciles there was even statistically significant protection from gliomas! So the data did not support a clear dose-response relationship. And the qualitative shift from protection to elevated risk raised concerns that the study's findings were fallacious.

Despite the problem of bouncing associations between dose and risk from one decile to the next, what if the risk level at the highest decile were reliable? Furthermore, although it was a case-control study and, therefore, potentially subject to various types of biases, it did not appear to be subject to any recall bias. This is because the hours of cell phone use for each individual in the study were objectively assessed from their phone records, rather than by interviewing the patients about their phone use behavior. In the absence of any identifiable biases, the committee felt that it had no alternative but to take the 40% increased risk of glioma at the highest decile of cell phone use at face value and acknowledge that it was "possible" that radio waves cause cancer.

Forty percent sounds bad, but how bad is it, really? Let's calculate the NNH to give us a sense of the level of risk, using the conservative assumption that the high-dose findings of the INTERPHONE study are real and that every cell phone user has the same alleged 40% risk as those in the highest dose decile. To put it another way, let's ask the following question: What would a 40% overall increase in risk mean to those 420,095 Danish cell phone subscribers from the cohort study if all of them, not just those with the most exposure, were at 40% increased brain cancer risk from their cell phones?

In any group of 420,095 people, about 2,500 will develop a brain cancer (or other nervous system cancer) at some point during their lifetime.[6] If all of these people were lifelong cell phone users, and their cell phone use really does increase their risk of brain cancer by 40%, we would expect to see 3,500 cancer cases instead of 2,500 within this group (i.e., 1,000 more cases). By dividing 420,095 people by 1,000 additional cases, we get an NNH of 420; this means that among 420 lifetime cell phone users, only one person might be expected to develop a brain tumor due to his or her phone use. Still, given the large number of cell phone users in the United States (91% of adults and 78% of teenagers),[7] it suggests that brain cancers produced by cell phones could be generating a substantial increase in brain cancers nationwide—if it were, in fact, true that cell phones cause brain cancer.

HABEAS CORPUS

If cell phones are in fact producing all of these cancers, where are the cancer patients? The fact is that the brain cancer incidence in the United States has remained relatively unchanged for 40 years—a period of time over which cell phone use has increased among adults from 0% in 1980 to 91% as of 2013, with the most rapid increase occurring between 1995 and 2005. There are now more mobile devices in the United States than there are people. When cigarette smoking started to become popular and the cigarette market exploded, there was a huge increase in lung cancers shortly afterward and the elevated rates of lung cancer didn't begin to subside until cigarette smoking started to drop. Why hasn't this pattern happened with cell phones? We have to ask: "If cell phones are truly killing us with brain cancer, where are the bodies?"

The data suggest that we do not have an increasing trend for glioma incidence in either men or women, for any age group.[8] We have the same glioma rates now, after decades of cell phone use, as we did before cell phones were invented. So even if it is theoretically possible that cell phones could cause brain cancer, they don't seem to be doing so in any great numbers.

It could be argued that if only the most highly exposed people were at significant risk of glioma, it might not translate into a noticeable increase in overall glioma rates. That is, if only the top 10% were affected, we might not be able to see an increase across the whole population. Point taken. If that's the case, however, it still suggests that radio wave-induced glioma is not a significant public health threat for the majority of people, and few (or none) of the glioma cases that occur annually can be attributed to cell phones. In any case, gliomas are not a public health threat that is on the rise, so there appears to be no reason to believe that we are facing a surge of brain cancers caused by cell phones or, for that matter, any other technologies introduced over the last 40 years.

PRELIMINARY HEARING

Some people are concerned that it just may be too soon to tell whether cell phones cause cancer. As we know from studies with ionizing radiation, there is a latency period between exposure and the occurrence of cancer. But latency is more related to cancer type, rather than the type of carcinogen. Ionizing radiation-induced brain tumors began to appear between five and ten years after the atomic bombings in Japan. So we would expect the same latency for cell-phone-induced brain tumors, if they were to occur. We haven't seen rates increase yet, but perhaps we just haven't waited long enough . . . perhaps. However, radio waves are not exclusive to cell phones. They have been used in telecommunications since the time of Marconi, and the early radio-wave workers were exposed to massive doses because Marconi erroneously assumed that the only way to increase transmission distance was to boost power output to extreme levels. As we've already learned, workers are usually the first to bear the consequences of exposure to hazardous agents because they typically experience the highest doses for the longest periods of time. The history of radio waves over the last 100 years suggests that those workers are not at increased risk of cancer. In contrast, the increases in cancer among ionizing radiation workers were very evident within a few years after x-rays were discovered. Cell phones emit just a minuscule fraction of radio-wave energy that early radio workers were exposed to, many of whom worked on transmitting towers while the towers were transmitting, with their heads right in the path of the radio waves. So how can cell phones be producing cancer among the general public when we don't see cancers among the workers? Good question.

A VERY HIGH BAR

Epidemiology has always faced the same dilemma. It is a science of measuring *associations*, but people want it to measure *causation*. Proving that one thing causes another is extremely difficult, even when causation seems obvious. It is easy to show that plant growth and sun exposure are associated. But try to prove that sunlight causes plants to grow. That is a trickier matter.

Of course, the problem of determining cause and effect casts a broad net; it affects more than epidemiology specifically, or even health science in general. But it is a particular dilemma for the field of public health, because it's very hard to solve a health problem when you don't know what's causing it. The infectious disease scientists took the first serious crack at solving the causation problem, and the best they could come up with was a list of criteria that, when met, strongly supported the causation of an illness by a particular suspect infectious agent. These criteria, known as *Koch's postulates*, were developed by Robert Koch (1843–1910) in 1884 to help identify the microorganisms that cause tuberculosis, cholera, and anthrax, and have been used by scientists to identify the causes of infectious disease ever since. More recently, Koch's postulates were employed to establish that a particular virus, now known as human immunodeficiency virus (HIV), actually causes the disease AIDS—a contention originally doubted by many.

In 1965, Austin Bradford Hill (1897–1991), a prominent British epidemiologist, developed a set of criteria that could be used to make the case for disease causation by an environmental agent, similar to how Koch's postulates are used for infectious agents.[9] There are a total of nine *Hill's criteria of causation*, and only rarely are all nine met. It is generally accepted, though, that the more criteria met, the stronger the evidence for causation. We cannot review all the criteria here, but the first five are considered the most important and are the most relevant to cell phones and cancer.

These are (1) the magnitude of association should be large; (2) the association should show consistency across studies; (3) there should be temporality, meaning that the exposure should precede the disease by a reasonable length of time; (4) there should be a coherence of explanation or, in other words, it should not contradict other known facts; and (5) there should be a dose-response relationship.

When epidemiology studies fulfill these five criteria, the case becomes very strong that the suspect environmental agent causes the disease. In the case of cell phones, we aren't even close to fulfilling the first five of Hill's criteria:

1. The magnitude of the association between cell phone and cancer is not large. The alleged 40% increase in risk is a relatively weak association. For example, the association between smoking and lung cancer is on the order of a 2,000% increase.[10]

2. The association between cell phones and cancer is not consistent between studies. Sometimes there is increased risk, sometimes no risk, and sometimes even protection.

3. Temporality is often taken to mean simply that the exposure must precede the disease, a requirement that seems intuitively obvious. But, actually, its definition can be extended to mean that the appearance of the disease must be sufficiently delayed after exposure to be consistent with its known latency period. For cancers, latencies are at least five years. In the case of the INTERPHONE study, the alleged association between cell phone use and cancer doesn't appear until after seven years of exposure, so this would be consistent with the temporality criterion. However, temporality has never been explicitly investigated, and individuals with tumors that are deemed to have arrived too early are often just dropped from the studies.

4. Radio waves are a type of nonionizing radiation. As we've learned, nonionizing radiations do not have sufficient energy to break chemical bonds, which is the means by which

DNA is damaged. The most they can do is produce heat. That is why the chief health concern for nonionizing radiations, such as microwaves, radar, or radio waves is burns from excessive heating. The prospect that nonionizing radio waves might cause cancer like x-rays, gamma rays, neutrons, alpha particles, and other ionizing radiations do, is counter to the known mechanism by which environmental agents cause cancer (i.e., by damaging cellular DNA). Radio waves from cell phones are too weak to cause even heat damage, let alone break chemical bonds. (Videos on the Internet allegedly showing cell phones cooking eggs are bogus.) Random reports that claim radio waves can damage DNA have not been reproducible and are, therefore, not considered credible.[11] If radio waves cause cancer, it would have to be through their own unique and unknown mechanism that somehow does not involve any damage to DNA. Is that within the realm of possibility? Of course it is. But if it's true, it would cause a complete paradigm shift away from everything we have learned about environmental carcinogenesis since 1915, when scientists Koichi Ichikawa and Katsusaburō Yamagiwa of Hokkaido University, in Japan, first started painting coal tar on rabbit ears to explore why chimney sweeps got scrotal cancer; this is the research that ultimately led to the discovery that chimney soot contains DNA-damaging chemicals.[12] In short, we have no coherent explanation of how radio waves might cause cancer that would be consistent with our 100 years of evidence that DNA is the target for carcinogenesis.

5. There is no clear dose-response relationship. In fact, the very study that defines a 40% increase as the high-end risk also suggests 20% protection at lower doses. In order for us to accept that this represents a true dose-response relationship, we would also have to accept that small doses of radio waves are actually beneficial to health, a contention that almost no one is willing to adopt at this point.[13]

It's clear that cell phones fail miserably as a cause for cancer based on the Hill's criteria. Does that mean it is impossible for cell phones to cause cancer? No. It simply means that cell phones don't meet even the minimum conditions that we would expect to see in epidemiology studies, if it were true.

In short, epidemiology alone can never prove causation, it can just make a very strong case that an environmental agent causes a particular disease. Using Hill's criteria, the case for cell phones causing cancer is extremely weak, suggesting that they do not cause cancer. But if you think showing causation is difficult, it's nothing compared to trying to prove that something does not cause disease. Few scientists would even try. It's noble to take on challenging tasks, but foolish to attempt the impossible.

INNOCENT BEYOND A REASONABLE DOUBT

The problem with the IARC classification system for potential carcinogens is that complete exoneration is not possible. How can anyone say with certitude that it is impossible for some particular environmental agent to cause cancer? That's an unreasonable standard to meet. In fact, the IARC carcinogen categories don't even allow for that. The best a compound can do is to be designated as "probably not carcinogenic to humans" (group 4), and only one agent out of 970 evaluated in the history of IARC panels has gotten that designation. Caprolactam, a chemical used in the manufacture of nylon, is the sole agent that has made it into group 4.

There are actually five possible groups for a potential carcinogen to be assigned to under the IARC classification system:

Group 1	Carcinogenic to humans	(113 agents)
Group 2A	Probably carcinogenic to humans	(66 agents)
Group 2B	Possibly carcinogenic to humans	(285 agents)
Group 3	Not classifiable as to its carcinogenicity	(505 agents)
Group 4	Probably not carcinogenic to humans	(1 agent)

In contrast to lonely caprolactam, the newly assigned radio waves have a lot of company in their category (group 2B) of "possibly carcinogenic to humans." The 285 agents in group 2B include aloe vera leaf extract, coffee, gasoline engine exhaust, talcum powder, pickled vegetables, and even the profession of carpentry. Just about every individual is exposed to one of these agents, and they are just as culpable of being a cancer-causing agent as radio waves.

The point is this: The news about radio waves should not have been that they were listed as "possibly carcinogenic" (group 2B). Rather, the news should have been that they were not listed as carcinogenic (group 1) or even "probably carcinogenic" (group 2A).[14]

WEIGHT OF EVIDENCE

The only thing more controversial than the risks of cell phones are the benefits. Cell phones are both glorified and demonized. Few people would dispute that a cell phone is extremely useful for calling in assistance when your car slides off an icy road and into a marsh, as Mr. Day worries about. But the value of talking on the phone about weekend parties while driving to high school, not so much. We'll leave it to you to assign a value to your cell phone.

On the risk side, you can choose to accept the contention that cell phones cause cancer and go with the worst-case estimate of a 40% increase in brain cancer risk, making the NNH for cell phones equal to 420. If you decide adopting this worst-case scenario is the most reasonable approach, then just ponder that number (420) for a while and decide for yourself whether this level of cancer risk is worth the benefit to you.

But the beauty of cell phones is that the means to control your personal risk is literally in the palm of your hand. If you're uncomfortable with this risk level, you can virtually eliminate your risk while still retaining the benefits. As we've already mentioned, it's the phones themselves and not the cell phone towers that are giving us the dose, and it's our heads that are taking the brunt of it. Fortunately, cell phones can be used with headsets as an alternative to holding the phone next to our head. By using headsets, the dose to the head can be reduced to nearly nothing, and if the cell phone isn't consistently pressed against some other part of the body the rest of your organs are spared too. Even using hands-free Bluetooth audio devices will largely eliminate the alleged threat because Bluetooth signals are notably weaker than the cell phone signals themselves. So just use headphones or Bluetooth and keep talking on your cell phone without worrying about cancer.

Alternatively, you can decide that there is little credible evidence that cell phones cause cancer, keep your cell phone, and move on to worrying about some of your other exposures, where evidence of an adverse health effects is well established. Either way, don't wait to make a decision until definitive proof of cell phone safety comes along, because that day will never come. It's your call, and there is no sense in waiting to decide. You might as well make a decision today. Science will never be able to guarantee safety. To quote Clint Eastwood, "If you want a guarantee, buy a toaster."

CIRCUMSTANTIAL EVIDENCE: NEUROLOGICAL SYMPTOMS

There is a small group of people who complain that radio waves produce headaches and other neurological symptoms, apart from any cancer they might cause. They claim to have a condition called *idiopathic environmental intolerance to electromagnetic fields*, or *IEI-EMF*. The medical community does not recognize IEI-EMF as a legitimate disease and suggests that the symptoms, if real, are either due to another cause or are psychosomatic in origin. Regardless, some of these people have sought complete refuge from radio waves by immigrating into the small town of Green Bank, West Virginia, where they claim to find relief. This town is in the National Radio Quiet Zone, an area of about 13,000 square miles (34,000 square kilometers) in a very rural and mountainous region straddling the Virginia and West Virginia border.[15] Personal cell phones and any other devices that transmit radio waves are banned in the Quiet Zone because they cause background noise at the National Radio Astronomy Observatory in Green Bank, which houses a radio telescope that measures low-level radio-wave radiation emanating from deep space.[16]

The Quiet Zone has been in existence since 1958. Longtime residents claim not to miss cell phones because they have never known them, and they like the fact that the technological constraints of the Quiet Zone help ensure the area will remain rural. The IEI-EMF sufferers say living in Green Bank has provided them relief from their radio wave-induced symptoms and are happy to trade modern wireless conveniences for good health.

Unfortunately, IEI-EMF, if it does truly exist, must be classified as an orphan disease because not enough people suffer from it to attract the attention of the medical community. There have been seven double-blind clinical trials, but they have all been negative, meaning no association between radio waves and symptoms was found.[17] *Double-blind studies* are to clinical trials what cohort studies are to population epidemiology. They are the gold standard for proving the existence of a clinical condition or a treatment effect. In a double-blind study, everything is concealed from the patients as well as their doctors—so both are "blinded" as to who's getting the real treatment and who's getting the placebo. This prevents any biases from being introduced. The code is cracked on the data only at the end of the study, at which point the treatment's test and placebo patients are identified and the results analyzed.

Since the double-blind studies failed to validate IEI-EMF, implying no association between the claimed neurological symptoms and radio waves, can we rule out that radio waves cause neurological symptoms? Just as we've seen for cancer, we cannot. We cannot dismiss causation simply because we lack any evidence for it. All we can say regarding the potential of cell phones to cause IEI-EMF is that there is a complete lack of any high-quality clinical evidence to support it, leaving only anecdotal testimonies and unreliable non-blinded studies as the alleged proof of its existence. Again, Hill's criteria of causation are not met. But in terms of coherence (criterion 4), perhaps there is a better case to be made for radio waves causing IEI-EMF than cancer.

In contrast to cancer, where DNA damage would presumably be a requirement for causation, the notion that neurological symptoms could be caused by radio waves seems more tenable, and more coherent, with the known mechanism of the nervous system. The human brain is the center of a complex network of neurons that transmits and receives electrical signals throughout the body, and radio waves can interact with electrical systems. Are radio waves interfering with the neuronal circuitry of the bodies of these people? If so, what is unique about their bodies that they should be so highly predisposed to symptoms when the vast majority of us is not affected?

These questions may never be answered. There are too few people claiming to be affected to warrant large-scale studies, and there is no current evidence that their alleged symptoms are real. A cancer is a cancer, and cancers are definitely not psychosomatic. But a headache or other pain is a self-reported symptom that is hard to document or authenticate. Until some quantifiable neurophysiological endpoint surfaces that scientists can measure and validate independent of the self-reported complaints of the patients, it is unlikely that IEI-EMF will get any serious attention from the medical community. This means that, for IEI-EMF suffers, permanent residence in Green Bank will likely remain their only option for relief.

SEX OFFENSE: INFERTILITY

There is one other frequent question about cell phones that seems to reflect widespread concern: Can cell phones carried in pants pockets cause infertility? Ever since male soldiers started telling their female companions that radar duty had rendered them sterile, the false rumor that radio waves produce male infertility has persisted. Hermann Muller directly addressed the question in his fruit fly model in the 1940s and found no credibility to it.[18] Atomic bomb studies and experiments with animals tell us that large numbers of spermatogonia cells—the cells that give rise to sperm—must be killed in order to produce infertility, and even then the condition is typically temporary. Radio waves do not kill cells, so the contention that radio waves produce infertility doesn't seem plausible. And sporadic reports that radio waves do induce infertility have not been reproduced or validated. Whether this alleged infertility should be considered a risk or a benefit of cell phones all depends upon your perspective. But an unplanned pregnancy cannot be blamed on a dead cell phone battery.

CHAPTER 15

HOT TUNA: RADIOACTIVITY IN FOOD

According to the Global Research Report, recent tests in California have unearthed contaminated bluefin tuna in nearby coastal waters. . . . In addition, the cesium-137 level has also gone up along the California, Oregon and Washington coasts. . . . In Alaska, the sockeye salmon population has declined apparently due to radionuclides as well.

—*Marisa Corley (reporter for* Liberty Voice*)*

Do not tell fish stories where the people know you; but particularly, don't tell them where they know fish.

—*Mark Twain*

ON THE RIGHT TRACK

Daniel Madigan is a recreational angler and a professional marine biologist. Fish intrigue him, particularly the big ones, like marlin, sailfish, and tuna. "It's black water, and all of a sudden you have this huge animal," he exclaims. He thinks such encounters with these enormous pelagic fish in the open ocean are enchanted moments, and they make him wonder: "Why now, and why here?"[1]

These giants of the sea lead mysterious lives that include migrations of thousands of miles, but the details of their migratory journeys are poorly understood. Madigan hopes to change that. He did his doctoral studies at the Hopkins Marine Station of Stanford University, a place where migratory behavior is studied for a number of ocean species, including great white sharks, leatherback turtles, black-footed albatross, and bluefin tuna. Most of this tracking is done with electronic transmitters that require individual animals to be captured, tagged, and released unharmed. No easy task, particularly when the animals are very large.

During his doctoral research, Madigan was exploring another approach. He was trying to determine whether you could tell the location from which a fish had migrated by tracking the specific combination of the nonradioactive (stable) isotopes of carbon and nitrogen in their flesh.[2] To put it simply, did particular regions of the ocean convey upon their marine residents a specific isotopic signature that scientists could use to determine an animal's original location? It seemed like a reasonable idea. In fact, anthropologists have been using a similar strategy to track ancient migration patterns of humans by characterizing the isotopic composition of the teeth from their remains. For example, anthropologists are able to tell from the isotopic signature in teeth that a person whose body is unearthed in southern England actually grew up in Scandinavia, suggesting he was likely a Viking invader rather than a local Brit.

Such isotopic signature investigations are a type of *tracer study*. In tracer studies, materials of interest, be they living or dead, are molecularly marked in some way, as though they had a molecular bar code. Information about movement and location of the material can then be obtained by following that marking. In Madigan's case, a unique combination of stable isotopes that characterized a specific ocean location was the marking he followed. But detecting and measuring stable isotopes is not easy. Certainly, it is much simpler to follow radioactive isotopes because they emit distinctive types of radiation that can be readily tracked, identified, and measured with radiation detectors. This is why scientists prefer to perform tracer studies with radioisotopes, rather than stable isotopes, whenever possible.

Radioisotope tracer studies can be done on either a microscopic or a grand scale. We've already seen a microscopic radioisotope tracer study. You'll recall that labeling of viral DNA and protein with different radioisotopes allowed virologists Hershey and Chase to trace the migration of the DNA, rather than protein, into the cells of the virus's host, suggesting that DNA, and not protein, was the substance of genes. Fortunately for Hershey and Chase, it was easy to tag viruses with radioisotopes in the laboratory. Unfortunately for Madigan, it is not easy to radioactively tag fish in the open ocean; hence, his nonradioisotope tracer approach was the next best alternative.

But Madigan had a thought. Suppose, somehow, a characteristic combination of radioisotopes entered the ocean at some specific location. Would that radioactivity be taken up by local sea life, and could tracing the radioactivity in such animals be used to follow their migrations to other regions of the ocean? Perhaps.

One day in August 2011, fishermen were unloading their local catch of Pacific bluefin tuna (*Thunnus orientalis*) on the San Diego docks, and Madigan was there. His knowledge of bluefin growth rates allowed him to identify by size alone those fish in the one- to two-year age class. This cohort of fish would have just completed their first transpacific crossing. If Madigan were correct, these would be the most recent immigrants from the bluefin spawning grounds in the waters off Japan.

ORTOLAN OF THE SEA

Pacific bluefin tuna are big animals. They can reach weights in excess of 1,000 pounds (450 kilograms) at maturity, but few ever reach full size due to overfishing. The northern Pacific population spawns off the coast of Japan and remains there until, as juveniles (one to two years old), a subgroup of the fish begin an eastward ocean migration that ends on the west coast of the United States; here they feed on schools of baitfish, particularly the California anchovy. Their transoceanic migration takes one to four months.

FIGURE 15.1. COMMERCIAL FISHERMAN WITH A BLUEFIN TUNA. These large fish travel great distances in the ocean, consuming local marine life as they go, and incorporating into their bodies whatever stable and radioactive isotopes happen to be in the local waters. When humans consume such fish, they also incorporate these isotopes into their own bodies, where they tend to congregate in specific organs. If these isotopes happen to be radioactive, consuming the fish will result in a radiation dose to those organs.

While the tuna are in American waters, local commercial fishermen catch them and, ironically, their carcasses end up back in Japan within 24 hours, this time making their transpacific trip in the cargo hold of commercial aircraft. Most of them are sold at the Tokyo fish market. This is the same market that bought the radioactive catch of the *Lucky Dragon No. 5* back in 1954, and very close to the site where that radioactive catch remains buried to this day.

The Japanese have an insatiable appetite for bluefin tuna, consuming over 90% of the world's catch. They consider bluefin the premier fish for sushi and that appetite drives worldwide demand. In 2013, a single 489-pound (222-kilogram) tuna fetched a record $1.7 million ($3,500 per pound) in the Tokyo market. With prices like that on their heads, no wonder the stocks of bluefin tuna in the Pacific Ocean are only at 5% of their historic levels.[3]

Madigan collected steaks from 15 different bluefin. He then contacted fellow scientist Nick Fisher, an expert on marine radioactivity at the School of Marine and Atmospheric Sciences at Stony Brook University in New York State. Would Fisher test the fish for radioactivity? Fisher told Madigan that he would, but really didn't expect to find anything. Soon the results were in and, to Fisher's surprise, all 15 fish were contaminated with manmade radioactivity. But not just any radioactivity. They were all contaminated with cesium-134 and cesium-137 radioactivity, a signature of recent nuclear power plant waste. The implication was clear. The tuna were contaminated with radioactivity they had likely picked up that spring in the waters off Fukushima, Japan, the site of a major nuclear reactor meltdown.

We'll learn a lot more about the Fukushima nuclear disaster of March 11, 2011, in the next chapter, but for now, all we need to know is that this nuclear accident resulted in a lot of radioactive waste escaping into the Pacific Ocean, and some of it got into sea life. These tuna had obviously been in the vicinity of Fukushima at the time of the accident and they carried that reactor radioactivity within their bodies across the entire Pacific Ocean. "This was just nature being amazing!" Fisher gushed. "Now, potentially, we had a very useful tool for understanding these animals."

Madigan and Fisher realized their findings were important and rushed to publish them in the prestigious *Proceedings of the National Academy of Sciences*.[4] In their paper, they concluded, "These results reveal tools to trace migration origin . . . and potentially migration timing . . . in highly migratory species in the Pacific Ocean." But nobody gave a damn about that. The press and the public had a more pressing question: Are the tuna safe to eat?

To answer any question about safety of radiation-contaminated food, we need to know two things: (1) the exact radioisotopes involved; and (2) the amount eaten. Knowing exactly which radioisotopes are implicated tells us whether or not they will become concentrated in specific human organs, and will inform us as to the specific types and energies of the radiation they emit. This information will allow radiation dosimetrists to calculate the dose received by different human organs from eating various amounts of the contaminated food. Knowing those organ doses, we can then determine how much is safe to eat.

We already recognize, from the radium dial painter experience, that radioisotopes can chemically mimic dietary nutrients and become concentrated specifically in the organs that use those nutrients. In the case of radium, its chemical similarity with calcium, the mineral constituent of bone, causes ingested radium to be incorporated into bone. Likewise, as we know from the Bikini Islanders' experience with nuclear bomb fallout, strontium seeks bone because it, too, is located in column two on the periodic table of elements, along with the calcium and radium. It was strontium that posed the major threat to the Bikini Islanders who ate coconut crabs following the hydrogen bomb tests, and it was their bones that received the bulk of the dose from the ingested strontium. Health data from the radium girls was directly applicable to the Bikini Islanders because radium and strontium behave very similarly in the body.

The Fukushima radioactivity found in the San Diego fish flesh came from two isotopes of cesium, specifically cesium-134 and cesium-137. With the exception of cesium-133, which is stable, all other isotopes of cesium are radioactive and their half-lives are relatively short in comparison to the age of planet Earth. So all the radioactive cesium produced by the supernova that created Earth is long gone. Thus, any radioactive cesium that we now find in our environment must have come from manmade sources, either nuclear power plants or nuclear weapons testing. Cesium radioactivity produced by bomb testing is ubiquitous, having spread all over the planet, and currently contributes slightly (~0.02 mSv) to the annual background radiation exposure for everyone in the world. So how was it that scientists were able to tell the cesium in the tuna specifically came from Fukushima and not from a nuclear bomb test or another nuclear power plant leak?

It turns out that cesium-134 and cesium-137 are produced in roughly equal amounts by the fission of uranium-235 (i.e., nuclear power plant fuel). But the 134 radioisotope has a half-life of just 2 years, which is very short in contrast to the 137's half-life of 30 years. So the 134 disappears from the environment relatively quickly while the 137 persists. That's why it's the 137 and not the 134 radioisotope that currently contributes to our background dose from atmospheric nuclear bomb testing years ago.[5] The cesium-134 decayed away long ago, so there are no currently detectible amounts of cesium-134 in our environment. Consequently, when you do find cesium-134 somewhere in the environment, you know that it was produced by a relatively recent fission reaction. But how recent?

As already mentioned, cesium-134 and 137 are produced by fission reactions in relatively equal amounts. In fact, the cesium released into the Pacific Ocean at Fukushima had almost exactly a one-to-one ratio (134 to 137, respectively). Since 134 has a 2-year half-life and 137 has a 30-year half-life, it can be readily seen that two years following a release to the environment only half the 134 would have remained, but the 137 would have hardly decayed away at all. Thus, at two years, the ratio of 134 to 137 should be about 0.5 to 1.

You may recall that the ratio of carbon-14 to carbon-12 in biological material can be used to accurately date archeological finds (radiocarbon dating; see chapter 3). The same approach can be used here to date the cesium found in the fish. So that's what Madigan and Fisher did. They determined the ratio of 134 to 137 in the fish to find out how long ago the cesium had been produced. They discovered it had been produced five months before the fish were caught. Since the fish were caught in August 2011, they must have picked up their cesium contamination around March 2011, the month of the Fukushima accident. Given that time frame, combined with the fact that the fish are known to migrate into Californian waters from Japan, the findings are pretty strong evidence that the cesium contamination in the fish came specifically from Fukushima.

Now, having identified cesium-134 and 137 as the Fukushima radioisotopes being ingested when eating bluefin tuna, how big a risk does eating the fish pose to sushi eaters?

Cesium is not in column two of the periodic table (i.e., the column with calcium, radium, and strontium). Rather, it resides in column one, which is inhabited by the biological nutrients sodium (element Na) and potassium (element K).[6] As mentioned earlier when we discussed nuclear fallout, sodium and potassium are electrolytes critical to cellular function and muscle activity (including cardiac muscle). As might therefore be expected, the body's sodium and potassium supplies tend to be somewhat enriched in muscle. And, yes, as you may have guessed, cesium is enriched in muscle as well. So cesium consumed as fish flesh (muscle) ends up in human muscle because it is mistaken for sodium or potassium by the body's physiological processes. Thus, cesium is more likely to be found in muscle, rather than bone. But there are other important differences between cesium and radium.

The turnover of the minerals in bone is extremely slow. So radium incorporated into bone stays there a very long time. Sometimes scientists express the residence time of an element in the body in a manner similar to how they express the decay of radioactivity. They say that the element has a certain *biological half-life*, and the interpretation of this value is the same as for a radiological half-life. For example, a biological half-life of one year for element X in a specific organ would mean that after one year the original amount of X in the organ would have decreased to half of the original amount through biological turnover processes and body excretion. For radium, its radiological half-life is 1,600 years, and its biological half-life in the human body is about 28 years.[7]

So for radium, the bones get a double whammy. The radium decays away slowly, plus it stays in the bone for a long time. In other words, the radium is going to give a whopping dose to the bones because the exposure time is so very long. In contrast, for radioisotopes with either a short radiological half-life or a short biological half-life, the exposure time is going to be too short to deliver a significant radiation dose (all else being equal). Fortunately, cesium falls in this latter category of radioisotopes. Its radiological half-lives, 2 years and 30 years (cesium-134 and 137, respectively) are much shorter than radium's (1,600 years). More importantly, cesium's biological half-life is just 70 days (0.20 years),[8] as opposed to radium's 28 years. Of these two determinants, its very short biological half-life, rather than its moderately long radiological half-lives, is the primary determinant of cesium's low radiation dose. In effect, most of the cesium radioisotopes are excreted from the body before they get the chance to decay.

To get to our goal of determining dose, we now need to employ some logic extrapolated from what we've already learned. For any ingested radioisotope, knowing its rate of uptake into the body, its target organs, its radiological half-life, its biological half-life, and the energies of its decay emissions (i.e., the energies of its alpha particles, beta particles, and gamma rays, etc.), dosimetrists are able to calculate the radiation dose rate to any particular target organ. These calculations involve calculus, but we need not get into any of that math here. Nevertheless, the strategy is simply to calculate the average length of time that one atom of the radioisotope spends in the organ, using its radiological and biological half-lives, and then determine the average radiation dose that a single atom would deliver during its residence time in the organ. Once that is known, you can simply multiply the dose delivered by a single radioisotope atom by the total number of radioisotope atoms ingested, and you have the total dose.

What the dosimetrist has actually calculated is an *organ dose*, which is not unlike a partial-body dose from a diagnostic x-ray, as we've already discussed for broken arm x-rays and mammography. So we now ask the dosimetrist to convert the organ dose (partial-body dose) from the radioisotope, into an effective dose (a whole-body dose equivalent), just like we previously did for partial-body external x-rays. She does the calculation, gives us the number, and we send her on her way, because effective dose is all we need to make our risk estimate. How is that? Because, as you remember from our earlier discussions of diagnostic x-rays, effective dose allows us to calculate our personal risk using the atomic bomb survivor findings.

Even if the details seem complicated, the take-home message is simple and short: With a little knowledge of the metabolic physiology of the elements and the emission characteristics of their radioisotopes, the effective dose for any ingested radioisotope can be directly calculated, and that's very important because effective dose allows us to predict cancer risk.

Before we leave all this internal dosimetry stuff, we should acknowledge one more thing. In this chapter, we're focusing on ingested radioisotopes. But the same approach works for inhaled radioisotopes (e.g., radon), and for radioisotopes absorbed through the skin. The approach works exactly the same, and it works well. How well we'll get to shortly. But now, let's get back to the tuna.

The scientists found that the San Diego tuna contained 4 and 6 becquerel per kilogram (Bq per kg) of Fukushima radioactivity, from cesium-134 and cesium-137, respectively.[9] To keep this simple, let's just combine these numbers and say that there were 10 Bq per kg of Fukushima cesium in the tuna. Is that a lot of radioactivity?

Everything is relative, so let's take that value of 10 Bq per kg and make some comparisons. In terms of health regulations, it is not high. The US Food and Drug Administration (FDA) allows up to 1,200 Bq per kg of cesium in fish, so the bluefin are at less than 1% of the regulatory limit. How do the folks at the FDA determine the regulatory limit? They simply perform the effective dose calculations we described above and then they make a worst-case assumption about a typical American's fish consumption. That is, they assume that people are eating a lot of the contaminated fish. In this case, they assume that the typical American eats one pound (0.45 kg) of fish per week and all of that fish is cesium-contaminated bluefin tuna. Then they determine how high the cesium radioactivity concentration would need to be in the fish before an individual's effective dose would exceed the US Nuclear Regulatory Commission (NRC) radiation-dose safety limits for the general public, and presto, they have a regulatory limit for consumption of cesium in fish. Basically, this is where the 1,200 Bq per kg regulatory limit comes from.[10]

What does all this mean in terms of public health safety? It means that it is virtually impossible for Americans to eat enough bluefin to exceed federal regulatory limits for effective dose, as long as the cesium radioactivity concentrations of the fish are <1,200 Bq per kg, even if they are eating extremely large amounts. However, the reality is that no one in the United States is eating that much bluefin since it is not readily available on the American market. Thus, the FDA's assumptions about Americans' bluefin consumption, like the EPA's assumptions about the time Americans spend in radon-contaminated homes, don't pass the smell test. Even for the sushi-eating Japanese, it would be extremely hard to eat enough cesium-contaminated bluefin sushi to get themselves into trouble, even if they could afford to do so.[11]

Now let's look at this health question in a slightly different way and make another comparison. Cesium is a manmade radioisotope, and people tend to worry more about manmade contaminants. But there is also a lot of natural radioactivity in food; not just in seafood, but in all foods. Again using the bluefin as an example, we'll now turn our attention to the natural radioactivity that the scientists found in the bluefin samples.

Potassium-40 (K-40) is a natural radioisotope, not manmade, and it was also found in all the bluefin samples. We already know that potassium is enriched in muscle. There was an average of 347 Bq per kg of K-40 radioactivity in the bluefin muscle (35 times the cesium radioactivity levels). Humans have less natural K-40 radioactivity in their muscles (<100 Bq per kg) than the fish, owing to a lower overall potassium concentration in our muscles compared to bluefin tuna.[12] Our body burden of natural K-40 contributes about 0.15 mSv per year to our annual effective background dose. (You'll recall total annual background effective doses for Americans are typically about 3.0 mSv per year, so about 5% of our annual background dose is due to the K-40.)

All the potassium on Earth has a radioactive component because 0.012% of natural potassium is the radioisotope K-40 (the rest being mostly the stable isotopes K-39 and K-41). Therefore, all of our bodies contain K-40 radioactivity. No point worrying about it. We can't get the radioactivity out of the natural potassium and we can't live without potassium. In fact, the human body is so dependent upon potassium that it keeps its potassium levels tightly regulated. Too much or too little and cardiac death soon follows.

The body doesn't store potassium so it needs a consistent dietary input to fulfill its needs. Fish and meat (i.e., animal muscle) are good sources of dietary potassium. For vegetarians, certain vegetables and fruits, particularly bananas, are a major source of their dietary potassium. So aren't we getting a lot of radiation dose from K-40 radioactivity by eating fish or bananas? And what additional radiation dose are we going to receive if we start eating more fish or start putting more bananas on our cereal every morning? Oddly, the answer is none! But how can that be?

Despite what you may have heard, you get no additional radiation dose from eating those naturally radioactive bananas. Why? Because the body must keep a constant level of potassium, so when you eat 10 excess milligrams of potassium, you pee out 10 milligrams of potassium. Hence, the body burden for K-40 remains steady and unchanged no matter how many bananas you eat, and thus your annual effective dose from K-40 also remains unchanged.[13]

This physiological balancing act of potassium results in an interesting paradox for our cesium-contaminated bluefin. Being a natural radioisotope, K-40 provides a uniform background dose in our bodies that cannot be changed by altering our potassium consumption, so the K-40 in the bluefin has no effect on our annual radiation dose. Oddly, even though there is 35 times more K-40 radioactivity than cesium radioactivity in the bluefin, only the much smaller amount of cesium radioactivity can actually increase our radiation dose if we ingest the tuna. This is because cesium is more radioactive than K-40, molecule for molecule, and cesium is an additional radioisotope that would not be part of our natural environment, were it not for atomic bomb testing and nuclear reactor leaks. Thus, this manmade radioisotope adds to our natural radioactivity.

At the end of the day, however, effective dose is effective dose. So what is the effective dose delivered from eating the cesium-containing bluefin, and how does it compare to what we're getting from natural potassium simply by virtue of having potassium-filled tissues? If every week for one year, someone were to eat one pound (0.45 kg) of bluefin with 100 Bq per kg of cesium (i.e., 10 times what Madigan and Fisher found in their fish samples), that person would have received an effective dose of 0.001 mSv from the cesium.[14] Compare that to the 0.15 mSv annual effective dose we receive from our bodies' natural potassium (a 150-fold higher dose than from the cesium). And further compare it to the total background effective dose that we normally receive in a year, which is 3.0 mSv (3,000 times more than the tuna cesium dose).[15] So, the radiation doses we receive from eating these supposedly radioactive tuna are trivial in comparison to the background radiation doses that we all receive, apart from the tuna.

At these extremely low levels of dose, cancer risk is strictly theoretical. No one has ever proved that there is any cancer risk from radiation doses that are this low. Nevertheless, since we're playing the conservative assumption game, let's assume there are, in fact, risks, and calculate the lifetime fatal cancer risk from the cesium in the tuna based on the atomic bomb survivor data:

$$0.001 \text{ mSv (effective dose)} \times 0.005\% \text{ per mSv} = 0.000005\% \text{ (or odds of 1 in 20,000,000)}$$

This means that the NNH would be 20,000,000 people. Expressed in words, 20 million people would have to eat a pound (0.45 kg) of the contaminated bluefin every week for a full year before we might expect one of them to come down with cancer because of the radioactive cesium that it contains. There are about 14,000,000 people living in Tokyo, so even if everyone in Tokyo each ate 52 pounds of contaminated bluefin tuna over the course of a year, we wouldn't expect even one of them to contract cancer from the experience. Now compare this with the annual baseline cancer rate for Tokyo, which is 150,000 cases per year.[16]

Let's try one more comparison for good measure. We'll now look at the risk in yet another way. As we've discussed before, cancer is, unfortunately, a common disease. We can expect about 25% of people living today to ultimately die of cancer. Thus, the lifetime risk for any one individual dying from cancer, all else being equal, is also about 25%. This is the baseline risk of cancer death. The 0.000005% cancer risk calculated above adds to this baseline risk. So the added risk of cancer death after a year's worth of daily bluefin consumption moves from 25% up to 25.000005%.

Now we've looked at the risk from eating the cesium-contaminated tuna in three different ways: (1) we've compared the cesium dose to annual background doses; (2) we've compared dose from ingesting the manmade cesium to the dose we receive from a natural radioisotope (K-40); and (3) we've compared the expected increase in cancer incidence from eating the cesium-contaminated bluefin to the baseline cancer incidence. Although one of these three comparisons may speak to you more than the others, it doesn't matter which one you choose because all three are just different characterizations of the same risk level.

As for the benefits of eating the bluefin sushi; well, that's another matter. Bluefin certainly doesn't contain any nutritional benefit that cannot be obtained from other foods. But, nevertheless, based on the prices people are willing to pay for it, a good number of diners must place a very high value on its flavor. Even at just $50 per pound, the 365-day bluefin diet we described above would have an annual cost of $2,600. Armed with all this information, the consumer can do her own risk-benefit analysis (and a cost-benefit analysis as well). Then it's decision time. To eat, or not to eat? That is the question.

NOT SO FAST!

We need to wait a minute here; some of us are still skeptical. All these doses from ingested radioisotopes are calculated with statistical models because the dose from ingested radioisotopes can't be measured directly in tissue. Suppose these models are way off? Suppose the dose (and thus the risks) are actually much higher than the models predict? These are good questions.

What these questions really are getting at is how much *uncertainty* there is in the dose modeling. Uncertainty is the bane of all risk assessments. What if our model parameters, input data, and assumptions are not accurate or valid? As we all know, for all computer-based models, garbage in means garbage out.

The answer is that most dosimetric models have not, unfortunately, been rigorously validated. There are just too many radioisotopes and too many exposure scenarios to consider. Also, it isn't usually possible to ask people to voluntarily eat or breathe radioactivity just to validate a dosimetric model. Nevertheless, some scientists have made serious efforts to assess the accuracy of their models in real-life situations. Three examples address situations that we've already discussed.

The first example is a little gruesome. It has to do with our old friend Robley Evans, the MIT scientist who labored mightily to estimate the bone doses for people who ingested radium, and Eben Byers, the business magnate who had a penchant for Radithor. In 1965, 33 years after poor Mr. Byers had passed away from his Radithor habit, Evans secured permission to recover Byers's skeleton from its coffin in a Pittsburgh mausoleum to measure the actual amount of radium in his bones.[17] Evans measured the radium radioactivity and then returned the skeleton to its coffin, where it remains to this day.

He measured a total of 225,000 Bq (6.1 μCi) of radium radioactivity in the skeleton. He then compared the actual radium measurements from Byers's skeleton to the amount of radium that his model had predicted to be in the bones based on Byers's own account of how much Radithor he said he had drunk over his lifetime. The results showed that there was actually twice as much radium in Byers's bones than had been predicted by the model. But the question remained: Was the model wrong by twofold, or had Byers underreported his Radithor consumption by twofold? That is, what was the source of the uncertainty, a weak model or poor input data? You can see why uncertainty is so hard to nail down.

We'll look at another example, this one from our radon story. As we have already discussed, radon is one of the few radioisotopes for which we don't typically use effective dose (in mSv) as a dose metric. Why? Because all the exposure data comes in the form of air concentration levels in mines (i.e., in *working levels*; *WL*), and we want to limit people's lung dose by limiting the radon concentration in their home air. Therefore, epidemiologists find it more reliable to directly assess risk in terms of radon air concentrations, rather than go back and forth converting WLs to modeled lung radiation doses with dosimetry models that might introduce uncertainty. Nevertheless, it would be interesting to see whether the dosimetric models result in lung cancer risk levels comparable to the risks levels calculated directly from WLs.

We won't go into all the details here, but the findings for radon in the lungs were similar to those for radium in the bones. The WL risk levels were within twofold of the risk levels predicted by modeling the dose to the lungs.[18] But the question still remains: Are the WL risks off by twofold or is the model's mSv dose off by twofold. Who's to say?

The final example comes from the tuna. Madigan and Fisher were able to use their fish tissue concentration model to back calculate the cesium levels that the bluefin tuna should have had while they were in Japanese waters, back in March 2011, in order for them to have had the tissue concentrations that were later measured in California waters in August 2011 (i.e., 10 Bq per kg). Their model indicated that the cesium concentrations for the fish while in Japanese waters would have been about 150 Bq per kg. Compare this to what the Japanese government reported as the actual cesium concentration in bluefin caught in local waters at the time: 170 Bq per kg. This amounts to a discrepancy between modeled and the measured values of about 13%. Not bad. But again, is the model off by 13% or are the measurements off by 13%? Tough call.

So validation of dosimetric models is difficult and, even when successful, it's hard to interpret what the information means in terms of the accuracy of the model. Nevertheless, enough validation studies have been completed for the major dosimetric models to suggest that they may be off some, but they are not off by orders of magnitude (i.e., multiples of ten). Rather, if they are off at all, it's likely they're off by no more than two to threefold, a discrepancy that we can usually live with. What would a threefold underestimation of cesium dose mean for the bluefin risk estimate we calculated above?

It could mean that the baseline cancer rate for the tuna eaters might actually be 25.000015%, rather than the 25.000005% we had predicted. Or it may mean that one person out of the 14,000,000 people in Tokyo might contract cancer, rather than none. If these small differences in risk are going to be a game changer for the risk-benefit analysis you just completed, you'd better reconsider eating that tuna. But in most cases, such small differences are not going to upset the cart. The added risk from the cesium is still pretty low, even if the dosimetric model isn't perfect.

THE UPSIDE OF DEATH AND TAXES

In previous chapters, we've alluded to uncertainty a few times. But this is the first instance where we've discussed uncertainty in a methodical way, and the first time we've acknowledged that uncertainty can be a systemic problem for risk assessment. We're going to see more about uncertainty going forward, but let's just recognize here that uncertainly comes in two flavors. The first flavor is the *known unknowns*. We may not be certain how much Radithor Mr. Byers actually drank, but at least we know that we don't know it. That's why Mr. Beyers's Radithor consumption rate is a known unknown.

For known unknowns, we can usually estimate how big a problem we have on our hands.[19] Technically speaking, we can put bounds on the magnitude of the uncertainty. In the case of Mr. Byers, he might have reported just half of what he actually drank, but he probably did not drink much less than that; otherwise he wouldn't have suffered the health problems that he did.

In contrast, the real problem for risk assessment is the *unknown unknowns*; that is, the things that we don't even know that we don't know. For example, suppose we are wrong about DNA damage being required for cancer induction. Let's say, out of the blue, some scientist conclusively proves to everyone's satisfaction that the gravitational pull from the moon—that same force that drives the ocean's tides—causes cancer in humans even though it doesn't produce DNA damage. That certainly would be a game changer across the board in our understanding of the mechanism of environmental cancer induction. And we might then have to go back and rethink the cancer risk from those cell phones; after all, we had considered the plausibility of their causing cancer suspect because radio waves don't damage DNA. But who would have thunk it? And it's precisely the "who would have thunk it" stuff that can cause the big problems. We must keep an eye out for these unknown unknowns.

Let's leave uncertainty to rest for a while. We'll get back to it shortly. But we've introduced the uncertainty concept now so we don't get too cocky with our risk percentages, our NNTs, our NNHs, and all that stuff. [20] There's only so far we can go with these risk metrics if the fundamental premises on which they are based are shaky. In situations where we have a lot of experience and tons of data collected over time, and the conditions in the future seem little different from those in the past, we can be fairly confident that such metrics are covering all the bases. In these cases, we can feel assured that our risk-benefit analyses are valid. But when data are scarce, the situation is new or different, or our experience is limited, uncertainty rears its ugly head. Do we really know that all swans are white, or have we just not looked at enough swans to be aware that *black swans* exist?[21] Hard to say. Are death and taxes, as Benjamin Franklin claimed, the only truly certain things in life?

THE LATEST BUZZ

The year 2011 is gone. The risk of eating bluefin back then is of historical interest, but what is the current situation? Scientists have asked that same thing.

In 2013, Eric Norman, a physicist at the University of California at Berkeley, was perplexed by ongoing reports on the Internet and elsewhere saying that foods were still contaminated with unsafe levels of radioactivity, two years after the Fukushima accident. Was this even possible? He thought not. When he went looking for data that supported these claims he could find none, so he decided to gather his own.

Norman had his students collect samples of plants, milk, fish, seawater, and salt from various locations in the Pacific Ocean, particularly coastal regions. Included were samples of seafood products from Hawaii, the Philippines, all over the California coastal area, and even Japan. Norman then recruited the assistance of Al Smith and Keenan Thomas, fellow faculty members and experts on radioactivity counting, to assess the samples. They looked for radioactivity in everything, but focused particularly on seaweed samples because seaweed concentrates potassium, and thus cesium. The search found nothing.

"We looked very hard," said Norman. Nevertheless, none of his samples showed any radiation that could be linked to Fukushima. This is likely due to two major factors. First, much of the radiation has simply decayed away and is continuing to do so. But the second factor is probably the more important. That is, the radioactivity has been diluted to such a large extent after mixing in the sea for two years that it is no longer readily detectible. "There's a lot of water in the Pacific," says Norman. "Whatever gets dumped in the ocean will get diluted by enormous factors."[22]

The window of risk from Fukushima radiation in seafood has apparently closed. The large amount of radioactivity dumped into the Pacific in March 2011 has decayed and dissipated to trace levels. Any ongoing leaks from the Fukushima nuclear power plant site are apparently too minimal to result in significant levels of radioactivity in the world's seafood.

How do we know that future radiation leaks won't change things for the worse? We don't. But we have a watchdog on duty. He is Ken Buesseler, a marine chemist at the Woods Hole Oceanographic Institute on Cape Cod, Massachusetts. Buesseler directs the Center for Marine and Environmental Radioactivity (CMER), which actively monitors radioactivity in the ocean. He fills a void that no government organization currently occupies, and he does it all on a shoestring budget funded largely by public donations. According to Buesseler, "Whether or not you agree that levels of radiation along the Pacific coast of North America are too low to impact fisheries and marine life, we can all agree that radiation should be monitored." He analyzes seawater samples sent to him by the public from coastal communities along the Pacific coast and, in the interest of transparency, he posts the results for all to see on the CMER website.[23] If any new source of radioactivity crops up in the Pacific, either from Fukushima or elsewhere, Buesseler's team should be able to detect it and warn us of the threat. So far he's found only trace amounts, but he intends to keep looking, as long as crowd funding continues to support his efforts.[24]

All this news is good for everyone, except for Madigan . . . and the bluefin tuna. For Madigan, the large reduction in radioactivity in the ocean from Fukushima also means the window of opportunity has closed for tracking the migration of fish using Fukushima radioisotopes as a tracer. He'll need to return to his stable isotope methods, with their many limitations, if he wants to keep following ocean fish migration patterns. As for the bluefin, their clean bill of health means that sushi gourmets can resume their bluefin indulgence without any worries about the risk of eating Fukushima radioactivity. As such, the Tokyo fish market can continue selling its bluefin tuna at astronomical prices and the carnage of these magnificent fish can proceed as usual . . . until they're all gone.[25]

CHAPTER 16

BLUE MOON: NUCLEAR POWER PLANT ACCIDENTS

When you want to know how things really work, study them when they're coming apart.

—*William Gibson*

There is no such thing as an accident; it is fate misnamed.

—*Napoleon Bonaparte*

ON SHAKY GROUND

At 2:46 p.m. on March 11, 2011, American engineer Carl Pillitteri was working with his Japanese crew inside Reactor Unit One at the Fukushima Daiichi Nuclear Power Plant, situated 150 miles (240 kilometers) northeast of Tokyo on the east coast of Japan. Suddenly it seemed like an enormous sledgehammer was pounding the foundation of the building. The concrete walls and floor began to crack, and everyone in the crew immediately knew this was an earthquake. After a few seconds the lights went out, leaving the crew in pitch darkness and unable to see anything. Panic ensued. Pillitteri suddenly felt his body being clutched tightly by the two young men standing on either side of him and the entwined trio impulsively began chanting prayers in a mixed cacophony of English and Japanese, like a bilingual foghorn blaring forlornly into a moonless night.

As if to shout down their prayers, earsplitting demonic sounds started to emanate from the direction of the steam-driven turbine. Pillitteri didn't know whether the sounds were coming from the turbine itself or from the building being flexed and twisted all around it. But he knew the sounds meant bad things were happening. Suddenly, he started to feel himself surrender to the idea that he and his crew were going to die. And a scream in the dark from a hysterical coworker, proclaiming that the turbine was about to blow, didn't help matters. At that point all Pillitteri asked of God was to "make it quick."[1]

Steam turbines are devices that look like jet engines. They take energy in the form of steam derived from boiling water and convert that energy into rotary motion. The rotary motion can be used to propel a vehicle, as in the case of old steam-powered railroad locomotives, or to drive a generator to produce electrical power. Coal power plants and nuclear power plants are based on the same thermodynamic strategy; heat boils water to make steam, which then turns turbines to generate electrical power. Simple as that. The heat can come from either burning coal or it can come from nuclear fission. Regardless of the source of the heat, once the water boils, the turbines start spitting out electricity just the same. Steam turbines harness a considerable amount of energy, so it's not good to be around when one "blows."

The turbine did not blow and, after about five minutes, the shaking subsided and a few lights came back on thanks to a diesel emergency power generator that had survived the quake. There weren't many working lights, but there were just enough for Pillitteri and his crew to find their way to the door leading outside. Miraculously, all of the crew made it out alive.

What Pillitteri had experienced was a magnitude 9.0 earthquake. A 9.0 earthquake is an enormous earthquake, rarely ever experienced. There have only been three larger than 9.0 since people began measuring earthquakes: Chile (9.5 in 1960), Alaska (9.2 in 1964), and Sumatra (9.1 in 2004).

Earthquakes are currently measured using the *moment magnitude scale*, which is more accurate that the older Richter scale that it replaced. An increase of just one unit on the moment magnitude scale represents a 32-fold increase in energy, not just 10-fold as many people mistakenly believe.[2] This means that even tenths of a point differences in the magnitudes of earthquakes can represent huge differences in damage.

Once outside, the crew members soon regained their composure, but they were uncertain what to do next. Pillitteri decided it probably best not to linger near the damaged reactor building. The building happened to be on sloping ground. To remove himself from the scene, Pillitteri turned uphill, facing away from the harbor below, and began to climb, soon arriving up at the employee parking lot on the crest of the hill. From his elevated perch he surveyed the situation below. There were damaged buildings and rubble everywhere. The damage extended well beyond the power plant grounds. Most of the buildings surrounding the plant had collapsed.[3] This was way too much damage for any local rescue crews to handle. Surely many would die before outside assistance could arrive. The nuclear power plant where he had worked since 2008 was in shambles, and many of his friends and colleagues among the 6,415 workers were likely dead. Slowly he began to absorb the enormity of the situation and he was beset with grief. How could this happen, and why couldn't it have been foreseen?

Earthquakes are notoriously hard to predict, although it's not from lack of trying. In fact, the Japanese were among the first to attempt it when, in the ninth century, they claimed that earthquakes could be predicted from the behavior of catfish. That didn't work out so well. Nevertheless, nearly twelve centuries later, we aren't doing much better than the catfish. The best that can be done is to make forecasts in the form of broad probabilistic statements concerning the chances of an earthquake happening in a certain region of the world over a certain number of decades.[4] If that's all you need, then geologists can help you. But if you want to know whether a major earthquake is going to occur at a given location in the coming year, you'll receive more guidance from the *Farmers' Almanac* than the US Geological Survey.

Furthermore, despite what you may have heard, earthquakes cannot be overdue. It makes intuitive sense that if a typically active seismic region hasn't had an earthquake for a long while, one is overdue to arrive and the probability of having one soon should, therefore, be higher. We can even imagine the pressure building up between the tectonic plates during the quiescent time, destined to be released in one calamitous burst. But this is an oversimplification of the highly complicated dynamics of fault lines and their tributaries. In reality, it's a safer bet to assume that the annual earthquake risk for the region remains constant no matter how long it's been since the last earthquake. Even the foreshocks reported to sometimes occur before major quakes are not consistently present, and sometimes they can occur without any large subsequent quake. In fact, they technically aren't even foreshocks unless they are followed by a major quake.[5]

There is some good news, however, at least for the Japanese. Japan has a very dense network of tremor detectors throughout the county that can give a warning seconds to minutes before destruction hits. It turns out that fast moving, but nondamaging, P waves emanate from the epicenter of the quake faster than the slower moving S waves, which produce the damage. (P and S stand for primary and secondary, respectively.) Depending on the distance from the epicenter, the P waves may precede the S waves by up to two minutes.[6] Two minutes is not a lot of time, but a nationwide broadcast warning would allow someone to climb down from a ladder, turn off a stove, get into an earthquake shelter . . . or shut down a nuclear reactor.

The Fukushima plant was connected to the P wave warning system. And, just as planned, its three currently operating reactor units (One, Two, and Three) were automatically shut down just before the earthquake hit. Its other three reactor units (Four, Five, and Six) were already shut down for refueling and maintenance, so all power generation at the plant immediately halted prior to any plant damage.

The Fukushima Daiichi plant is owned by Tokyo Electric Power Company (TEPCO). In 2011, the plant, with its six functional nuclear reactors, was one of the largest nuclear reactor sites in the world. Japan relies heavily on nuclear power because of its relative shortage of fossil fuels. Before the Fukushima accident, there were 54 operating nuclear reactors in Japan, providing 40% of the country's energy needs. (For comparison, 20% of electric power generation in the United States is nuclear.)

Fukushima's first reactor was purchased from General Electric, Thomas Edison's old company. General Electric had beaten out a competing bid from its longstanding rival, the Westinghouse Electric Company, which had been offering a substantially larger and more complicated reactor. Westinghouse may have won the AC/DC wars, making AC current the standard for transmission of electrical power worldwide, but Edison's General Electric Company became the standard nuclear power reactor at Fukushima. When TEPCO bought more reactors for Fukushima Daiichi, they bought them all from General Electric.

The earthquake was detected by its P waves, the reactors were shut down before the damaging S waves arrived, and a catastrophe was averted . . . right? Not quite. Shutting down a nuclear reactor does not eliminate the need for cooling it. It still has a lot of residual heat that needs to be dissipated. In addition, the radioactive fission products—the byproducts of nuclear fission—further contribute to the heat, even when the reactor is not running. For these reasons, reactor core cooling must be continually maintained, and for that you need an intact cooling system. The cooling water intake pipes for the six reactors were damaged somewhat by the earthquake, and that was a potential problem. But water is not the only thing you need to cool a reactor.

SEA SICKNESS

From the parking lot, Pillitteri had a clear view of the harbor below and he noticed the crew of a freighter frantically trying to get their ship out of harbor and onto the open sea. He suddenly realized that they were acting in anticipation of the coming tsunami, a response that had been drilled into Japanese seamen so much that they scarcely needed to think about it. Sure enough, just as he watched the ship leave harbor and head east toward the open ocean, he saw a huge swell of water heading in. And then he saw the ship safely ride over it.

Tsunamis are tidal waves of enormous size that are produced by earthquakes originating under the sea. A fault line demarcating the intersection of the Eurasia, Pacific, and Philippine tectonic sea plates lies just off the east coast of Japan. The plates slide over and under each other in fits of starts and stops, rather than in a continuous smooth slide. Each time there is an abrupt movement of the plates, an earthquake tremor follows. These tremors vary in size in proportion to the amount of plate slippage. In Japan, small movements and small quakes happen on a daily basis. (Japan has about 1,500 earthquakes per year.) But it's the big movements, all at one time, that cause the problems. On March 11, 2011, the tectonic plates made a big slip and a big earthquake followed, moving the entire main island of Japan 13 feet (4 m) further east than it had been on March 10.[7]

When one plate slides underneath the other during an earthquake, it's as though the bottom had dropped out of all the ocean water above that lower plate. The sudden drop of the enormous water column—the fault line is 19 miles below the water surface—causes the energy produced to emanate in all directions in the form of waves. In the open ocean, where the water is very deep, the waves take the form of large swells on the surface, but as those undulating swells move toward the coastline, the water quickly grows shallow and waves begin to crest, forming a roiling curl of great height that breaks onto the shore.

This is why the seamen needed to take their freighter straight out to sea in anticipation of the tsunami that would soon hit the harbor. Meeting the wave head on in the deeper water of the open ocean, while it's still in the form of a swell instead of a ferocious curl, allowed the boat to ride over the swell rather than slam into the curl. It may seem crazy, but if a tsunami is chasing your ship, it's better to make a 180-degree turn and head straight at it because you're never going to outrun the thing, and delaying the inevitable encounter by running toward land will only make things worse.

Pillitteri and his crew had survived the shockwave of an earthquake, but others in town were not so fortunate. The manner of death from the shockwave of an earthquake is not unlike that from the shockwave of a bomb. Buildings collapse, crushing people and possibly starting secondary fires as cooking stoves, heating devices, broken gas lines, and live electrical wires ignite the building materials that have fallen on them. Trauma and possible burning is therefore the first wave of death from an earthquake. But those who survive the shockwave from an earthquake near the sea cannot afford to sit and lick their wounds. They must drag themselves to higher ground as quickly as possible because very soon their likely mode of death is going to change from crushing to drowning.

Depending on how far the earthquake's epicenter is from shore, the time until the tsunami hits will vary. For Fukushima, the epicenter of the quake was 60 miles at sea, which meant that the tsunami wouldn't arrive for 40 minutes . . . 40 precious minutes. Those people skeptical that the tsunami would come, and those that thought they were too far from the shoreline for the water to reach them, were sadly mistaken and would pay with their lives.

Ironically, the indication that a tsunami is imminent is that the water at the shoreline rapidly recedes, leaving harbors dry and boats resting on the sea floor below their high and dry docks. This happens because the tsunami's cresting wave is created from the water in front of it, producing a deep forward trough. The trough quickly empties that harbor of water, resembling a fast-forwarding video of a receding tide. The occurrence of water rapidly retreating from harbors and beaches can be considered the start of the five-minute warning. It means that the crest of the tsunami will typically arrive within five minutes. In situations where offshore earthquakes are too far away to be felt on the coastline, this rapidly receding water may be the only warning that people get. And five minutes gives little time to get out of the wave's path.

Whether or not the harbor below had gone dry, Pillitteri hadn't noticed. His gaze was further out to sea, watching the wall of water that the freighter had just climbed move toward land. At 3:26 p.m., a wave 13 feet (4 meters) high hit the 18.7-foot (5.7-meter) seawall protecting the harbor and was successfully stopped. But the relatively small first wave was followed by two waves closer to 49 feet (15 meters) high. (The precise height of these two waves will never be known, since the wave height meter at the plant wasn't able to measure anything over 24.6 feet [7.5 meters].) Both of these waves plowed over the wall as though it were merely a speed bump in the road. They hit the shore, moved inland, and ripped buildings from foundations. Cars that happened to have their windows closed trapped air briefly and bobbed around on the surface for a short while. Presently, the cars took on water, capsized, and sank, drowning anyone inside. People who happened to be outdoors at the time were soon slammed with seawater thick with flotsam that included the cars and even whole buildings. If not killed immediately, they swirled around in the cold March seawater groping for some higher structure to pull themselves onto and out of the fray.

As Pillitteri watched, he suddenly realized that the water would soon make its way onto the power plant grounds, perhaps even make it as high as the parking lot where he stood. But by this time he had become completely fatalistic, and his fear was replaced by awe of the natural forces he was witnessing. Forces much grander than any nuclear power plant were at play, making human efforts to harvest a little natural energy from nuclear fission appear feeble.[8] Pillitteri was mesmerized by the sight.

Pillitteri dodged another bullet. The water never made it to the parking lot where he stood. But it did reach the reactor building from which he had just escaped.

The fact that the water had breached the seawall meant big trouble for the nuclear reactors. Having lost power from the outside electrical grid due to the earthquake and unable to generate their own power due to the automatic core shutdowns, the cooling units were reliant on diesel backup generators. They had enough diesel fuel to operate for days and, even if fuel should run out, there were backup batteries to keep the electric cooling pumps running for at least eight more hours, buying still more time for the reactors to cool down to a safe temperature. The trouble was, the generators and the batteries were in the basements of the reactor buildings and the basements were about to get flooded.

At 3:37 p.m., seawater entered the basement of Reactor Unit One, the building that Pillitteri and his crew had just vacated, and all power needed to control the electrically operated valves for the cooling systems was lost. But fortunately, there was a backup cooling system in Reactor One that didn't require power. It was a completely gravity operated condensation cooling system. Basically, water vapor from boiling water in the reactor rises through pipes to the roof were it runs through a condenser coil to return it to liquid water, and the condensed water then returns to the reactor via gravity to commence the cycle again. Furthermore, the system can be replenished manually with water from a fire-truck hose, so that any water vapor leak does not run the reactor vessel dry. This is a brilliantly simple system that requires no electricity once it's operating, but its activation was electronically controlled. Unfortunately, when the water from the tsunami flooded the basement of Reactor Unit One, the resulting loss of power caused the condenser cooling system to shut down because of a flaw in its fail-safe logic.[9]

To be *fail-safe* implies that a machine, instrument, or procedure has been designed in such a way that if it should fail, it will fail in a way that will not result in harm to property or personnel. You'll recall that Louis Slotin defied fail-safe principles when he unwisely performed his criticality experiments by lowering components onto a nuclear core (see chapter 5). Fail-safe rules required that he should have always raised the components up toward the core. The fail-safe logic of this safety rule was that, if something should slip, gravity would pull it away, rather than toward a critical mass situation. But Slotin, as we already know, ignored this fail-safe rule and lowered a neutron-reflecting brick onto a core. When the brick did slip and fall onto the core, there was a criticality burst that resulted in Slotin receiving a lethal dose of radiation, instead of just a broken toe.

The flaw in fail-safe logic that killed the cooling condenser system had to do with the circuit that controlled the condenser's valves. The circuit was erroneously programmed to interpret a loss of power as an indication of a leak somewhere in the cooling pipes.[10] Thus, the circuit commanded all the valves of the condenser system to close down in order to stem the alleged leak. In this case, however, there was no leak, and closing the valves just meant stopping the coolant flow. Now there was no means to keep the reactor from overheating. In most reactor accidents, there always seems to be some type of perilous mistake that causes things to spiral out of control.[11] At Reactor Unit One, this was it.

The other Reactor Units at the site eventually experienced their own electrical failures, each ultimately resulting in loss of their reactor cooling.[12] All the reactor operators could do was sit and watch the reactor dials by flashlight, without the ability to intervene. Their hands were tied by lack of power. Meanwhile, two operators in Reactor Unit Four were already dead. They had not been as fortunate as Pillitteri's crew. They became trapped in their turbine building when the water rose in the basement and drowned.[13]

At 6:45 a.m. on March 12, the morning after the earthquake, TEPCO officials announced that radioactivity *may* have leaked off the plant grounds. A curious announcement given the fact that it could easily have been confirmed. People didn't know what to make of it. Was this TEPCO's oblique way of saying that the reactor cores had melted down? Even if a reactor meltdown hadn't yet claimed any lives, there apparently was at least one casualty already; that was timely and reliable information from TEPCO.

Meltdown means that the reactor fuel gets so hot that it turns into molten metal, similar to hot lava. This molten radioactive fuel can burn holes in its container, releasing the reactor's radioactive fission products into the environment. If a meltdown has been caused by uncontrolled criticality, the melting process may actually help to stop the accident because the melted fuel tends to spread out into a geometric configuration that is not conducive to sustaining criticality.

But the Fukushima reactors were not experiencing uncontrolled criticality, because they were successfully shut down. The cores were at risk because their cooling systems failed. In the end, all three of the operating reactors would completely melt down.

TOWER OF BABEL

Anyone who's ever had a flooded basement in their home might ask: Why put essential electrical systems in a basement where the possibility of flooding always exists? The answer is because a very high seawall has eliminated the flooding threat. Really?

The seawall the TEPCO built to protect Fukushima's harbor was 18.7 feet (5.7 meters) high. Why this height and why not lower or higher? It turns out that the seawall height was based on the Chilean 1960 earthquake, one of the more recent of the large quakes that have sent tsunamis to Japan. This was a massive 9.5 quake, even larger than the 2011 quake, but its epicenter was on the other side of the Pacific Ocean. The tidal wave it produced arrived hours later and it took 142 Japanese lives. This 1960 disaster was fresh on the minds of those who built the Fukushima sea wall in 1967. However, the worst Japanese tsunami in recorded history was actually produced by the Meiji Sanriku earthquake of 1896, well before most people had been born. Although it registered a magnitude of only 7.2, its waves reached heights of 100 feet (30 meters) and killed 22,000 people. But it would have been cost prohibitive to build 100-foot-high seawalls, so the Chilean 1960 tsunami became the standard for seawall height along the east coast of Japan.[14] The height of 18.7 feet (5.7 meters) would have been good enough for the 1960 tsunami, but totally inadequate for 1896, or for 2011.[15]

IMPERIAL WISDOM

As faith in TEPCO's ability to contain the situation waned, fear and anxiety mounted back in Tokyo, fed by a spreading rumor that the emperor had fled his Tokyo palace and gone to Kyoto, a city about 230 miles (370 kilometers) further away from Fukushima.[16] Had their emperor abandoned them? His father, Hirohito, had remained in Tokyo even during the Allied Forces' firebombing of the city. Was the Fukushima accident a bigger threat to Tokyo than that? The rumor turned out to be false, but it had a chilling effect on people nonetheless. Something needed to be done to restore calm.

After suffering five days of bad news, an increasingly distrustful and hostile populace needed a reassuring voice, and they weren't hearing one from TEPCO. So, on March 16, the grandfatherly emperor, Akihito, addressed his nation in a televised broadcast urging his people to stay calm:

> I am deeply hurt by the grievous situation in the affected areas. The number of deceased and missing increases by the day and we cannot know how many victims there will be. My hope is that as many people as possible are found safe. I hope from the bottom of my heart that the people will, hand in hand, treat each other with compassion and overcome these difficult times [and not] abandon hope.[17]

SKETCHY NUMBERS

That earthquakes and tsunamis pose a threat to nuclear power plants is well known, but the magnitude of the risk is hard to assess (i.e., earthquake prediction). At any rate, earthquakes and their resulting tsunamis are known unknowns. We know they are a problem and we have at least one tenable solution (i.e., seawalls), but we're not exactly sure how big the problem is (i.e., height of the wave), so we don't know how big our solution needs to be (i.e., height of the seawall).

Norman C. Rasmussen (1928–2003), a professor of nuclear engineering at MIT, is considered the father of assessing nuclear power reactor safety using statistical models. In the early 1970s, he headed a federal committee charged with determining the risk of a nuclear reactor core accident occurring. The committee's approach was to identify the series of failure events that would all need to occur for a core accident to happen (termed *fault trees*), estimate the probabilities of individual failures for each branching point on the tree, and then multiply those probabilities by one another to get the overall probability that multiple system failures would simultaneous occur in a way that could result in a core accident. In 1975, the committee issued its report, which was officially known as WASH-1400, but unofficially called the Rasmussen Report.

The report received worldwide attention, as did Professor Rasmussen, because its conclusion was that the risks posed by nuclear power plants were extremely small. The report contended that the odds of an accident involving damage to the reactor core at a commercial nuclear power plant was just 1 chance in 20,000 operation years (i.e., odds of 1:20,000).

An operation year is analogous to the working month that we previously used for assessing the lung cancer risk of mine workers. By a working month, we meant one miner working a mine for one month. In this case, an operation year means one year of power generation for a single reactor. We're using it here as a measure of the total time that a reactor is actually generating power, which is the period when a core accident is most likely to occur.

But remarkably, in March 1979, just four years after the Rasmussen Report was published and when the American nuclear industry had less than 500 operational years of experience under its belt, the Three Mile Island reactor accident occurred on the Susquehanna River in Pennsylvania. In that case, a valve failed to shut properly and vented reactor coolant to the environment. The situation initially went unnoticed by reactor operators due to instrument failures, and then confusion about the situation led them to override an automated cooling system because they mistakenly thought that the coolant level in the reactor was too high when it was actually too low. This unfortunate sequence of events cascaded into a partial meltdown.[18] Luckily, most all of the radioactivity was contained on site so there were no health consequences, but it was a very close call and justifiably got people's attention. Was this just extremely bad luck, or was the risk of a core accident actually closer to 1:500 than 1:20,000?

The US Nuclear Regulatory Commission decided to revisit the risk issue with another committee that was charged with auditing the Rasmussen committee's report. The second committee, headed by Harold W. Lewis (1923–2011), a professor of physics at the University of California at Santa Barbara, endorsed Rasmussen's probabilistic approach to assessing risk, particularly with regard to the use of fault trees to track how problems might spread through a plant. But it also found flaws in the Rasmussen model, specifically in that Rasmussen had neglected to consider some forms of risk, most notably fires, and only considered a limited number of reactor designs to the exclusion of others. In fact, Three Mile Island's reactor design had been excluded. Additionally, the report was severely criticized in that it failed to adequately acknowledge the large uncertainty associated with its risk estimates. Nevertheless, this second committee did exonerate the Rasmussen committee on one of its omissions. Noting that Rasmussen's committee had not considered the possible risks of sabotage, the Lewis committee noted, "The omission was deliberate, and proper, because it recognized that the probability of sabotage of a nuclear power plant cannot be estimated with any degree of certainty." At least there was something both committees agreed on.

Rasmussen had not, however, neglected to consider tsunamis. His committee recognized that tsunamis and hurricanes posed risks, but thought those risks were vanishingly small. In the words of the report:

> Some plants are located on the seashore where the possibility of tidal waves [tsunamis] and high water levels due to hurricanes exists. The plant design in these cases must accommodate the largest waves and water levels that can be expected. Such events were assessed to represent negligible risks.[19]

The Lewis committee concluded that the Rasmussen committee failed to consider a number of factors that could increase risk, and suggested that the true risk might be much higher than the Rasmussen Report had depicted. But this second committee fell short of offering its own risk estimate.[20]

It has been 36 years since the Lewis committee issued their report. We now have more than 15,500 operational years of commercial nuclear reactor experience worldwide (as of 2014).[21] According to the International Atomic Energy Agency (IAEA), which tracks and rates nuclear incidents on a severity scale from 1 (low) to 7 (high), there have been a total of ten nuclear power plant accidents of varying severity—levels 4 (accident with local consequences) through 7 (major accident)—during that period (1952 to 2014). Ten accidents out of a total of 15,500 operational years translates into a rate of one accident per 1,550 operational years, or the odds of 1:1,550.

Currently (2014), there are 430 operational nuclear reactors in the world.[22] A risk level of 1:1,550 would, therefore, suggest that we might expect a significant reactor core accident to occur among one of these 430 reactors once every three to four years (3 years × 430 operational reactors = 1,290 operational years; 4 years × 430 operational reactors = 1,720 operational years). But the 1:1,550 odds of an accident was our risk rate of the past. A moving target counterargument would posit that we've learned our safety lessons and that more modern reactors have much safer designs, making the accident rate of the past an overestimation of future risk. Although that may be true, it is also true that many existing reactors are very old and therefore are more likely to have maintenance-related failures. So it's very hard to tell what the overall effect on the historic reactor accident rate will be. If our future resembles our past experience, then 1:1,550 would remain our best risk estimate going forward.

Whether it's 1:20,000 (Rasmussen), 1:1,550 (historic), or something else, the risk is definitely not zero, and the list of assumptions and uncertainties that go into these risk calculations is long. As a rule, when uncertainties are high, using our simple NNH calculation to judge our personal risk doesn't really hold water. For example, we could take the 1:1,550 odds and convert them to NNH = 1,550 operational years, and interpret it to mean that we would need to live 1,550 years next to a single-core local nuclear power plant to expect to experience one core-related accident. Since none of us will live anywhere near that long, we might judge our personal risk to be slight. This would be true . . . if our neighborhood nuclear power plant is a typical nuclear power plant . . . if the plant is following all safety regulations . . . if there are no undetected maintenance issues . . . if there are no engineering design flaws . . . if there are no operator errors . . . if there are no severe earthquakes . . . and the list goes on. The point is not whether the risk estimate is right or wrong, or whether the NNH logic is flawed when it comes to nuclear power plants. Rather, the point is that this apparently precise risk estimate comes with an enormous grain of salt; that is, it comes with a high level of uncertainty. Some people have high salt tolerance, while others prefer to keep their food bland. Everyone needs to make her own decision about the amount of salt she can handle in her diet.

Another issue is that the situation with nuclear power plants may be more complex than simply calculating statistical probabilities. Some have argued that complicated systems, such as nuclear power plants, have an inherent characteristic called *interactive complexity*, and that failures can occur due to interactions between different components of the highly integrated systems. Furthermore, these interactions are nearly impossible to foresee.[23] Interactive complexity presents new types of risks that never occurred to the designers of the systems (i.e., an unknown unknown). This allows catastrophes to happen when two or more systems fail at the same time in unexpected ways. Furthermore, if the complex system is also tightly coupled, such that one event quickly proceeds to the next, there is little time for operators to intervene and fix the problem. More worrisome still is that because there is no time for operators to learn to understand the nature of the novel interactive problem, they may misread it and intervene in the wrong way, making the situation worse than if things had been allowed to run their course.

It may seem illogical, but piling backup safety systems on top of backup safety systems can actually exacerbate problems because it increases interactive complexity and introduces more unknown unknowns. Viewed in this way, we are fighting ourselves when we add too many safety redundancies into a system, no matter how well intentioned, because we are simultaneously adding interaction nodes that are new targets for failure.

The implication is that, if we try to remedy risky situations with complex solutions, we can actually invite more risk, potentially increasing the level of risk to the point that an accident becomes inevitable. In such a scenario, accidents become the norm rather the exception and these types of incidents have, therefore, been called *normal accidents*.[24] Whether the Fukushima Daiichi power plant accident should be classified as a normal accident or just really bad luck is a matter for debate. But certainly, having six different nuclear reactors in one location at common risk from external forces, and having them share electrical systems, steam vents, cooling systems, and so forth, results in a highly complex operating environment that's primed for a domino effect of failures that could lead to a catastrophe.[25] The Fukushima Daiichi plant was one of the oldest, largest, and most complex reactor plants in Japan. It is often overlooked that there were 17 other newer, smaller, and simpler nuclear power plants that were also affected by the 2011 earthquake, and four of them were even hit with the same tsunami, yet only the Fukushima Daiichi plant failed and it failed in a big way (i.e., three meltdowns).[26] That says something for the value of simplicity.

There is no way to eliminate unknown unknowns, but studies have shown that by simplifying systems you can greatly reduce sources of systems risk, and lessen the likelihood of interaction-driven failures that are nearly impossible to anticipate. Furthermore, if we loosen the "tightness" of internal couplings within a complex system, we can further lower the level of this type of risk. The modifications need not just be technical. Organizational simplification can also have a positive impact, as has already been demonstrated in other highly complex systems, such as air traffic control.[27] Thus, managers of complex technological systems can learn much from studying the safety improvements within other complex systems. We may not be able to precisely measure the risk of a nuclear core accident, but there are definitely ways to minimize it.

The Lewis committee was convened because the occurrence of the Three Mile Island nuclear accident suggested that risks of a nuclear core accident were actually significantly higher than predicted by Rasmussen's statistical models. In that sense, the Three Mile Island accident was fortuitous because it caused us to revisit Rasmussen's risk models with a more critical eye. In the process, we learned a great deal about the weaknesses of predicting catastrophic accidents with statistical models, particularly with regard to uncertainty. Fortunately, not one life was lost at Three Mile Island. Had the Three Mile Island accident not occurred, we most likely would not have revisited Rasmussen's work and would have remained completely naive with regard to reactor safety until eventually some other, possibly devastating, nuclear reactor accident happened.

It is important to remember that just because an accident has never occurred before, doesn't mean that it is unlikely to occur in the future. To think that it does amounts to submitting to the black swan fallacy,[28] as Professor Lewis explained:

> The general argument that the fact that one has operated safely for a finite period of time proves that the safety level is adequate is just not statistically right. . . . [It is] a psychological trap to

believe that because something has not [yet] happened, you are doing just fine.[29]

PULLING NO PUNCHES

Given that complex systems have high levels of uncertainty that can thwart attempts to precisely quantify the risk of failure, how then should we judge their safety? In such cases, we can sometimes gain insight by just assuming the catastrophe has happened and predicting its health consequences. In other words, if the dreaded event eventually occurs, no matter how great the odds against it, what exactly are the health consequences that we can anticipate? Fortunately, this question is easier to answer than predicting the rate of nuclear core accidents. We can do it by studying similar, worst-case accidents. The value of characterizing the consequences of such a worst-case scenario is that it provides us with two things: (1) we can judge the direness of the worst-case event; and (2) we can determine what type of emergency responses we need to have in place.

In terms of a worst-case scenario for a nuclear power plant accident, Chernobyl offers us an excellent model. The Chernobyl nuclear power plant accident of 1986 was, by far, the worst nuclear reactor accident of all time.[30] It resulted from a series of operator misadventures set within an environment of incompetence that was also cloaked in secrecy. The causes of the accident have been studied in detail, and make a good playbook for what not to do at a nuclear power plant, but that's not our concern here.[31] We're more interested in the public health consequences of the Chernobyl accident.

Although Chernobyl and Fukushima Daiichi both received the highest accident rating (level 7) on the IAEA scale of nuclear accidents, Chernobyl was far worse, affecting far more people. Over 572,000,000 people among 40 different countries received at least some exposure from Chernobyl radioactivity. (Neither the United States nor Japan were among the exposed countries.) It took over 20 years to fully assess the cancer consequences. Finally, in 2006, an international team of scientists completed an analysis of the dose and health data and reported on the cancer deaths that could be attributed to Chernobyl radioactivity.[32] Their detailed analysis included countrywide estimates of individual radiation doses in all 40 exposed countries, and region-wide estimates for the most highly contaminated regions (Gomel, Mogilev, Bryansk, Tula, Kiev, Rivno, and Zhytomir) within the most highly contaminated countries (Belarus, Russian Federation, and Ukraine).

Here's what they found. Even in the most highly contaminated regions, the effective doses the public received from the radioactivity averaged only 6.1 mSv (equal to two years of natural background), and the average dose for all exposed people was just 0.5 mSv (equal to two months of natural background). Using statistical models that accounted for age, gender, and other demographics, the scientists predicted a total of 22,800 radiation-induced cancers (excluding thyroid cancer) among this group of 570,000,000 people. That is, 22,800 cancers in addition to the approximately 194,000,000 cancer cases that would normally be expected in the population even if there had been no Chernobyl accident (i.e., 194,000,000 versus 194,022,800), or a 0.01% increase in the cancer rate.

For people receiving an average dose, we can translate the 0.01% cancer risk (i.e., odds of radiation-induced cancer = 1:10,000) into an NNH of 10,000. Of course, for those in the highest doses group (6.1 mSv), the risk is proportionately larger, but only 2% of the exposed people fell into this higher dose group. That said, on average it would require 10,000 people exposed to Chernobyl radioactivity to expect even one of them to have cancer as a result of their exposure. This number is too small to have any measurable impact on the cancer incidence rates for any cancer registries, so the predicted values will likely remain theoretical. Simply put, the study concluded that there was an immeasurably small number of excess cancer cases that could be attributed to Chernobyl radioactivity in the environment . . . with the exception of thyroid cancer.

Unfortunately, the one type of cancer that could have easily been prevented was not. Having apparently learned nothing from the

Marshall Islanders' unfortunate experience with iodine-131 (half-life = 8 days), the population surrounding Chernobyl was not warned about the possibility of iodine-131 in milk and other locally produced agricultural products. Had they avoided eating these foods for just three months, nearly all (>99%) of the radiation-induced thyroid cancers could have been prevented. But that did not happen. The people were not warned, iodine-131 contaminated food was eaten, and thyroid cancers resulted. For Chernobyl, even the cancer rates from iodine-131 were a worst-case scenario because much of the local population had an iodine-deficient diet, meaning that their starved thyroids sucked up any iodine that became available, an extreme situation that would not have happened in countries such as the United States or Japan, where diets are richer in iodine. Studies suggest that there will be a total of about 16,000 thyroid cancers produced as a result of Chernobyl iodine-131 exposure.[33] Since thyroid cancer is one of the rarer cancers (i.e., it has a low background incidence), excess thyroid cancers due to iodine-131 can more readily be seen as elevated thyroid cancer rates in cancer registries. In fact, a spike in cases was seen within a few years following the accident. Fortunately, thyroid cancers are among the more successfully treated cancers. Of the 16,000 thyroid cancers produced, it is anticipated that at least two-thirds will be cured with standard therapy.[34]

At Fukushima, there was much less iodine-131 released. Compared to Chernobyl, the thyroid doses from fallout were extremely low, the population exposed was smaller, and people were advised to avoid local dairy products and drinking water due to possible iodine-131 contamination. Iodine-131 uptake into the thyroids of exposed people was later measured and the doses estimated to average just 4.2 and 3.5 mSv, for childern and adults, respectively.[35] Compare this to Chernobyl where millions of people received thyroid doses in excess of 200 mSv (i.e., more than 50 times as much)—high enough to see appreciable amounts of excess thyroid cancer. The Fukushima thyroid doses, in contrast, were very close to annual background levels, too low to expect to see a significant increase in clinical thyroid cancer.

At Chernobyl, 127 reactor workers, firemen, and emergency personnel sustained doses sufficient to cause radiation sickness (>1,000 mSv), and some received lethal doses (>5,000 mSv). Over the following 6 months, 54 died from their radiation exposure.[36] And it's been estimated that 22 of the 110,645 cleanup workers may have contracted fatal leukemias over the 20 years following the cleanup.[37]

As for radiation sickness at Fukushima, there was none, not even among the reactor workers. Two workers who had leaky respirators received effective doses of 590 mSv and 640 mSv.[38] These doses are above the Japanese occupational limit for a lifesaving event (250 mSv), but still below the threshold for radiation sickness (>1,000 mSv). Their lifetime cancer risk will increase about 3% (i.e., from the 25% background rate to 28%) due to this exposure, but they are unlikely to have any other health consequences; nor are their future children.

These two radiation workers and others like them are the heroes of Fukushima. They stood their ground and did their jobs despite the risks, and likely saved the lives of others. Radiation workers acknowledge that their profession exposes them to more radiation risk than the general public and voluntarily agree to sustain exposures up to the regulatory limits set for employees, which are routinely 10 times higher than those set for the general public.[39] They understand that their profession, like many others, entails some added risk of death, and they accept that risk as a condition of their employment.

Their heroism, however, may come at a very high personal cost, even if their health risks are actually limited to a slightly elevated cancer risk. The biggest fear for young Kai Watanabe, a radiation worker who fought to save the Fukushima plant, is not the cancer risk but rather that his radiation exposure will cost him a normal life: "Let's say I tell a woman about my past, that I've absorbed all this radiation and may get sick or father children that are deformed, so we shouldn't have children. Is there a woman out there that will accept me? . . . Neither of us would be happy with a situation like that. So I think it's best for me to stay single."[40] It's a shame that Watanabe would deprive himself of the joys of marriage and fatherhood because of health risks that are actually quite low. This is a tragic example of how exaggerated fears of radiation can damage lives.

With regard to radiation sickness, it is safe to say that the risk from a nuclear core accident is borne almost exclusively by people on the reactor grounds, most notably by the radiation workers servicing the core. It is hard to imagine how anyone off the grounds could get a dose high enough to cause radiation sickness. We have seen, in the case of the *Lucky Dragon No. 5* fishermen, that fallout can cause radiation sickness even at great distances from a nuclear detonation, but the fallout levels from a hydrogen bomb are orders of magnitude higher than what can occur from a nuclear reactor accident. And there is no possibility of a nuclear reactor exploding like a hydrogen (fusion) or even an atomic (fission) bomb,[41] so the fallout potential is much lower, as we now know from both the Chernobyl and Fukushima experience.

ASHES TO ASHES

At the time of the Fukushima accident, the radiation threat was out of the minds of earthquake and tsunami survivors. That was a worry for tomorrow. Today had enough problems of its own, and among them was how to account for the missing.

Setsuko Uwabe was hunting for her husband Takuya, but he couldn't be found at any of the evacuation centers. Setsuko, a cook, worked in the cafeteria of a local nursery school situated on a hillside overlooking Rikuzenataka, a small town up the coast from Fukushima Daiichi. She had spotted the tsunami coming toward shore from the elevated kitchen window of the school, and was able to evacuate herself along with many of the school children to still higher ground just before the tsunami hit. They were all saved. But her husband Takuya, a municipal worker, was working in the town below. She did not see how he could have easily escaped and no one at the evacuation centers had seen him. If not among the living, perhaps he was among the dead. She decided to volunteer at the death registry desk. As long as he didn't appear on the death registry, she could maintain hope that he was still alive. But beyond gathering information for death certificates, volunteers at the death registry were also expected to help with arranging cremations, a job for which she was not emotionally prepared.

Among the buildings that were destroyed in the Fukushima Prefecture were the crematoria. And even those crematoria that survived didn't have the capacity to satisfy the heavy demand for cremations. Typically, crematoria are designed to handle less than a dozen bodies per day, while the need was for hundreds of bodies per day. Setsuko scheduled as many cremations as she could, but the fortunate families that were able to secure a cremation time were expected to prepare the body themselves. They were instructed that the corpse needed to be tied to a burnable wooden plank, and wrapped securely with a blanket. Then they would have to find their own transportation to get the body to the crematorium. Also, urns for the ashes were in short supply. They would need to find some type of container among the rubble if they wanted to keep the ashes. For the deceased who had no one to claim them, or had families that couldn't comply with the cremation requirements, the body was dumped into a mass burial pit.

After a couple of days, Takuya's body was found near the river, along with the body of his best friend since boyhood. Setsuko went down to the makeshift morgue at the local gym to view Takuya's remains. He appeared to be sleeping. There was no visible damage to any of his body parts. It was as though life had simply been sucked out of him. With the help of their son and daughter, Setsuko prepared his body as instructed, and transported it to the crematorium. Luckily, they were able to secure a flower vase as a substitute for an urn, and took as much of Takuya's ashes home with them as would fit in the vase.[42]

Takuya was just one of many who died. In the end, the death toll from the earthquake and tsunami reached 15,900. In addition, there are 2,600 still missing and presumed dead. No deaths were attributed to radiation exposure, and no significant increases for any type of radiation-induced cancer or inheritable mutations are expected to occur.

THE NEW NORMAL

There were 340,000 people displaced by the 2011 Fukushima disaster. Will it ever be safe for them to go back home? Will things ever return to normal in the Fukushima Prefecture? Good questions.

The radioactivity levels in the Fukushima Prefecture will never drop down to their prior levels within any of the survivors' lifetimes. The area is too contaminated with radioactive fission products for that to occur. The reality is that, although radioactivity will decay away and dissipate with time, it will always be elevated compared with other areas of Japan. This is an unfortunate fact.

When it will be safe to return is entirely dependent upon how you define safe. As we've learned, the term *safe* really means low risk, and not everyone agrees on what low risk means. The Japanese government has set an annual effective dose limit to the public of 20 mSv as its remediation goal for the Fukushima Prefecture.[43] Prior to 2011, one mSv above background was Japan's regulatory limit for the public. To the Japanese people, this raising of the safety limit appears like the government is backpedaling on its safety tolerance because it knows it can't deliver on any commitment to reduce the annual effective dose below 20 mSv. This, of course, is true. If they could deliver 5 mSv, then 5 mSv would have been their new safety standard. This is the problem with moving regulatory dose limits after the fact to accommodate inconvenient circumstances. It breeds distrust.

Perhaps the Japanese government would be better off to just explain what the risks are at the various dose levels and let the people decide for themselves if they want to go back. For example, 20 mSv annually is not unlike our earlier situation with an annual whole body spiral CT scan, which also delivers an effective dose of 20 mSv. As you recall, we calculated a lifetime risk of cancer from a whole body spiral CT scan as 0.1% (or odds of 1:1,000). If you lived for five years at that level, it would thus be 0.5% (or 1:200).[44] Alternatively, if the effective dose were 10 mSv per year, then the risk would be half of that from 20 mSv. Yes, it's true that children may be at somewhat higher risk and older people at lower risk, so you may want to take that into account in your personal situation, but overall, these are the typical risk levels. Now ask yourself: Would it be worth it to me to go back to my home knowing I was facing this level of cancer risk?

This is why transparent risk characterization is better than setting opaque safety limits, particularly if you're going to start moving those limits around to suit the circumstances and, thereby, destroy people's confidence. The above risk estimates were the same before the accident as they were after the accident. The risk per unit dose doesn't change with the circumstances, but the "safe dose limit" apparently can. People who think that regulatory limits represent thresholds for safety are quite mistaken. The limits are merely arbitrary lines that are drawn in the sand by some regulatory body, marking the fuzzy border between the dose levels that entail "acceptable" versus "unacceptable" amounts of risk. If you don't like where that line has been drawn, pick up a stick and draw a different line for yourself. When it comes to risk tolerance, there will always be different lines for different people.

These are the issues people from the Fukushima Prefecture are now facing. It's not necessary that all of them arrive at the same conclusion about their personal safety. Whether or not to return should be an individual choice, and people can make different decisions, all equally valid, as long as they have the facts they need to make a credible assessment of the situation.

BETWEEN A ROCK AND A WET PLACE

Dotting the hills along the eastern coastline of Japan are hundreds of ancient stone monuments resembling large gravestones, inscribed in an archaic dialect that today's inhabitants cannot read.[45] Nevertheless, scholars have deciphered most all of them. The stones turn out to be neither religious shrines nor gravestones, but rather

FIGURE 16.1. THE OLDEST TSUNAMI WARNING STONE IN JAPAN. Although the inscription is worn away, the stone in the foreground is believed to be a tsunami warning tablet erected after the great Jōgan tsunami of 869 AD. On the day of the March 11, 2011, earthquake, the Miyato-jima Islanders who heeded the warning of this ancient tablet evacuated to higher ground and were saved from the subsequent flood. In gratitude, they installed a new tablet next to the old stone acknowledging their ancestors' sage advice. The new tablet reads: "Following the tradition of the tablet, thousands of residents could evacuate. We are grateful for the Jōgan tablet." (*Source*: Photograph source: www.megalithic.co.uk. Generously provided by the courtesy of Mr. Hatsuki Nishio.)

warning signs. The specific phrasing of the inscriptions varies from stone to stone, but the underlying message is always the same. The stone located in the village of Aneyoshi states its warning quite clearly and succinctly: "High dwellings are the peace and harmony of our descendants. Remember the calamities of the great tsunamis. Do not build your homes below this point!"[46]

It turns out the stones mark the high water lines of previous tsunamis, some dating back as far as 869 AD. Although none of the stones is believed to be quite that old, it is thought that many of the current stones are replacements for much older predecessors; literally a monumental effort by subsequent generations to keep the lifesaving message alive in perpetuity.

Modern Japanese ignore these stones at their own peril. The people of Aneyoshi listened to their stone and their village survived the March 2011 tsunami. But people in some other villages thought the construction of modern seawalls provided sufficient protection and that warnings on the stones had become obsolete. The 2011 tsunami, however, crested most seawalls, and the villages with buildings below their stone paid a heavy price for ignoring their ancestors.

Sometimes risk management reduces to simply heeding the lessons of your own history. No complex engineering nor elaborate statistics are required, just some simple timeless wisdom.

CHAPTER 17

THE THINGS THEY CARRIED: GEOPOLITICAL RADIATION THREATS

A man who carries a cat by the tail learns something he can learn in no other way.

—*Mark Twain*

"But I don't want to go among mad people," said Alice.
"Oh, you can't help that," said the cat. "We're all mad here."

—*Lewis Carroll,* The Adventures of Alice in Wonderland

STOPPING BY THE WOODS ON A SNOWY EVENING

Far from the ocean, in the remote hills of western Maryland, 115 miles (185 kilometers) from Washington, DC, rests a stone monument, alone by itself in the woods. The monument is topped by a simple Christian cross and is inscribed in plain English with the name of Major Robert E. Townley. It shows his birth date as May 3, 1921, and his death date as January 13, 1964. Although it resembles a grave marker, Townley's body is not buried under the monument. His body is buried in his home state of Alabama. Rather, the monument marks the impact site of Buzz One Four, a massive B-52 bomber that carried Townley to this spot and crashed. He died alone in a fiery blaze, the only crew member that was not able to eject from the plane before impact. That January night, coming a mere seven weeks after the assassination of President John F. Kennedy, was one of the darkest and coldest nights of the Cold War. The air temperature in the mountains was 10°F (–12°C), and there were three feet (1 meter) of snow on the ground when the plane hit. By sunrise there would be another eight inches (0.2 meters).

The plane had been en route from Westover Air Force Base (AFB) in Massachusetts to its home at Turner AFB in Georgia when it hit severe turbulence. Changing altitude did not improve things and soon a part of the tail section—a known structural weak spot of B-52 aircraft that had already contributed to three previous fatal crashes—was torn off the plane. The pilot then lost control and instructed the crew to eject. But Townley, the bombardier, who had survived his military service in both World War II and the Korean War, had trouble with his seat restraints and couldn't free himself in time. He went down with the aircraft and its cargo, which included two hydrogen bombs.

Buzz One Four was assigned to the 484th Bombardment Wing. The two weapons it carried were Mark 53 hydrogen bombs. The Mark 53 (later known as the B53) was one of the most enduring nuclear weapons deployed by the United States. It remained a key component of US nuclear deterrence until 1997.[1] It could be carried by B-47, B-52 and B-58 aircraft, and its warhead was later adapted for the Titan II intercontinental ballistic missile (ICBM).

The Mark 53 warhead (9,000 KT or 9 MT) is 600 times more powerful than the Hiroshima atomic bomb (15 KT). Since a stick of dynamite (TNT) weighs five pounds, the explosive power of a Mark 53 warhead is equivalent to more than 10 sticks of dynamite for every man, woman, and child in the United States (i.e., 50 pounds of TNT per person). Given this level of explosive power, letting a Mark 53 hydrogen bomb drop in free fall from your plane, and living to tell about it, as Tibbets's crew had done with their "mini" atomic bomb, is impossible. No upwind approach and precise angle of departure will solve the problem of escaping alive when it comes to dropping hydrogen bombs. But you don't need to drop a hydrogen (fusion) bomb with the same precision as an atomic (fission) bomb. No need to aim at a specific ground structure, like the T-shaped bridge in Hiroshima, in order to maximize enemy casualties. The destructive power of a hydrogen bomb is so massive that dropping one anywhere within a city will completely destroy the whole city regardless of what specific ground target you happen to hit. So this feature of hydrogen bombs allowed the engineers to come up with a simple solution to protect the bomber. The Mark 53 was equipped with five parachutes: one 5-foot pilot chute, one 16-foot chute, and three 48-foot chutes. These parachutes let the bomb drift down into the approximate target area, while also greatly slowing its descent to the ground, so the bomber that releases it has ample time to depart the scene to safety before the bomb blows.

While the air force scrambled to get men and equipment to the scene of the crash and recover the bombs, the townspeople of nearby Grantsville (population 503) came out in force to find the missing crew members who had parachuted down miles from one another into dense forest. The town opened its doors gladly for the airmen and civilians alike who were assisting in the rescue effort. The women of St. John's Lutheran Church served 1,500 dinners to the rescue workers. The American Legion Hall, the fire house, and the elementary school together set up enough bunk beds to sleep nearly 500 people. Over a colossal five-day search through the snowy woods, two crew members were recovered alive and two were found dead from exposure to the cold.

While the search for crew members was going on, bomb experts were focusing on to how to get the hydrogen bombs out of the woods. They were three quarters of a mile from the nearest road, so a bulldozer was appropriated to cut a path through the forest,

FIGURE 17.1. CRASH OF BUZZ ONE FOUR. Armed military guards stand watch over the wreckage of a downed B-52 bomber plane that crashed in the woods of western Maryland on January 13, 1964, carrying two hydrogen bombs. Salvage workers soon cut a road to the crash scene and carried off the two bombs, along with all of the plane debris. More crashes of American Air Force B-52s carrying hydrogen bombs would soon follow. (*Source*: AP Photo/William A. Smith; image used with licensed permission from AP Images)

while the bomb disposal crew dismantled enough of the bombs to ensure they wouldn't explode during transport out.

Lifting them was another problem; they each weighed over four tons (3,600 kilograms). Local resident Ray Giconi who owned a quarry nearby, offered the use of his huge forklift and two dump trucks. So the rescue workers picked up the bombs with the forklift and lowered each one separately into its own truck. Meanwhile, the radiation protection personnel surveyed the crash site for any radioactivity that may have leaked out of the bombs. There was none. The warheads had remained intact.

After thanking the Grantsville residents for their patriotic services, the military drove off with the bombs to parts unknown. The residents then returned to their homes with an unforgettable story that they would pass on to their grandchildren.[2]

Some would argue that this western Maryland incident underscores how vulnerable we are to our own nuclear bombs, and that we could have nearly blown ourselves to bits, all due to a little bad weather. Yet, others would point out that all planes have an inherent risk of crashing that we can reduce but not eliminate. They would contend that the crash rate for B-52s was much lower than for most other types of aircraft. Crashes, they would argue, are a known part of the risk equation. That's why bombs have multiple safety mechanism to prevent detonation in the event of a crash. They would say the western Maryland incident is testimony to the fact that flying with hydrogen bombs is safe because the risk of their detonation, even in a crash situation, is negligible. In western Maryland, no one was even exposed to radiation, let alone a nuclear detonation, and not a single civilian had died. The system had worked, thus proving the safety of the system. Right?

Not really. The bombs had actually been in tactical ferry configuration (electrically and mechanically deactivated).[3] They were not on a military mission at the time of the crash, but were rather just being ferried back to their home base at Turner, Georgia, because their previous two missions had both been scrubbed due to problems. On the first mission (January 7–8, 1964), engine trouble had required a layover for repairs at an AFB in Spain. On the second (January 11–12, 1964), bad weather over the Atlantic had forced a flight plan diversion and an unscheduled layover at Westover AFB in Massachusetts. In Westover, the bombs were disarmed so that they could be safely ferried back to their home base in Georgia.

Therefore, the crash really did not amount to an unplanned test of the bombs' safety systems, since they were partially dismantled, and there was no chance that they would have detonated their nuclear warheads during the crash. The only real radiation threat was that the accidental detonation of their conventional explosives might have compromised the integrity of the warhead and spread radioactive material across the countryside. This potential spread of radioactivity has some parallels with dirty bombs in terms of the possible consequences to the surrounding population.

A *dirty bomb* is a weapon that couples radioactive material with a conventional explosive for the main purpose of spreading radioactivity and thereby simulating nuclear fallout. They have no value to the military, only to terrorists. Dirty bombs are really designed to exploit people's heightened fear of even small amounts of radioactivity, and thus induce terror. As with nuclear bombs, their percussive effects pose the biggest health threat, since it is difficult to pack enough radioactivity into a dirty bomb to cause radiation sickness. The larger the explosive component, the further the radioactivity is spread. When the radioactivity is spread over a greater area, however, its concentration is greatly diluted and its potential health impact is, therefore, greatly lessened as well. Nevertheless, terrorists may see widespread dispersal as an advantage, because it also means that the area requiring radioactivity cleanup is increased and fear is, likewise, spread to more people.

CASH ON DELIVERY

Let's fast-forward four decades to another nuclear transport. This time it is a ship carrying the nuclear material, specifically a cargo freighter that's carrying depleted uranium in its hold. The freighter encounters no weather issues and arrives safely at its destination port, New York City.

The day was September 11, 2002, and ABC News decided to commemorate the one-year anniversary of the 9/11 attack on the

World Trade Center in New York City by concocting a stunt to demonstrate America's ongoing vulnerability to terrorists. So they packaged a cylinder containing 15 pounds (6.8 kg) of depleted uranium (about the size of a beer can) and loaded it onto a cargo freighter heading out of Istanbul and destined for New York. In New York, the uranium passed through US Customs without being detected, at which point ABC News revealed to their television viewers what they had done, thereby infuriating government officials and scaring everyone half to death. The news anchors asked their viewers: "If we could do it, why not the terrorists?" Good question.

The antiterrorism experts were not as concerned about ABC's stunt as the public was. Some experts pointed out that customs security is just one layer of a multilayered systems approach to protecting us from nuclear terrorists.[4] They also correctly noted that you need a lot more than 15 pounds of uranium to make an atomic bomb, particularly if it's depleted uranium you're talking about. To make an atomic bomb you need enriched uranium, not depleted uranium. Enriched uranium has a higher concentration of the uranium-235 radioisotope, which is what's required to achieve supercriticality. (Depleted uranium is the waste that's produced during the uranium-235 enrichment processes.) There are a whole series of highly technical steps that must take place before a functional atomic bomb could be made from scratch, the first being the need to obtain a large quantity of highly enriched uranium (HEU). Even if this particular security layer had failed, surely all layers would not have. So here again, we should breathe easier. Right?

Again, not really. Since we have not yet experienced any nuclear bomb attacks by terrorists, we have no information with which to calculate the attack rate for terrorist nuclear bomb incidents, and thus no means to estimate risk that is based on real data. We must, therefore, resort to theoretical models to predict the risk of being attacked by nuclear terrorists wielding a homemade bomb. Fault tree logic is sometimes employed to assess such theoretical risks, as was done for nuclear power plant accidents prior to the Three Mile Island incident. The thought is that there must be sequential failures of all the multiple layers of deterrents, rather than just a single layer, in order for terrorists to ultimately achieve success and make a functional bomb. This is why the uranium sting that ABC News perpetrated wasn't as crippling an indictment of the overall effectiveness of the United States' antiterrorism system as might first appear. For example, if you wanted to make your friend a cake for her birthday and someone was keeping you from buying eggs, stopping you from bringing flour into your kitchen, sabotaging your electric mixer, and cutting the power to your oven, it would be extremely difficult to get the job done. Even if one layer of deterrence (e.g., the broken electric mixer) was imperfect (e.g., you alternatively mixed the batter with a spoon), it could still contribute to the overall deterrence strategy in a significant way. This is because the individual deterrence success rates, even if low, are multiplicative with one another and quickly compound to a substantial deterrence level, just as compound bank interest can change pennies saved into a major nest egg. The compounding of many points of deterrence makes the overall probability of terrorists foiling all the security barriers and producing their own bomb extremely small. At least that's the theory.

The problem is that we can't really measure the failure rate at each node in the fault tree of a nuclear deterrence network, at least not in the same way that Rasmussen did for mechanical failures at nuclear power plants. In his case, the failure rates for certain components of the reactor core had been tested and real data were available to calculate the risk. As for predicting the failure rates for individual deterrent measures, however, the only way to do it is to make mock challenges, as ABC News did, and then assess the percentage of time that deterrence point fails. This is very difficult to do for every individual deterrence step, so many of our deterrence procedures live on, untested as to their effectiveness.

Exactly one year later, on September 11, 2003, the second anniversary of the 9/11 attack, ABC News repeated their stunt at another port. This time they used a ship headed to Long Beach, California, from Jakarta, Indonesia. Again, the ship contained the same cylinder of depleted uranium and, again, the uranium made it through US Customs unchallenged. Allowing that the sample size for this challenge of port security is only two (i.e., $n = 2$), one might say that ABC News's data suggests that the failure rate for this layer of the nuclear terrorist deterrence system is 100%. Not good.

PLAN B

If the overall effectiveness of multiple layers of deterrence is reassuring to us, it can also be very discouraging to terrorists who want to beat the system. Discouragement itself might have its own deterrent effect and thus channel terrorists into lower-tech options for demonstrating their hatred. It could make them settle for just getting some radioactive material and making a dirty bomb by mixing it with conventional explosives; this would not be anywhere close to the same lethality as detonating a nuclear bomb, but would still be pretty frightening given the public's hysterical fear of anything radioactive. Or it may cause terrorists to seek nuclear bomb alternatives that fall short of actually making their own bomb.

One option terrorists have is to wait until some unscrupulous sovereign state, such as North Korea, Pakistan, or some nuclear wannabe nation, makes nuclear weapons for its own use, and then simply negotiate a clandestine deal to buy one from them.[5] This alternative seems quicker and more likely to succeed than trying to make their own bomb. Not only that, sovereign states have more financial resources, manpower, and technical expertise, and could probably, therefore, make more efficient, smaller, and devastating bombs than terrorists could make themselves. Thus, buying one would potentially have a greater return on investment for terrorists than making a homemade nuclear bomb.

Still, there are some significant obstacles that could frustrate this approach for obtaining nuclear weapons. Since bomb materials are so hard to come by, shady governments with nuclear munitions may not be willing to share their nuclear weapons even if they and the terrorists have a common enemy. Those governments may even worry that their terrorist "friends" would turn against them once they have secured a weapon. They might further worry that once the terrorists use the weapon against their common enemy, that enemy might find it easier to retaliate against the seller of the bomb (i.e., the sovereign state) than the buyer (i.e., the terrorist organization). All these issues could be a major disincentive to sovereign states entertaining any thoughts about selling their bombs to terrorists, no matter how much they sympathize with their goals. For these reasons, the terrorists may need to infiltrate the sovereign state that has the desired weapons, and bribe or steal their way to a nuclear bomb. As opposed to garnering the technical expertise required to construct their own nuclear bomb, bribing and stealing are skill sets that terrorists abound in. Such a plan would, therefore, play to their strengths.

At this point, you may be appalled that this book seems to be revealing viable nuclear bomb acquisition strategies to terrorists. Don't be. These avenues for obtaining nuclear weapons are only news to you. The terrorists are well aware of them. In fact, various documents seized from Al Qaeda operatives in Afghanistan and elsewhere in the wake of the New York 9/11 attack suggested that the organization was actively exploring all of these options.[6]

FEAR OF FLYING

It would have been easier to accept the Buzz One Four accident as a demonstration of the safety of air transport of hydrogen bombs had it been the only example of a B-52 bomber accident involving hydrogen bombs on United States soil, but it was not. In fact, there had been two previous accidents with B-52 bombers carrying hydrogen bombs that had brought the United States a little closer to nuclear catastrophe.

The first happened in Goldsboro, North Carolina. On January 24, 1961, just three years prior to the Buzz One Four crash, a B-52 bomber had broken up in midair, due to a major fuel leak in the right wing that caused a weight imbalance; this stressed the wing to the point that it ripped from the plane. When the fuselage broke open, its two hydrogen bombs dropped out and landed in a farmer's field.[7] In this instance, six crew members bailed from the plane and two died in the crash. Obviously neither bomb detonated, because the state of North Carolina still exists, but it had a chilling effect nonetheless. Rather than just being ferried to a new location, these bombs were on a mission when the plane went down, and were ready for action. At the time, it was assumed that redundant safety mechanisms in the bombs had all worked as planned by the bomb designers and catastrophe had been avoided according to script. But further investigation revealed that the arming mechanisms on both bombs had been inexplicably activated, and gone through all but one of the seven steps needed for detonation, because the safety rods of the bombs had somehow been removed. Technically they were still not armed because of the one remaining safety step, but some saw that distinction as semantics. As nuclear historian Chuck Hansen said: "It was like a fully loaded pistol with the safety off, and the hammer cocked—it is not armed until the final safety mechanism, the trigger, is pulled."[8]

There was speculation that centrifugal forces, acting on the safety rods when the plane started into its downward spin, had caused the rods to extract from the bombs. The safety rods were intended to prevent the bombs from detonating while still in the plane. The rods were designed to extract only when the bombs were released from the bomb bay. Their removal enables all steps of the bomb's automated activation sequence to take place. Unfortunately, the effect of centrifugal forces on the safety rods had not been considered by bomb designers. (It was an unknown unknown.) Thus, the one thing stopping detonation of the bombs was a manual safety switch that remained in the "SAFE," rather than "ARM" position. Only this single safeguard, a switch controlled solely by the flight crew, prevented detonation.

Bomb experts were able to disassemble and remove one of the bombs from the crash site, but the other had penetrated too deeply into the mud of the waterlogged field to allow its extraction (42 feet; 13 meters). It was deactivated in place, covered back up with dirt, and left in the field. It remains buried at its site of impact to this day.[9]

A few months later, on March 14, 1961, another B-52 with two hydrogen bombs ran out of fuel during flight and crashed. A sudden loss of cabin pressure had forced the pilot to fly at a lower altitude. The low altitude flying caused increased fuel consumption, and an inflight refueling attempt was only partially successful. The crew then inexplicably refused an offer of an additional inflight refueling, bypassed a potential emergency landing site, and continued flying until the fuel tanks were empty. They ejected safely, and the plane crashed 15 miles west of Yuba City, California. A subsequent investigation found that the crew had been using Dexedrine, a stimulant, to mitigate their fatigue, and this was surmised to be a contributory factor to the accident.[10] The people had failed the system, but the technology had not. The bombs did not detonate, thanks to those redundant safety mechanisms.

One might be curious as to why warplanes loaded with multiple hydrogen bombs were flying all around the United States during peacetime. They were all part of operation Chrome Dome.[11] This was the strategy used by the Strategic Air Command (SAC), the component of the Air Force that was responsible for maintaining the peace through displays of massive nuclear force.[12] Rather than have the bombers and their bombs in fixed, vulnerable ground locations, the thought was to keep one dozen nuclear bombers in the air at all times, fully armed with hydrogen bombs, and ready to attack the Soviet Union on a moment's notice. This was considered the best way to deal with the hypothetical threat of a preemptive Soviet attack on ground-based bombers. The twelve bombers were continually flying mock bombing missions, with real bombs directed at the Soviet Union. They would approach Soviet airspace, but then veer off before crossing the border, thus maintaining a perpetual game of chicken that was intended to keep the Soviet government constantly intimidated by an ever-present threat of total annihilation.

Even after the Buzz One Four crash in western Maryland, Chrome Dome missions proceeded as usual; this entailed twelve planes with multiple hydrogen bombs flying 24/7. And B-52 crashes continued to occur.

In 1966, a B-52 collided with its refueling plane and crashed in Spain, taking its four Mark 28 hydrogen bombs down with it.[13] Three of the bombs landed in the vicinity of the fishing village of Palomares on the Mediterranean coast. The parachutes for two of the three bombs failed and their conventional explosives detonated, but did not cause a fission detonation,[14] making a subsequent fusion explosion impossible. (Recall that fission explosions are required to trigger fusion explosions.) Instead, plutonium from the core of the warhead was spread over a 0.75 square mile (2 square kilometers) area, requiring extensive cleanup. The fourth bomb went missing for nearly three months, until it was finally found at the bottom of the sea and recovered.

Then, less than two years later, on January 21, 1968, a cabin fire caused by a seat cushion that was innocently stowed in front of a heating duct caused a B-52 to crash near Thule AFB in Greenland, a territory of Denmark.[15] The plane crashed with four MK-39 hydrogen bombs inside. Nothing detonated, but the impact of the crash compromised the integrity of the warheads, and the raging fuel fire helped disseminate the radioactivity over 3 square miles (8 square kilometers). Again, a massive radioactivity cleanup effort, this time hampered by extremely cold temperatures, was required. It took nine months to clean up the site. All of the radioactive material recovered was removed from Greenland and shipped back to the United States at the insistence of the Danish government.[16]

Since these last two accidents took place on foreign soil, they became international incidents, and the United States had to confess to its allies that it had been routinely flying hydrogen bombs over their countries for years, something it had promised it would not do. Good luck was being pushed to its limits and political pressure was bearing down on SAC.

These five crashes, three in the United States and two on the territory of allies, caused the United States to reevaluate the wisdom of a nuclear strategy based on airborne hydrogen bombs that put its own homeland and that of its allies at risk of accidental destruction. Consequently, the United States abandoned use of airborne nuclear bombers and began to rely on a nuclear defense strategy based on a combination of both ground-based intercontinental ballistic missiles (ICBMs) hardened against preemptive attack within subterranean silos, and submarine-based ICBMs that eluded preemptive attack by constantly moving around under the oceans of the world, like an aquatic version of Chrome Dome. This complex and tightly coupled defense system is designed to keep us safe by the perpetual threat of immediately launching a devastating nuclear counterattack on our enemies as soon as we detect that a launch has been targeted on us.

Now that Chrome Dome is history and ICBMs are the basis for our strategic defense against other nuclear powers, are we safer from accidentally annihilating ourselves? Hard to say. Again, we have to contend with the moving target argument. The systems are completely different now than they were back in the days of Chrome Dome, so our experience with that is irrelevant to our current risk assessment. Nevertheless, declassification of government records regarding our early experience with ICBMs shows that the program had growing pains similar to Chrome Dome's, the most notable being an explosion in an ICBM silo in Damascus, Arkansas, in 1980, that blew the hydrogen bomb warhead off a Titan II missile (a Mark 53 warhead similar to those in Buzz One Four's two bombs) and clear out of the silo into the countryside.[17] Again, no detonation occurred, but recurring mishaps are certainly a precarious way to keep reassuring ourselves that our redundant fail-safe mechanisms are effective.

The Damascus incident happened in 1980. We might ask, are ICBMs safer now than they were back in 1980? After all, we now have 35 more years of experience with ICBMs since the Damascus incident. No one can say for sure. Specific information about our current missile defense system is classified. But we do know that as recently as January 2014, eleven ICBM air force officers were charged with illegal drug use and 31 were charged with cheating on ICBM proficiency exams. Later that same year, two air force commanders were fired and a third disciplined for leadership lapses and misbehavior. These commanders were all in senior positions at three different nuclear missile bases that collectively control over 300 of the US Air Force's estimated 450 Minuteman 3 ICBMs. The most senior officer was dismissed specifically because of "a loss of trust and confidence in his leadership abilities."[18] It does not speak well for the safety of the ICBM system when the military officers with their fingers on the nuclear buttons dabble in recreational drugs, cheat on exams, and cannot be trusted to lead.

ON A ROLL

Despite all the unknowns, surely decades of experience handling tens of thousands of hydrogen bombs without an accidental detonation indicates that the risk is negligible or even nonexistent, doesn't it? What this question is really asking is whether the rule of succession applies.

The *rule of succession* was the brainchild of mathematician Pierre-Simon Laplace (1749–1827). This statistical principle is based on the notion that the longer that something has not occurred, the less likely it is that it will ever occur.[19] Laplace even came up with an equation to calculate the probability of the event's occurrence with increasing time:

$$P = 1 - [(n + 1)/(n + 2)]$$

where P is the probability of an accident, and n is the number of days that have passed without an accident happening. By inspection of the equation, you can see that as time passes the probability keeps getting smaller and smaller toward a limit of zero. Stated simply, if something hasn't happened in many days, it is extremely unlikely that it will happen tomorrow. Given that more than six decades have passed without an accidental detonation of a nuclear bomb, the rule of succession suggests that the probability of such an occurrence happening tomorrow is virtually zero. It seems logical, and reassuring. But is it true?

Just as the occurrence of the Three Mile Island accident underscored that Rasmussen's faith in fault trees to capture all nuclear power plant risk was misplaced, the surprise occurrence of other events thought to be extremely improbable has led statisticians to question the validity of the rule of succession. It turns out that there was an unappreciated fallacy of logic underlying the rule of succession that is sometimes referred to as the *turkey illusion*, after a story that statistician and risk analyst Nassim Taleb tells to illustrate the fallacy. Taleb says that we are deceived into accepting the validity of the rule of succession because we see the world as through the eyes of a turkey.

Here's how a turkey sees the world. A turkey hatches on a farm, and the farmer takes it from its mother and feeds it. The young turkey wonders if the farmer will return the next day to feed it again. The farmer does. The turkey wonders yet again about its food prospects for tomorrow, and tomorrow again brings the farmer bearing more food. This goes on and on for many days and the turkey eventually comes to the conclusion that its prospects for getting fed the next day are virtually certain, so it stops worrying about starving. But one day the farmer shows up without any food and kills the turkey, because that day happens to be Thanksgiving. The circumstances had changed, and the turkey was unaware that Thanksgiving Day was a day unlike any it had experienced before. This is the turkey illusion. It is the fallacy of logic that assumes circumstances remain static.

The world of today is very different from the world during the time of Chrome Dome (1960–1968). The technology has changed, the weapons have changed, their delivery methods have changed, the military has changed, and even our enemies have changed dramatically. We can gain no comfort from the fact that an accidental detonation of a nuclear weapon has not, so far, occurred. The rule of succession has been debunked. The probability of an accidental nuclear detonation occurring tomorrow might be nearly zero or it might be 100%; we just don't know what world events will bring us tomorrow. Times have changed, our useful data for predicting accidental detonation risk are practically nonexistent, and our uncertainty level is extremely high. We need to be ever mindful that Thanksgiving Day may be coming.

QUICK COUNT

We have already seen, in the case of nuclear power plants, that when uncertainty is high, we are sometimes better served to move away from risk calculations and instead try to assess the direness of the worst-case outcome. As early as 1966, an attempt was made to do this so we could prepare ourselves for the aftermath of the worst possible event in human history—a total nuclear attack on the United States by the Soviet Union, followed by an American nuclear retaliation on the Soviets. The US Defense Department commissioned a study with the RAND Corporation, a private think tank that the government heavily used for consulting services during that period, to assess the consequences of a massive nuclear weapon bombing of the United States and predict the health outcomes. The report was entitled *The Postattack Population of the United States*.[20]

The study took a novel approach. It relied on a computer program called QUICK COUNT, which was designed by RAND as an attack simulator and damage-assessment model. The program took every potential hydrogen bomb target site in the United States (all cities with >50,000 people), and drew a circular perimeter around their population centers that demarcated the limit of percussion waves exceeding 5 psi. It then drew an outer circle, beyond the 5 psi line, where lethal levels of fallout might be expected to occur.[21] The program then made the reasonable assumption that everyone within these two circles would die from either the shockwave or radiation sickness. By overlaying these circles on top of 1960 census maps, it was able to provide an estimate of the death toll. The program allowed the user to input various attack scenarios (e.g., number and size of bombs, and their hypothetical targets). It would then do a "quick count" of deaths and spit out the fatality statistics that you could reasonably anticipate from such an attack. Quite a handy tool if you need to deal with the aftermath of the ultimate war.

The study envisaged several different lines of attack that the Soviets might employ, using different combinations of various types of nuclear weapons, with each delivering from 0.3 to 100.0 MT of explosive power. It assumed that as many as 1,200 weapons would be launched, and all of the hypothetical launch scenarios were based on the premise that there would be "a brief nuclear exchange terminated by a peace settlement." The analysis focused on the condition of the United States population after the attack. The study's report is mostly concerned with postattack demographics, particularly with regard to age, race, and sex, and serves as a fascinating window into the mindset of Americans in the 1960s.

Noting that the young (<15 years) and the elderly (>65 years) would be less likely to survive the attack than those in the 16 to 64 age range, a bright spot was noted. Since the young and the old would have been a drag on recovery anyway, since they "contribute little to the social product and draw heavily on social resources," their loss was actually anticipated to speed recovery because the surviving population thus would have been culled down to the most highly productive individuals (16–64 year olds).

Regarding race, there was concern that postattack America might have an altered racial balance. Since blacks tended to be concentrated in cities, an attack strategy that focused on cities would disproportionately kill blacks. But if the Soviet attack strategy was focused primarily against military installations, which tended to be populated mostly by whites, it might tip the racial balance in postattack America in favor of blacks.

Lastly, there was a concern that women would be killed off in higher proportion to men. The thought was, since men worked outside the home and tended to occupy office buildings and other commercial structures that provide a measure of shielding, they would be better able to survive than women who tended to spend their time in flimsy residential buildings or retail establishments with lesser shielding properties. Thus, women were more likely to be exposed to blast and radiation effects. Also, allegedly being less robust than men, women were thought to be more likely to die from their injuries. The combination of greater exposure and relative weakness of women suggested that there would be a shortage of females in postattack America. This could decrease the birthrate, hinder population recovery, and "even place a strain on the social norms governing sexual liaisons." It's odd what people choose to worry about when faced with total annihilation.

The report's conclusion, based on the 1960 population data maps, was that up to 62% of the American population would likely die within the first three months following an attack. That was the bad news. The good news was that as bomb sizes increase, local geographic variations in racial and gender distribution (the main factors that drive the death disparities) tended to become less pronounced in determining the composition of the postattack population. To put it simply, when the bombs approach sizes where most everything is destroyed, it doesn't much matter exactly where you happened to be located at the time of the bombing, because your goose is equally cooked. So everyone has about the same risk of dying, regardless of race or gender. Viewed in this way, large hydrogen bombs are egalitarian, blindly killing without regard to race, gender, or any other social demographic.

GLOBAL WARMING

Since the end of the Cold War and the dissolution of the Soviet Union, the prospect of a total nuclear exchange is thought to have greatly lessened. At least progress in nuclear disarmament through weapons reduction treaties has greatly diminished the number of nuclear weapons that we have trained on each other. Presumably, that alone has lowered the risk, although we still rely heavily on a policy of *mutually assured destruction* (*MAD*) to lessen the prospect that there is anything to be achieved by starting a nuclear war. MAD is, therefore, considered to be a major deterrent to preemptive attacks on us from our sane enemies.[22] But we are not addressing the risks and benefits of total nuclear war here. How close we are to total nuclear apocalypse with our adversaries is a topic for another day. Rather, we just want to know here what the consequences of a single nuclear weapon detonation would be on us personally, whether detonated by an accident of our own military or through an intentional terrorist attack.

Again, the government has provided us with a report: "The Effects of Nuclear Weapons," first published in the 1960s, but periodically revised through 1977.[23] This report comes with a handy circular slide rule that allows a reader to determine the effects of nuclear detonation of various sizes, at various distances. Knowing your distance from some likely bomb target, and making an assumption about the size of the bomb, you can quickly determine what types of damage you might expect at your location. It's hard to get your hands on one of these original slide rules today, but a virtual replication of the original has been made into a tablet app called "Nuke Effects."[24] Better still, Alex Wellerstein, a professor at the Stevens Institute of Technology in New Jersey, has produced a free web-based program called NUKEMAP that overlays damage circles for various sized nuclear weapons on Google Maps at any desired geographic location, and immediately generates a body count based on the latest census data.[25] It is, in effect, a more modern and sophisticated version of QUICK COUNT that can be used to precisely map nuclear bomb casualty statistics of personal relevance to individuals living at any particular location.[26]

If you are like most Americans, these handy calculation tools will probably reveal you're living or working near enough a potential nuclear target to be well within the immediate death zone of a hydrogen bomb (assuming a 10-MT fusion device); but you're likely outside of the death zone of an atomic bomb (assuming a 10-KT fission device), because the casualty radii of fission bombs is much more limited. Your likelihood of encountering lethal fallout, in contrast, is more serendipitous since wind patterns are a stronger determinant of lethal doses than distance from the epicenter. Fallout can travel quite far on the winds, and settle long distances from the epicenter, as we know from the *Lucky Dragon No. 5* and Bikini Island bomb testing experience. Scary stuff indeed. Given our high vulnerability, what are we to do?

In the early days of atomic bombs, civil defense was quite popular, and may have had some legitimate benefit, though slight.[27] It was based on the idea that by taking a few precautions, and doing some simple training exercises, we could gain protection from the consequences of an atomic bombing, with an expectation of life soon returning to normal. In the advent of hydrogen bombs, however, we've stopped kidding ourselves. We no longer promote "duck and cover" as a way for school children to survive a bombing, we no longer build bomb shelters in our homes, and we no longer buy into the fantasy that hydrogen bombings are survivable.[28] We just go about our lives in ignorant bliss and hope that we never see a nuclear bombing within our lifetimes.

The old adage that an ounce of prevention is worth of pound of cure was never truer than for nuclear weapons, except that it might be more accurate to say that an ounce of prevention is worth a megaton of cure. Since a megaton of cure will always be beyond our capabilities, we are left with prevention as our sole option for survival. Have we done enough prevention?

On August 29, 2007, 39 years after operation Chrome Dome had ended and 6 years after terrorists had hijacked 4 planes and flown 2 of them into the World Trade Center, a B-52 bomber, prophetically nicknamed Doom 99, was mistakenly loaded with 6 cruise missiles armed with nuclear warheads at Minot AFB in North Dakota. The plane was left unguarded, parked on a runway overnight. In the morning, it was flown to Barksdale AFB in Louisiana, crossing South Dakota, Nebraska, Kansas, and Oklahoma, in violation of current regulations that prohibit nuclear weapons from being flown over United States soil. At Barksdale, the plane was again left unguarded on the runway for nine more hours. Later in the day, air force maintenance workers accidently discovered the armed nuclear weapons. All the while, no one back at Minot AFB had noticed that six nuclear bombs were missing. This was a security breach in which ABC News had played no part.

A task force was convened by the air force to investigate the incident. They reported their findings to the Pentagon:

> Since the end of the Cold War, there has been a marked decline in the level and intensity of focus on the nuclear enterprise and the nuclear mission [within the US Air Force]. . . . The decline in focus has been more pronounced than realized and too extreme to be acceptable. The decline is characterized by embedding nuclear mission forces in non-nuclear organizations, markedly reduced levels of leadership whose daily focus is the nuclear enter-

> prise, and a general devaluation of the nuclear mission and those who perform the mission.[29]

In other words, nuclear weapon duty had lost its glamour. To rectify this situation, the task force made 16 specific recommendations to improve nuclear security and restore nuclear weapon responsibilities to their former luster. This included nuclear training for B-52 pilots, because existing basic and advanced B-52 training courses "largely ignore the nuclear mission," and "there are no flying sorties devoted to the nuclear mission in either course." No nuclear training for B-52 bomber pilots? Yes indeed, times had definitely changed!

Some people were concerned, however, that as bad as it sounded, the task force actually had understated the problem, so a further investigation was commissioned with the USAF Counterproliferation Center with the goal to "provide a deeper understanding of the context of internal and external forces that led to the unauthorized movement of nuclear weapons."[30] The findings of the investigation, published in 2012, were quite damning to the air force:

> Through our research we have found the problems to be far more systemic than the Air Force leadership admitted in 2008. The problems were at all levels of the Air Force and institutionalized through years of change at the strategic, operational and tactical levels. . . . We see the problem in three areas: leadership, management, and expertise. Each of these elements is critical to the other and without improvement in all three the nuclear mission will likely fail again.[31]

Likely fail again? We certainly have come a long way from the air force's assertion in the early 1960s that they had everything under control and the risk of losing nuclear weapons was low to nonexistent. Most disturbing was the report's final comment that "it was clear to the investigators that the Air Force no longer valued its nuclear role and mission." Nuclear weapons had gone completely out of fashion.

Unfortunately, terrorists are less concerned that nuclear weapons are no longer fashionable. In fact, terrorists have a contrarian view.

They believe that possession of nuclear weapons actually enhances their swagger, and would be glad to take any unfashionable nuclear weapons off our hands. No need to bother delivering the weapons, they are happy to stop by and pick them up . . . or rather, they'd probably prefer to skip the heavy lifting altogether and just detonate the weapons in place. There's likely a worthwhile target somewhere nearby. Happy Thanksgiving!

EPILOGUE

N-RAYS

> The most interesting information comes from children, for they tell us all they know and then stop.
>
> —*Mark Twain*

This is a good place to stop. The story of radiation is far from over, but we have nearly reached the limit of what history can tell us. To go any further would require considerable speculation about what the future might bring. Although futurists have no qualms about telling us what will happen tomorrow, speculation is a slippery slope that we best not climb here. Nevertheless, we've learned a lot of things about radiation that should serve us well as we move into the future, if we have the wisdom to heed the lessons from our past.

In this book, we have specifically focused on the properties of radiation that drive its health effects, and have learned how the risk of suffering adverse consequences is largely determined by dose. We've also examined how scientists measure risk. We've even taken a stab at characterizing radiation's benefits and have explored how they might be measured as well. We've learned to weigh both the risks and benefits when making any decision regarding radiation, and we've warned ourselves about the ever-present threat that uncertainty poses to the validity of those decisions.

Armed with this information, we now know what questions to ask and of whom to ask them. We realize that we can look to the experts, but we need to hold them to high standards and critically consider all that they tell us, because some may have biased viewpoints.

Speaking of biases, there is one story about radiation that we've skipped over. But it's a story that happens to speak to our very concern about deception. Let's end this book now with that story. It's a little story with a big lesson. It's the story of N-rays.[1]

EUREKA . . . AGAIN?

In 1903, French physicist and distinguished member of the French Academy of Science Prosper-René Blondlot (1849–1930) may have been trailing in the field of Crookes tube research, but he was determined to catch up. Roentgen had made a big splash with his discovery that Crookes tubes emitted x-rays (1895), but Blondlot was confident that there were still things to be discovered.

Roentgen had painstakingly shown, in the eight years since their discovery, that x-rays share many properties with light, their neighbor on the wavelength spectrum. But there was at least one property that they do not share. Unlike light, x-rays cannot be refracted by a prism. Roentgen had been first to show that, and others had corroborated his finding. They found that x-rays just go straight through a prism, as though it were made of air, or water, . . . or human flesh. But what about polarization? Are x-rays polarized like rays of light are? That's what Blondlot wanted to know, and he wanted to know it before Roentgen did. Answering the question would be an important contribution to science and perhaps even win him a Nobel Prize.

Science builds on what has come before. And new discoveries beget new questions. Blondlot's polarization question was just that, a consequence of Roentgen's discovery of x-rays. This was an opportunity for Blondlot to build on what Roentgen had done before him and add his unique contribution to mankind. He was in the right place at the right time to answer the polarization question; that is, if he worked fast.

Saying an electromagnetic wave is *polarized* means that it has a specific orientation, similar to how the spinning Earth has an orientation that is defined by the axis of rotation around its north and south poles. The physics of wave polarization is quite complex and we don't need to know anything more about it here, but let's just say that light waves, when viewed head on, come at you with either a horizontal orientation, a vertical orientation, or something in between. It's similar to staring down the point of a knife blade; you are able to clearly see whether the blade is oriented horizontally or vertically. A wave of light comes at you just like the blade of a knife—it approaches with a specific orientation.

Polarization is a fundamental property of electromagnetic waves, just like wavelength is, so you might expect x-rays to be polarized just like light. But are they, and how would you prove it to be so?

Blondlot had an idea. As everyone well knew by 1903, electrons jumping through space, from one electrode to the other, produced x-rays. Roentgen had clearly shown that with the Crookes tube. Furthermore, the more electrons that jumped the electrode gap (i.e., the higher the electrical current), the more x-rays produced. Yet, Blondlot wondered if the reverse might be true. If you aimed an x-ray beam at the electrodes, might the x-rays alter the number of electrons jumping? Blondlot further reasoned that, since the jumping electrons took the form of a visible spark, changes in brightness of the spark should indicate changes in the numbers of electrons jumping due to the impinging x-rays.

So Blondlot set up an experimental apparatus to beam x-rays at sparking electrodes. When the x-rays were turned on, the spark brightness did seem to change, supporting Blondlot's hypothesis. He then reasoned that, if x-rays are polarized, the brightness of the spark should be dependent on the orientation of the electrodes in relationship to the x-ray beam. So he modified the apparatus in order to rotate the paired electrodes' orientation in relation to the direction of the beam, and called his new instrument a *spark gap detector*. As he rotated the electrodes through a 360-degree turn, yes indeed, the brightness of the spark did seem to vary. Orientation of the electrodes in relation to the beam is important! X-rays are polarized!

But wait. If the changes in spark brightness were really due to x-rays, then blocking the beam with cardboard should have no effect on the sparking phenomenon he was studying, since x-rays readily penetrate cardboard. He attempted to block the beam with cardboard, and he got the same results, seemingly confirming that it was x-rays producing the effect on the spark. He tried some other materials that Roentgen had shown to be transparent to x-rays and they, too, did not alter his spark findings. But then he tried a prism.[2] A prism placed within the beam seemed to refract the beam and modify the spark brightness effect. But what did that mean? Prisms don't refract x-rays!

It was at this point that exuberance overtook reason. Blondlot jumped to the conclusion that the prism results proved that the changes in spark brightness were due to some type of invisible rays other than x-rays; these rays were some unique form of penetrating radiation that could be refracted by prisms. Blondlot decided that this was a new type of ray that he had discovered and, therefore, he had the right to name, just as Roentgen had named his x-rays and Willie Bragg had named his cuttlefish. Blondlot chose to call them *N-rays* after his home institution, the University of Nancy. He rushed to publish.

It wasn't long after publication, however, before Blondlot's claim to be the first to discover N-rays was challenged by other French scientists who maintained that they had actually been the first to discover them. They petitioned the French Academy of Sciences, asking the Academy to rule on the matter of primacy in the discovery of N-rays. After all, a Nobel Prize was on the line. After an investigation, and to Blondlot's chagrin, the Academy sided with one of his challengers, Augustin Charpentier (1852–1916), who claimed to have first discovered N-rays emanating from living organisms. This claim was corroborated by a spiritualist who said he made a similar observation about N-rays coming from animals, and another who claimed that he had even detected N-rays coming from human cadavers. Nevertheless, the academy recognized that Blondlot's spark gap experiments had been a major achievement and bestowed on him a cash award of 50,000 francs. The Academy asserted it made the award to Blondlot in recognition of his body of work that included investigations into N-rays, without including any statement regarding his primacy in discovering them. Thus, they endorsed Blondlot's findings with N-rays, while cleverly sidestepping the issue of whether his work had been the first.

The scientific environment in 1903 was a completely different one than Roentgen had faced just a few years earlier in 1895. By 1903, scientists did not need to fret about whether they would be classified as kooks if they made outrageous claims about invisible rays. Roentgen had liberated them from that concern. In the decade following Roentgen's discovery, people were completely receptive to the idea of various invisible waves flying all around us and doing all sorts of amazing things. No longer was Roentgen's rigor a prerequisite for validating such a claim. Rigor just slowed down the pace of discovery.

Yet, the tried and true methods of scientific inquiry could not be bypassed without consequence. What was true in the past was true today. What was learned from past experience was best not forgotten.

Some of the more traditional physicists had a hard time swallowing the N-ray story, particularly the optical physicists, who were well versed in Newton's experiments with prisms. The story about the N-rays and the prism did not sit well with these scientists who knew their prisms. One of these skeptics was Robert W. Wood (1868–1955), a physics professor at the Johns Hopkins University. To him, what Blondlot was claiming smelled fishy. In the same Johns Hopkins tradition of skepticism that had motivated fly geneticist Thomas Morgan to test whether Gregor Mendel's genetic principles of inheritance were as true for fruit flies as they were for peas, Wood decided to test Blondlot's N-ray claims. As Wood suspected, he could not reproduce Blondlot's finding in his own laboratory, lamenting that he had "wasted a whole morning" playing with sparks.

When Sir William Crookes and Lord Kelvin also announced that they couldn't get the spark experiments to work either, things got serious.[3] Wood decided that the issue was important enough for him to travel to France to visit Blondlot's laboratory and actually see the evidence for himself. So off from Baltimore to Nancy he went. Quite a journey in 1903.

At a demonstration in Blondlot's laboratory, Wood failed to see the differences in spark brightness that Blondlot and his colleagues asserted were obvious. The tense situation then became a parody of the story, "The Emperor's New Clothes." Since Wood couldn't see the changes in spark brightness, Blondlot insisted that there must be something wrong with his eyes (or possibly his brain). Wood then retaliated by exploiting the fact that the experiments needed to be done in a darkened room. When the lights were turned out, he surreptitiously removed the prism from the experimental apparatus, but Blondlot's group still claimed to see the prism-dependent variations in spark brightness even though the prism was absent. When the room lights were turned back on and the prism proved missing, the jig was up. Who had the bad eyes now? This and other control experiments insisted upon by Wood revealed all the experiments to be tainted, not with fraud, but with observer bias. Just a bit too much wishful thinking.

Wood returned to Baltimore and submitted a paper to the journal *Nature,* in which he carefully dissected Blondlot's various N-rays experiments one by one, pointing out their biases, and thus putting the final nails in the coffin of N-rays.[4] Or so he thought.

Wood's paper was not well received by everyone. Since their apparent discovery, over 20 scientists had published dozens of papers on the properties of N-rays. Jean Becquerel, son of radioactivity discoverer Antoine Henri Becquerel, was one of them. Jean had reported that he could stop N-ray emissions by "anesthetizing" them with chloroform! A lot of scientists now had their reputations on the line. They would look like fools if N-rays proved to be a mere fantasy.

Things quickly started getting ugly, with accusations flying back and forth, some of them racist. Certain N-ray critics sarcastically suggested that only the Latin races seemed to have sensory and intellectual capacities to detect N-rays. The Latins retorted that Anglo-Saxon scientists had their senses dulled by overexposure to fog and beer. Roentgen and the Curies wisely decided to remain quiet on the subject of N-rays and just let things run their course. Blondlot finally pleaded for peace between the warring factions: "Let each one form his personal opinion about N-rays, either from his own experiments or from those of others in whom he has confidence." But Wood had mortally wounded N-rays and they were destined to suffer a slow, but inevitable, death.

Surprisingly, some support for N-rays persisted as late as the 1940s, which was well into the nuclear age. But then, N-rays completely disappeared from the scientific literature and from physics textbooks, without any formal pronouncement of their demise. It was just as James Clerk Maxwell, the mathematician who produced the equations that foretold the existence of radio waves long before they were discovered, had once said regarding the old corpuscle (particle) theory of light: "There are two theories of the nature of light, the corpuscle theory and the wave theory; now we believe the wave theory because all those who believed in the corpuscle theory have died."[5]

This is why no one today ever gets an N-ray scan and we don't make cancer risk estimates for N-ray doses. But the reason for retelling the N-ray story here is not to posthumously embarrass the unfortunate Professor Blondlot, or those other scientists who bought into his N-ray tall tale. Rather, this story is told to underscore the point that all of us, including scientists, are human. We have biases and we make mistakes. And we don't always see things as they really are. Our impressions of reality are often colored by the events of the day, and we easily forget the lessons of our past. The moral of the N-ray story is that we need to remain open to new ideas, yet not deceive ourselves with wishful thinking or blindly accept trendy notions. We should be simultaneously inquisitive and yet suspicious, just as Roentgen, Morgan, and Wood were. We must be skeptical of all new claims, not to be obstructive to progress, but rather because the offspring of skepticism is rigor. It is rigorous inquiry that purges us of our biases and makes it harder for others to deceive us, and for us to deceive ourselves. Going forward into our future with radiation, let us embrace skepticism and insist on rigor.

By the way, in answer to Blondlot's original question, x-rays are indeed polarized. The discovery was quietly made in 1904, one year after the N-ray incident, by Charles Glover Barkla (1877–1944), a 27-year-old British physicist who was trained at the Cavendish Laboratory by J. J. Thomson. Barkla ignored Blondlot's sparks and used a completely different experimental method.

Barkla studied x-ray polarization using an approach that was similar to the way polarized sunglasses filter out glare (a population of light waves with mixed orientations). Polarized lenses have microscopic parallel slits that can only admit light waves that have the same orientation as the slits (horizontal), thereby filtering out any light waves not oriented with the slits (vertical). If one polarized sunglass lens is laid over another, and then the two are rotated with respect to each other, an orientation can be found in which no light at all will pass through the stacked pair. In effect, you've oriented the parallel slits of the two lenses perpendicular to each other (e.g., horizontal slits are overlaid with vertical slits), such that light passing through the first lens results in a population of waves with the wrong orientation to pass through the slits of the second lens. Thus, everything gets blocked. This, in essence, is the approach that Barkla took to show that x-rays were polarized. He pointed an x-ray beam through a series of apertures and showed that the x-rays passed through the apertures as though they were polarized.

Barkla also used small gas ionization chambers to actually measure the x-rays transmitted through the apertures, so he didn't need to rely on changes in brightness or any other visual clues. His experiments were quickly duplicated and accepted by other scientists.

Shortly thereafter, the conclusion that x-rays are polarized soon became a commonly accepted fact (at least among physicists), thanks to Barkla. Then Barkla went on to make more valuable discoveries about the properties of x-rays that eventually earned him the Nobel Prize in Physics in 1917. This was just two years after Willie Bragg had received his Nobel for definitively showing that x-rays can bounce off of crystals, and 16 years after Roentgen had won his prize for convincingly showing, to his own satisfaction and that of everyone else, that x-rays—those invisible electromagnetic waves of energy that can pass through solid matter as though it were air—are real. Just as real as the light from the sun, even though no one can see them.

ACKNOWLEDGMENTS

First and foremost, I want to thank all of my friends and colleagues who generously gave of their time to read and critique various chapters from this book. Merging story with science is a delicate and intricate dance. The lay readers made sure that the science never suppressed the narrative, and the scientific readers helped ensure that the science was never compromised for the sake of the story. I am indebted to all of them, many of whom caught errors that saved me from publishing inaccurate information. Nevertheless, if any errors remain, I accept full responsibility.

These wise and critical readers are listed here in alphabetical order, along with others individuals who made contributions to this book: Mary Katherine Atkins, Tyler Barnum, Gerald Beachy, Michael Braun, Richard Brown, Ken Buesseler, Marissa Bulger, John Campbell, Thomas Carty, Harry Cullings, Daniel Dean, Mary Ellen Estes, Matthew Estes, Alexander Faje, Gregory Gilak, Andrew Herr, John Jenkin, David Jonas, Anna Jorgensen, Helen Jorgensen, Matthew Jorgensen, Christopher Kelly, Manu Kohli, Allan L'Etoile, Ann Maria Laurenza, Paul Laurenza, Collin Leibold, Paul Locke, Ryan Maidenberg, Nell McCarty, Matthew McCormick, David McLoughlin, Jeanne Mendelblatt, Kenneth Mossman, Fred Palace, Jessica Pellien, Gary Phillips, Virginia Rowthorn, Cathleen Shannon, Karen Teber, Mark Watkins, Jonathan Weisgall, and Timothy Wisniewski.

I want to single out Paul Laurenza—a renaissance man whose wide knowledge base spans both the humanities and the sciences—for special thanks. His legal training gave him keen and critical eyes. He dissected every sentence of this book and purged the prose of all pesky ambiguities, forcing me to say exactly what I meant and mean exactly what I said. Sadly, due to his extreme literalness, he sometimes deprives himself of the joys of the double-entendre and fails to appreciate the humor inherent in irony. Nonetheless, he does fully recognize the dangers posed by poor syntax and whimsically placed commas. (I will never look at a comma the same way again!) I appreciate all that he did to elevate the quality of this book, and I greatly value his enduring friendship. Thanks, Paul!

I'd like to acknowledge that my personal understanding of radiation risks has benefited greatly though my associations, over the years, with some truly outstanding scientists and teachers, each of whom influenced my thinking in profound ways. I hope that I have done justice to their wise counsel, and that I have met the same high scientific standards that they themselves always maintained. I sincerely thank all of them for sharing their wisdom with me, and feel blessed for having had the privilege to know such remarkable professionals and count them among my friends. For those who have already passed, I hope that this book pays tribute to their enduring memory. These sages include: Michael Fry, Thomas G. Mitchell (1927–1998), Marko Moscovitch (1953–2013), Kenneth Mossman (1946–2014), Jonathan Samet, and Margo Schwab.

I especially want to thank Jessica Papin, my literary agent at Dystel & Goderich Literary Management. She pulled my book proposal from the "slush pile" and skillfully guided this rookie author all the way to the publishing finish line. I am likewise indebted to Ingrid Gnerlich, my editor at Princeton University Press, who came through on all of her promises and added a tremendous amount of literary value to the book. She was a true pleasure to work with. I also thank all of the editorial staff at Princeton University Press who worked so hard to put this book together and promote its sale, including: Eoghan Barry, Colleen Boyle, Eric Henney, Alexandra Leonard, Katie Lewis, Jessica Massabrook, Brigitte Pelner, Caroline Priday, Jenny Redhead, and Kimberley Williams.

I further want to thank Carole Sargent, Director of Scholarly Publications at Georgetown University, for providing wise advice during every step of the publishing process, and the members of my Georgetown writers' group, Jeremy Haft and Anne Ridder, for their support.

What little skill I have in writing in narrative style I owe to Raymond Schroth S.J., my journalism professor at Fordham University. He has very high writing standards that I always struggled to attain. I hope this book meets with his approval and does credit to his fine teaching. Thank you, Fr. Schroth.

On the home front, I want to thank my wife, Helen, and my children, Matthew and Anna. They graciously endured my daily dinnertime banter about the ups and downs of writing a book, and provided the encouragement I needed to get the job done. I value their unwavering support and want them to know that I love them very much.

Lastly, I want to thank Mark Twain for sharing his simple and ageless wisdom with the rest of us. We need more Mark Twains in the world.

NOTES AND CITATIONS

Some of the citations in the endnotes, as well as some references listed in the bibliography, include web pages with URLs that may expire after this book has been published. Such web pages may be recoverable using the Internet Archive Wayback Machine (http://www.archive.org/web/web.php).

PREFACE

1. Gottschall J. *The Storytelling Animal.*
2. Paulos J. A. *Innumeracy.*

CHAPTER 1: NUCLEAR JAGUARS

1. Slovik P. *The Perception of Risk.*
2. NCRP Report. *Ionizing Radiation Exposure of the Population of the United States.*

CHAPTER 2: NOW YOU SEE IT

1. In about 100 AD, glassblowers in Alexandria, Egypt, discovered that by adding manganese dioxide to their glass formulations, the glass became clear.
2. Newton is more famous for his contributions toward understanding gravity than for his work in optics. A popular, though false, story has him formulating his theory of gravity after an apple fell on his head.
3. Gleick J. *Isaac Newton*, 60–98.
4. Jonnes J. *Empires of Light*, 35.
5. Jonnes J. *Empires of Light*, 71.
6. Stross R. *The Wizard of Menlo Park*, 172–173.
7. Stross R. *The Wizard of Menlo Park*, 172.
8. Stross R. *The Wizard of Menlo Park*, 180.
9. With direct current, electrons flow continuously in the same directions

within the transmission wires. With alternating current, electrons flow in one direction for a short time and then in the opposite direction for a short time, in a continuous cycle.

10. Carlson W. B. *Tesla*, chapter 5.

11. Edison had once asked the SPCA to supply him with stray dogs for his electrocutions. The organization declined to comply (Stross R. *The Wizard of Menlo Park*, 176).

12. "Coney Elephant Killed," *New York Times*, January 5, 1903.

13. Stross R. *The Wizard of Menlo Park*, 178.

14. Stross R. *The Wizard of Menlo Park*, 18.

15. McNichol T. *AC/DC*, 143–154.

16. http://wn.com/category:1903_animal_deaths

17. Around this time, the board of directors of the electric company that Edison founded decided that his name wasn't as valuable to the masthead as they had once supposed, and they changed the name from Edison Electric to General Electric. (Stross R. *The Wizard of Menlo Park*, 184–186; Jonnes J. *Empires of Light*, 185–213; Essig M. *Edison & the Electric Chair*.) For the complete story of Edison's role in developing the electric chair, see Essig, M. *Edison & the Electric Chair*.

18. The song was recorded by the Brooklyn, New York, band Piñataland in 2001.

19. Stross R. *The Wizard of Menlo Park*, 124.

20. Wavelength is one of the parameters that characterizes an electromagnetic wave. We will define and discuss wavelengths more fully later. Just know for now that the wavelength of a form of radiation determines exactly what that radiation is able to do.

21. Larson E. *Thunderstruck*, 22.

22. This first transmitted message was uninformative. It was simply the letter "s" in Morse code (i.e., dot dot dot) repeated over and over again.

23. Larson E. *Thunderstruck*, 69.

24. Edwin H. Armstrong (1890–1954) and Lee DeForest (1873–1961) were notable contributors to the development of voice radio.

25. Watts are a unit of power (i.e., energy per unit time), named in honor of James Watt (1736–1839), a Scottish mechanical engineer who made major contributions toward the development of the steam engine. Electrical power is typically expressed in some multiple of watts (1 watt = 1 joule per second). For example, a milliwatt (one thousandth of a watt) is the power output of a laser pointer, while a lightning strike typically deposits at least a megawatt (one million watts) of power.

26. For a power comparison, a modern cell phone only uses three watts of power to communicate across the Atlantic. (Of course, the cell phone's power is only required to transmit the signal to the nearest cell tower, which in turn relays the message to satellites or land lines. The cell phone itself does not send its signal across the entire distance of the call.)

27. The shorter electromagnetic wavelengths (<200 meters) are better at skip propagation, a phenomenon in which the radio waves are reflected back to Earth from the ionosphere, thus allowing further transmission along the curvature of

Earth. Shortwave radio is now mostly used by amateur radio enthusiasts for two-way international communication between low-power transmitter/receivers that are able to broadcast signals over thousands of miles.

28. Larson E. *Thunderstruck*, 383.
29. Larson E. *Thunderstruck*, 125.
30. Larson E. *Thunderstruck*, 103.
31. Larson E. *Thunderstruck*, 98.
32. Larson E. *Thunderstruck*, 385.
33. Spelled *Röntgen* in German.
34. Berger H. *The Mystery of a New Kind of Rays*, 13.
35. Berger H. *The Mystery of a New Kind of Rays*, 13.
36. Electrons are charged particles and, like all charged particles, are pushed and pulled by magnetic fields.
37. Nikola Tesla nearly made the same discovery one year earlier. Tesla was using the fluorescent glow from Crookes tubes in photographic experiments. He even attempted to photograph his good friend Mark Twain solely using the eerie illumination from a Crookes tube. But the faint light required long exposure times, and his photographic plates were always mysteriously spoiled when developed. He got frustrated and moved on to other things. After hearing of Roentgen's discovery of x-rays, he went back and looked at the developed plates. Tesla then realized that shadows of the camera body's screws and the lens cap were visible in the photographic images, evidently the result of x-rays penetrating the camera's body and exposing the plate. He is said to have cursed himself: "Damned fool! I never saw it." The photographic evidence for this anecdote is lacking, however, because he allegedly smashed the plates in anger (Carlson W. B. *Tesla*, 221–222).
38. Roentgen W. C. "On a new kind of rays."
39. Berger H. *The Mystery of a New Kind of Rays*, 16.
40. Berger H. *The Mystery of a New Kind of Rays*, 17.
41. Berger H. *The Mystery of a New Kind of Rays*, 17.
42. Badash L. *Radioactivity in America*, 9.
43. Berger H. *The Mystery of a New Kind of Rays*, 29.
44. Berger H. *The Mystery of a New Kind of Rays*, 30.
45. Even more remarkable, although no one knew it at the time, by February 7, 1896, Chicago physician Emil H. Grubbe had already been using the newly discovered x-rays for over a week to treat breast cancer in a patient named Rose Lee (see chapter 6). Another milestone for x-rays was that the Montreal gunshot victim used the x-ray of his leg as evidence at his shooter's trial, making it the first use of x-ray evidence in the courtroom.
46. Perhaps Roentgen should have been less altruistic. World War I created tremendous financial difficulties for him, and he faced the end of his long and prestigious life in poverty. In fact, his entire estate was depleted and his university never received any of his Nobel Prize winnings.
47. Thompson was primarily known for his work in electrical engineering and his authorship of a popular series of science books on calculus, electricity, and magnetism. He also wrote biographies of the scientists Lord Kelvin and Michael Faraday. He can be regarded as the Carl Sagan of his day. Thompson repeated

Roentgen's x-ray experiments the day after hearing about them, and he found the results as astounding as Roentgen had asserted. The blessing from Thompson no doubt gave credence to the fantastic x-ray claims of the more obscure Roentgen, and helped popularize the story of his discovery. In contrast, Thompson thought little of Marconi. He described Marconi as "a mere adventurer [who] claims to be an original inventor" (Larson E. *Thunderstruck*, 222).

48. Berger H. *The Mystery of a New Kind of Rays*, 45.

49. Kean S. *The Disappearing Spoon*, 270–271.

50. To be perfectly accurate, the specific subtype of x-rays described here is called *Brehmstrahlung* (braking) radiation. Roentgen's x-rays were of the Brehmstrahlung type, but another type of x-rays, known as *characteristic* x-rays, also exists. See Lapp L. E., and H. L. Andrews. *Nuclear Radiation Physics*, 150–156.

51. Israel P. *Edison*, 309–310.

52. Berger H. *The Mystery of a New Kind of Rays*, 34.

53. Brown P. *American Martyrs to Science Through the Roentgen Rays*, 39–40.

54. Brown P. *American Martyrs to Science Through the Roentgen Rays*, 41.

55. Linton O. "Francis H. Williams and William H. Rollins."

56. Brown P. *American Martyrs to Science Through the Roentgen Rays*, 17.

57. Brown P. *American Martyrs to Science Through the Roentgen Rays*, 17–18.

58. Brown P. *American Martyrs to Science Through the Roentgen Rays*, 29–30.

59. Einstein A. *Relativity*.

60. A nanometer is a unit of length equal to one billionth of a meter. Since a metric meter is about the length of British yard, a nanometer is approximately one billionth of a yard.

61. Ultraviolet radiation is also sometimes called black light, because it can't be seen. (The purple glow from commercial black light bulbs used for special visual effects in nightclubs and other entertainment venues is due to the small amount of visible light in the violet range that contaminates the ultraviolet beam.)

62. An ion is an atom (or group of atoms) that has acquired an electric charge by gaining or losing some electrons.

CHAPTER 3: SEEK AND YOU SHALL FIND

1. Fluorescence and phosphorescence are two closely related luminescent phenomena that involve the storage and release of visible light by various chemicals. For the purposes of this book, we will use the term fluorescence to mean both.

2. Lindley D. *Uncertainty*, 33.

3. Uranium is a very heavy element (heavier than lead) that was discovered by M. H. Klaproth, a German scientist, in the 1790s. He gave it the name *uranium* in honor of planet Uranus, which had also just recently been discovered. Uranus was the ancient Greek god of the open sky.

4. If this coincidence had not been the case, it would likely have been many more years until radioactivity was discovered. But as fortune would have it, uranium sulfate happens to have both an unstable atomic nucleus (the uranium nucleus) and a photo-excitable electron shell. These atomic and chemical properties

are unrelated to each other. The only fluorescent mineral in Becquerel's collection that happened to have them both was uranium sulfate.

5. One becquerel (Bq) = one atomic disintegration per second.

6. Ore is any mined rock or earthen material that contains sufficient quantities of a mineral such that the mineral can feasibly be extracted for commercial purposes.

7. Zoellner T. *Uranium*, 130–179.

8. Badash L. *Radioactivity in America*, 11.

9. Clark C. *Radium Girls*, 43.

10. Badash L. *Radioactivity in America*, 11.

11. To be specific, this is a property of the radioisotope uranium-235. We will explore nuclear fission further in chapter 4.

12. This is generally true of the lighter elements. For the heavier elements, however, atoms tend to be most stable when there are somewhat more neutrons than protons.

13. Atoms can have unstable proton-neutron ratios for various reasons. Naturally occurring isotopes with unstable ratios are thought to be the remnants of a supernova that occurred at least six billion years ago. The supernova created our solar system and all of its isotopes in every conceivable proton-neutron ratio. Those isotopes with the most extreme ratios decayed away quickly. Those with ratios closer to one-to-one persisted, since they were more stable, and decayed much more slowly. Most long-lived natural radioisotopes (e.g., uranium) are supernova remnants. Alternatively, those natural radioisotopes with short lifetimes are usually produced from decay of a parent radioisotope with a longer lifetime (i.e., a radium parent decays to ultimately produce radon; chapter 5). Additionally, proton-neutron ratios can be artificially changed by bombardment with subatomic particles (see the splitting of the atom story in chapter 4, and the example of cobalt-60 production in chapter 6).

14. The emission of gamma rays from radioactive material was first discovered by Paul Villard, another Frenchman, in 1899 (Badash L. *Radioactivity in America*, 13).

15. Single photons cannot be identified as being either x- or gamma rays, but collectively they can be identified. This is because x-rays are emitted over a range of energies, while gamma-ray emissions are typically of a single energy (i.e., they are monoenergetic).

16. Willard Frank Libby (1908–1980) was awarded the Nobel Prize in Chemistry in 1960 for developing the technique of radiocarbon dating.

17. Marie Curie (née Maria Salomea Skłodowska) was ethnically Polish. She immigrated to France at the age of 24 to study at the Sorbonne (University of Paris).

18. Marie Curie alone was awarded another Nobel Prize in 1911 for the discovery of polonium and radium, because Pierre had died in 1906 and the Nobel Prize cannot be awarded posthumously.

19. The periodic table of elements is a tabular arrangement of all known elements, organized on the basis of their proton numbers, electronic configurations, and chemical properties. The table was invented by Russian scientist Dmitri Mendeleev (1834–1907) in about 1869, and remains the classic reference tool of

chemists. A major utility of the table is that elements with similar chemical properties (i.e., those within the same chemical family) are listed within the same column of the table. Thus, knowing the column of an element allows chemists to deduce its chemistry (Kean S. *The Disappearing Spoon*).

20. To this day, some of Marie Curie's scientific notebooks are too contaminated with radioactivity to be safely handled. They are currently in storage in lead-lined boxes in Paris.

21. Badash L. *Radioactivity in America*, 12.

22. The distance to the horizon at 1,000 feet above sea level is less than 100 miles.

23. Larson E. *Thunderstruck*, 116, 169, and 215.

24. Larson E. *Thunderstruck*, 312.

25. Larson E. *Thunderstruck*, 410.

26. The Heaviside layer is a stratum of ionized gas in Earth's atmosphere found roughly 56–93 miles (90–150 kilometers) above the ground.

27. Larson E. *Thunderstruck*, 383.

CHAPTER 4: SPLITTING HAIRS

1. In reality there aren't even plums in plum pudding. The recipe calls for raisins, which were commonly known as plums in pre-Victorian England, and plum pudding actually had the texture of a modern-day, moist and dense muffin, rather than the creamy consistency of American pudding.

2. Davis E. A., and I. J. Falconer. *J.J. Thomson and the Discovery of the Electron*.

3. Another apt analogy for the size of the nucleus within an atom is "a fly in a cathedral."

4. The historical accounts of the discoveries of scientists working at the Cavendish laboratory that appear in this chapter are largely drawn from the excellent book, *The Fly in the Cathedral* by Brian Cathcart. Biographical accounts of the life of Ernest Rutherford come from the authoritative scholarly biography, *Rutherford: Scientist Supreme* by John Campbell.

5. Kean S. *The Disappearing Spoon*, 301.

6. Reeves R. *A Force of Nature*, 31.

7. Cathcart B. *The Fly in the Cathedral*, 21.

8. Frederick Soddy would win the Nobel Prize in Chemistry in 1921 for work he performed at McGill on the radioactive decay and isotope theory.

9. Rutherford E. *Radio-Activity*.

10. The first use of the term in connection with radioactivity is actually attributed to Frederick Soddy. When Soddy first realized that radioactive decay was converting one element into another, he is said to have exclaimed, "Rutherford this is transmutation!" Rutherford retorted: "For Mike's sake, Soddy, don't call it transmutation. They'll have our heads off as alchemists." Later, Rutherford was proud to use the term in connection with his work.

11. Campbell J. *Rutherford*, 478–481.

12. As of 2014, 29 Cavendish researchers have won Nobel Prizes.

13. Crowther J. G. *The Cavendish Laboratory 1874–1974.*

14. The actual source of helium for balloons and other purposes is the pockets of gas that form in porous rock when radioactive minerals deep within Earth decay, releasing alpha particles. The alpha particles take on electrons and become helium gas. The gas is trapped and cannot escape to Earth's surface, so it accumulates in porous rock. Rutherford once attempted to date the Earth's age based on the helium gas content of its rocks and the known half-lives of the radioactive minerals within them. His data suggested that Earth was at least 40 million years old. His methods were inaccurate, however, and scientists now know that our planet is at least 4 billion years old.

15. Rutherford used gold because it was inert (i.e., it didn't chemically react with other things) and because it was easily malleable into thin foils, making it convenient for experimentation. Other elements didn't have these practical advantages, but they would have worked as well.

16. Lindley D. *Uncertainty*, 46–47.

17. Italian scientist Amedeo Avogadro (1776–1856) proposed in 1811 that the volume of a gas is proportional to the included number of atoms regardless of the type of gas. (Actually, he said that all gases at standard temperature and pressure will have equal numbers of molecules, not atoms, but since the chemists also knew the number of atoms per molecule, they could adjust their calculations for gas molecules that contained more than one atom.)

18. Since the very small electrons contribute negligible amounts to atomic mass, their masses can be disregarded in the calculation.

19. Neutrons also served to dilute the positive electrical charge of the nucleus, because a nucleus composed only of self-repulsing protons would blow itself apart.

20. Cathcart B. *The Fly in the Cathedral*, 208.

21. The nucleus of a hydrogen atom consists of a single proton.

22. An *electron volt* (eV) is a unit of energy equal to the work done when an electron is accelerated through a potential difference of one volt. Historically, the eV was developed as a unit because of its usefulness in particle accelerator sciences. But it is not part of the International System of Units (SI), which employs the unit of *joules* rather than eV.

23. Cathcart B. *The Fly in the Cathedral*, 141.

24. Badash L. *Radioactivity in America.*

25. Cathcart B. *The Fly in the Cathedral*, 185.

26. Radon has been a very important player in the story of radiation's health effects. We will learn much more about radon in chapter 12.

27. The now defunct Kelly Hospital was at that time the country's leading center for radiation therapy of gynecologic cancers. Named after its founder and head, the gynecologic surgeon Howard Atwood Kelly, the Kelly hospital staff probably felt that by giving the tubes to the Cavendish scientists they were indirectly returning a favor to Ernest Rutherford. It was Rutherford who had originally given Kelly, in about 1912, an apparatus to collect radon gas from radium (Aronowitz et al. "Howard Kelly establishes gynecologic brachytherapy").

28. Ironically, the word atom comes from the Greek ἄτομος (atomos), meaning

uncuttable or indivisible, as in something that can be divided no further. So the ancient Greeks would have considered a split atom the ultimate oxymoron.

29. This is just the opposite of what a Crookes tube does when it accelerates negatively charged electrons toward a positively charged anode (see chapter 2).

30. Poole M., et al. "Cockcroft's subatomic legacy."

31. Cathcart B. *The Fly in the Cathedral*, 66–84.

32. Cathcart B. *The Fly in the Cathedral*, 165.

33. Cockcroft and Walton won the 1951 Nobel Prize in physics for "Transmutation of atomic nuclei by artificially accelerated atomic particles." Translated into plain English: They split the atom. Strictly speaking, Rutherford had already done this in 1917 when he produced transmutations by alpha-particle bombardment of nitrogen, but the expression "splitting the atom" did not come into widespread use until later, and then more typically in connection with Cockcroft and Walton's proton accelerator cleavage of lithium than with Rutherford's alpha-particle transmutations of nitrogen.

34. Cathcart B. *The Fly in the Cathedral*, 158.

35. Cathcart B. *The Fly in the Cathedral*, 64, 106.

36. Cathcart B. *The Fly in the Cathedral*, 159.

37. To be accurate, we are talking about mass (kilograms), not weight (pounds). Weight is affected by gravity, while mass is not. For example, astronauts would weigh less (in pounds) on the moon than on Earth due to lower gravity, but their mass (in kilograms) would be unchanged. The difference between mass and weight is an important scientific distinction, but the terms are sometimes used interchangeably in common speech. For our purposes, we will use the noun "mass" and its unit kilograms, with the verb, "to weigh" (because the verb "to mass" does not exist).

38. The international standard unit of energy is the joule (J). One J is equal to: $(kg \times m^2)/s^2$.

39. Isaacson W. *Einstein*, 272.

40. Cathcart B. *The Fly in the Cathedral*, 250.

41. Kelly C. C. *The Manhattan Project.*

42. Fermi had also been supremely confident when he announced his discovery of transuranic elements. Nevertheless, after he had already been awarded the 1938 Nobel Prize in Physics for the discovery, he was proven wrong. His transuranic elements were actually just fission products. The true transuranic elements were discovered some time later by Edwin Mattison McMillan (1907–1991). Reluctant to admit its mistake, the Nobel committee awarded McMillan the 1951 Nobel in Chemistry, and allowed Fermi to retain his Nobel in Physics (Kean S. *The Disappearing Spoon*, 141–143).

43. The Chicago Pile is now buried in eternal rest in Red Gate Woods, a forest preserve within the Palos Division of the Forest Preserve District, Cook County, Illinois.

44. For a short and cogent explanation of how nuclear fusion releases energy, see Close F. *Particle Physics: A Very Short Introduction*, 107–111.

45. Cosmic radiation actually includes all radiation originating anywhere in space, not just from solar flares.

46. Brooks M. *13 Things That Don't Make Sense*, chapter 4.

47. A hydrogen gas explosion is not the same as a hydrogen bomb. Hydrogen gas (H_2) is highly reactive with oxygen gas (O_2) and can cause a chemical explosion that results in the production of water (H_2O), but this is not the same as a hydrogen nuclear fusion reaction. Hydrogen gas chemical explosions did occur at the Fukushima nuclear power plant accident in 2011, but it is not possible for a nuclear hydrogen fusion explosion to occur at a nuclear fission power plant.

CHAPTER 5: PAINTED INTO A CORNER

1. Schneeberg (Germany) and St. Joachimsthal (Czech Republic) are neighboring mining communities in the Ore Mountains—a mountain range that straddles the present-day border between Germany and the Czech Republic. The Ore region has a long history of mining various minerals, and the term "ore"—a naturally occurring solid material from which minerals can commercially be extracted—received its name from this place.
2. Schuttmann W. "Schneeberg lung disease and uranium mining."
3. Isaac Newton worked extensively with mercury (then called quicksilver). Samples of his hair were analyzed in 1979 and found to contain mercury at toxic levels. But it is not clear whether the mental illness that he suffered later in his life can be explained by the mercury (Gleick J. *Isaac Newton*, 99).
4. Clark C. *Radium Girls*, 20.
5. Rodricks J. V. *Calculated Risks*, 136–161.
6. It was often the practice of coal miners to bring caged canaries into the mines to detect the presence of methane gas, which is lethal. Canaries are extremely sensitive to methane. If a miner's canary died, it suggested the presence of the gas, and the miners evacuated before they succumbed to the methane fumes.
7. As we have already seen, the early researchers of the ionizing radiations (e.g., x-rays, gamma rays, beta particles, alpha particles, neutrons, etc.) often suffered skin irritations as a consequence of their research, and cancers sometimes later developed in the same body areas that had been irritated. Roentgen was aware of this, and he protected himself with lead shielding while experimenting with x-rays. The less cautious scientists and engineers, like the Curies and Edison, suffered significant health consequences due to their relatively high exposures. In contrast, Marconi and other pioneers of the nonionizing radiations (e.g., radio waves) suffered no apparent health effects even from massive exposures, suggesting that the ionizing radiations were potentially dangerous while the nonionizing radiations were not.
8. To be perfectly accurate, a rarer radioisotope of radon (radon-220) was first discovered in 1899, by Ernest Rutherford (Marshall J. L., and V. R. Marshall. "Ernest Rutherford").
9. An older term "daughter" is sometimes used instead of "progeny." (Although it may be politically incorrect, no decay products are ever called "sons.")
10. Lorenz E. "Radioactivity and lung cancer."
11. *Waterbury Observer*, "After glow."
12. The first person to actually invent a radium-based fluorescent paint was William J. Hammer in 1902, but he did not attempt to patent it because he

thought radium was too scarce and costly to have any practical use in paint. But George F. Kunz, an executive at the exclusive jewelry firm, Tiffany & Company, stole Hammer's idea, patented the paint in 1903, and used it to adorn several jewelry items that were sold to wealthy buyers in their Manhattan store. Radium paint would not come into widespread use until rising medical demands for radium greatly increased commercial radium production to the point that prices started to come down (see chapter 6).

13. Clark C. *Radium Girls*, 15.

14. Clark C. *Radium Girls*, 108–109.

15. Hacker B. C. *The Dragon's Tail*, 22.

16. The term *irradiate* is a verb meaning the act of exposing something to radiation.

17. Recall that carbon-14 (six protons) decays into nitrogen-14 (seven protons); see chapter 2.

18. NCRP Report. *Some Aspects of Strontium Radiobiology*.

19. The Waterbury Clock Company would ultimately become Timex.

20. Clark C. *Radium Girls*, 161.

21. Quinn S. *Marie Curie*, 411.

22. Quinn S. *Marie Curie*, 410.

23. Clark C. *Radium Girls*, 172–176.

24. Clark C. *Radium Girls*, 172.

25. Physicians were offered a discounted rate of $25.00 per case.

26. Clark C. *Radium Girls*, 176.

27. Lazarus-Barlow W. S. "On the disappearance of insoluble radium salts from the bodies of mice."

28. Engelmann W. "Radium emanation therapy."

29. Zueblin E. "Radioactive therapy in medicine."

30. Pierre Curie did something similar with the mouse, an animal model more relevant to humans. He found that just a few granules of radium placed near the backbone of a mouse produced paralysis in three hours (Mullner R. *Deadly Glow*, 13).

31. Mullner R. *Deadly Glow*, 15–17.

32. Clark C. *Radium Girls*, 174.

33. A total of 112 dial painters are thought to have died from radium poisoning, with the last radium-related death in 1988. (Mullner R. *Deadly Glow*, 5.) Several radium dial painters lived into their hundreds. The last known radium dial painter was Mabel Williams. She died at 104 years old in Olympia, WA, on July 23, 2015.

34. Buchholz M. A., and M. Cervera. *Radium Historical Items Catalog*, 1.

35. Clark C. *Radium Girls*, 202.

36. Clark C. *Radium Girls*, 201–202.

37. Upton A. C. "The first hundred years."

38. Yoshinaga S., et al. "Cancer risks among radiologists."

39. Upton A. C. "The first hundred years."

40. The roentgen was originally defined as a unit of radiation exposure equal to the quantity of ionizing radiation that will produce one electrostatic unit of elec-

tricity in one cubic centimeter of dry air at 0°C and standard atmospheric pressure.

41. Technically, a roentgen is a unit of exposure, not dose. So to define a unit of dose in roentgens is not appropriate. Nevertheless, at the time, the terms "exposure" and "dose" were often used interchangeably.

42. The limit was soon reduced to 0.1 roentgen per day.

43. Quinn S. *Marie Curie*, 415.

44. Quinn S. *Marie Curie*, 416.

45. Quinn S. *Marie Curie*, 413–414.

46. Greenwood V. "My Great-Great-Aunt Discovered Francium."

47. Quinn S. *Marie Curie*, 431–432.

48. *Nature*. "Curie laid to rest with France's heroes."

49. *Nature*. "X-rays, not radium, may have killed Curie."

50. Clark C. *Radium Girls*, 163.

51. Brothers Cecil and Philip Drinker took an interest in studying the health effects of radium dust in cats. Philip, an engineer, invented a machine to control the inhalation rates of cats. In 1928, the apparatus was scaled up to human size and used to assist polio victims in breathing. The machine, known as the *iron lung*, saved the lives of many polio patients (Clark C. *Radium Girls*, 207).

52. Radioactivity carried within the body is known as the body burden and is expressed in either units of radioactivity or mass, typically microcuries or micrograms, respectively. The *maximum permissible body burden* for any radioisotope is one that will result in an MTD.

53. Clark C. *Radium Girls*, 193.

54. Taylor died of complications from Alzheimer's disease in 2004, at the age of 102.

55. Oransky I. "Lauriston Taylor."

56. Frame P. W. "Tales from the atomic age."

57. This chapter highlights only some of the key events leading to the introduction of national and international radiation protection standards. Those seeking a complete and chronological history of radiation protection practices should refer to: Taylor L. S. *Organization for Radiation Protection*.

58. Isaacson W. *Einstein*, 471–478.

59. Zoellner T. *Uranium*.

60. Hacker B. C. *The Dragon's Tail*, 29.

61. The Manhattan Project began two years before the United States formally entered World War II (December 7, 1941), but the war had been waging in Europe since Germany invaded Poland on September 1, 1939. The atomic bomb's intended use was against Germany because the Germans were correctly suspected of having their own atomic bomb project, but the war in Europe ended before they had completed the task. After the Japanese surrendered, it was learned that they also had their own unsophisticated bomb project that was led by Yoshio Nishina (1890–1951), a former student of Ernest Rutherford. Some evidence suggests that Nishina may have sabotaged the project by assigning Japan's most gifted physicists to menial tasks and the less intelligent physicists to the critical work (Rotter A. J. *Hiroshima*, 66).

62. Mullner R. *Deadly Glow*, 125.

63. A radiation protection standard used in 1939 illustrates the crudeness of dental film dosimetry (i.e., the measurement of dose based on the level of exposure of photographic dental film) at that time: "If after a week of exposure and standard development, none of the films is so dark that, when it is put up to a printed page in good light, the letters cannot be distinguished through it, then protection is adequate" (Hacker B. C. *The Dragon's Tail*, 37).

64. Dose versus exposure: The term "dose" refers to what actually gets into your body due to a certain exposure. A good analogy is secondary cigarette smoke. If a group of people is in a room where someone is smoking, the nonsmokers are all exposed to the same smoke-contaminated air, but their doses will differ. If someone is holding her breath, she may receive no dose from the smoke exposure, while someone else with a fast respiration rate with deep breaths will receive a high dose. It's the dose that drives health effects, not the exposure. Although the two are highly correlated and are often confused, dose and exposure are not the same thing.

65. One mSv is equal to 100 mrem.

66. The older dose equivalent unit—the mrem—is still used by the US Nuclear Regulatory Commission (as of 2015) despite the fact that it has been replaced by the mSv in every other country in the world. In the author's opinion, the NRC's policy of sticking with mrem is a bad practice that unnecessarily confuses the public. This was particularly true during the nuclear power plant disaster in Japan in 2011, when the Japanese authorities were reporting to the media in mSv and the American authorities were using mrem. In this book, dose equivalents will be reported in mSv throughout.

67. Microorganisms can survive radiation doses hundreds of times higher than humans can.

68. Since nonionizing types of radiation, by definition, don't produce any ions, the mSv is not an appropriate unit for measuring their doses. Nonionizing radiation measurement requires metrics that quantify energy deposition based on other physical principles.

69. Hacker B. C. *The Dragon's Tail*, 43.

70. We will deal further with the exact meaning of *dose equivalent*, how it is measured, and its utility for predicting cancer risk in chapter 9.

71. This concept survives today as *ALARA*, an acronym for *as low as reasonably achievable*. Its use in the radiation work environment is now legally mandated by the regulations of the US Nuclear Regulatory Commission.

72. Neutrons release visible blue light as they move at extremely high speeds through matter. This blue light is called Cherenkov radiation, after its discoverer, but it is actually just light from the blue part of the visible wavelength spectrum. It has no health consequences of its own, but it does indicate the presence of high-speed particulate radiation, typically neutrons.

73. Conant J. *109 East Palace*, 339.

74. Conant J. *109 East Palace*, 340.

75. In fact, this is the exact safety approach that is used in nuclear power plants today. Control rods are mechanically raised, under electrical power, to initiate the nuclear fission reaction, so that accidental loss of electricity in the plant will cause

the rods to drop down under the force of gravity and, thereby, terminate the fission reaction.

76. Beryllium is a dull brittle metal that is prone to undergoing nuclear reactions that emit neutrons. It can absorb both alpha particles and gamma rays and emit neutrons in exchange. It was the alpha particle reaction of beryllium that Irène Curie and Frédéric Joliet first witnessed (see chapter 4). Slotin's team was using the beryllium to absorb gamma rays emitted from the plutonium core and convert them to neutrons, thus boosting the neutron yield of the core and pushing it closer to criticality. Because of its neutron producing characteristics, beryllium was widely used at the time by nuclear physicists who studied neutrons. Unfortunately, its brittleness causes it to produce a dust that is highly toxic to the respiratory system. Enrico Fermi was one victim of this dust. He died from beryllium poisoning, at the age of 53 (Kean S. *The Disappearing Spoon*, 192–193).

77. This is reminiscent of the alpha particle bounce-back experiments that allowed Rutherford to detect the atomic nucleus and measure its size (see chapter 4), but radar works on a much grander scale than alpha particles (i.e., bouncing off planes and ships versus nuclei).

78. Brown D. E. *Inventing Modern America*, 80–83.

CHAPTER 6: THE HIPPOCRATIC PARADOX

1. Hodges P. *The Life and Times of Emil H. Grubbe*, 7.

2. Hodges P. *The Life and Times of Emil H. Grubbe*, 21.

3. Radiation oncology is the medical specialty that uses radiation therapy to treat cancer. (Although radiation therapy is often called "radiotherapy," the author believes that this word is misleading because it implies that radio waves are used. Ironically, radio waves are the one type of radiation that is not used to treat cancer because its energies are too weak to kill cancer cells, or any other cells.)

4. Grubbe didn't graduate from medical school until March 1898, but maintained an active radiology practice at 2614 Cottage Grove Avenue in Chicago, starting in February 1896. Grubbe was criticized by some physicians in the neighborhood for being a quack because they were suspicious of x-rays and he was practicing medicine without a medical degree. Nevertheless, Grubbe carried on. At one point he made the claim that he and his staff were treating 70 patients per day for various ailments (not all of them cancer).

5. Hodges P. *The Life and Times of Emil H. Grubbe*, 25.

6. It was first known to Europeans as *Jesuit's bark* because it was introduced by Jesuit missionary Agostino Salumbrino (1561–1642), who brought it back from Lima, Peru, after observing the native Quechua using it to treat fever. The active ingredient was later found to be quinine. Quinine and its synthetic derivatives are still used to treat malaria today.

7. Upton A. C. "The first hundred years."

8. Regaud C., and Ferroux R. "Disordance des effets de rayons."

9. Radiation castration does not achieve the same effects as surgical removal of the testicles in one respect—hormone elimination. Radiation does not typically kill the testosterone-producing cells (Leydig cells) in the testicles since they, unlike

spermatogonia, are not dividing. Secondary sex characteristics, libido, and male aggressiveness are not typically affected by radiation doses that cause complete sterility. Since one of the purposes of livestock castration is to reduce testosterone-related male aggressiveness, radiation would not usually be a practical substitute for surgical castration (Mossman K. L., and W. A. Mills. *The Biological Basis of Radiation Protection*, 174–177).

10. Hodges P. *The Life and Times of Emil H. Grubbe*, 69.

11. A highly autobiographical book about the history of radiation therapy, written by Grubbe, is thought to be unreliable (Grubbe, E. *X-Ray Treatment*). Information regarding Grubbe's contributions to radiation therapy reported here was obtained largely from a biography of Grubbe written by Paul C. Hodges, *The Life and Times of Emil H. Grubbe*. In writing his book, Hodges actually verified Grubbe's essential claims of priority in radiation therapy with independent documents on deposit in the Smithsonian Institution in Washington, DC.

12. Mullner R. *Deadly Glow*, 31–33.

13. Implantation of radioactive "seeds" (usually palladium-103) in the prostate to treat prostate cancer is an example of modern brachytherapy. In the United States, more than 50,000 men receive prostate brachytherapy each year.

14. At the time, Austria owned the Ore Mountains, home of the Schneeberg and St. Joachimsthal radium mines, which were the only known source of pitchblende (radium ore).

15. Mullner R. *Deadly Glow*, 18 (quoted in the *New York Times*, July 17, 1904).

16. A stay bolt is a threaded metal rod that is used to connect opposing metal plates and hold them in a fixed position relative to each other.

17. Carnotite is a yellowish green mineral commonly found in the form of crusts or flakes in sandstone.

18. The site of the extraction plant in Canonsburg is contaminated with radium to this day. The federal government owns the site and the US Department of Energy (DOE) manages it. The contaminated area, estimated to contain 100 curies of radium, is confined within a containment cell designed to be impervious for 1,000 years. Its integrity is monitored continually by DOE to ensure that radioactivity does not leak into the water table and contaminate nearby Chartiers Creek. DOE's license to operate the site has no expiration date, as it is expected they will manage the site in perpetuity.

19. Mullner R. *Deadly Glow*, 25.

20. Mullner R. *Deadly Glow*, 27.

21. Purification of radium was complicated because carnotite ore from Paradox Valley also contains barium (a nonradioactive metal). Barium and radium are both alkaline earth metals (like calcium; see chapter 5) and thus share similar chemistries; these make them resistant to separation from each other by chemical means.

22. It is estimated the Standard Chemical Company purified a total of 180 grams of radium (about the same mass as a large bar of soap), and most all of it passed through the Flannery Building at 3530 Forbes Avenue, in Pittsburgh. The building went through a series of subsequent owners who mostly used it as office space; ultimately it housed a bank. When the bank tried to sell the building in the

1980s, the sale failed due to the discovery of radium contamination. Litigation ensued, and the property remained in limbo. Finally, the Pennsylvania Department of Environmental Protection oversaw the building's decontamination, which was completed in September 2003, and the building reopened for unrestricted use. Today the building houses commercial office space.

23. Standard Chemical ultimately entered the radium watch business, and created a division in Chicago known as the Radium Dial Company. This was a competitor to the fluorescent dial business of United States Radium Corporation in New Jersey (chapter 5).

24. In the years to come, the mines of the Belgian Congo would become of extreme military importance because they represented the only known source of high-grade uranium ore outside of Nazi control. Uranium ore was required in large quantities for the Manhattan Project (Zoellner T. *Uranium*, 45–46).

25. Robison R. F. "Howard Atwood Kelly."

26. After the Kelly Hospital closed in 1952, the building was found to be highly contaminated with radium. The switchboard from the first floor reception area was so covered with radioactivity that it was decided to encase it in concrete and dump it into the ocean. The building was eventually decontaminated and cleared for safe occupation by the US Atomic Energy Commission. (Mangrum, H. "1418 Eutaw Place.") Later, the building was abandoned and became occupied by homeless people when the neighborhood deteriorated. It was ultimately demolished and the property reduced to a vacant lot. The site is currently a Baltimore City park.

27. Most of the information in this chapter about Kelly's particular clinical procedures was obtained from: Aronowitz J. N., and R. F. Robison, "Howard Kelly establishes gynecologic brachytherapy."

28. Kelly H. A., and C. F. Burnam. "Three hundred and forty-seven cases of cancer of the uterus and vagina."

29. Rutherford had actually used a similar apparatus to discover radon. Working with Marie Curie's observation that pitchblende emitted a radioactive gas, he built an apparatus to collect bubbles of the gas into an inverted glass flask. Rutherford soon identified the captured gas as radon-220, a previously unknown element (Kean S. *The Disappearing Spoon*, 302). (The more naturally abundant isotope of radon, radon-222, was later identified by Friedrich Ernst Dorn.)

30. It was not the radon-222 itself that provided the penetrating gamma rays needed for therapy. Radon-222 primarily emits alpha particles that are poorly penetrating, only rarely emitting a gamma ray (510 KeV gamma; 0.076% abundance). Rather, it was the gamma rays emitted by radon-222's short-lived downstream progeny in the uranium-238 decay chain that were producing the tumor-penetrating gamma rays.

31. By 1914, Kelly was exclusively using radon encased in glass ampules, rather than radium itself, for his radiation therapy. These were the glass ampules of spent radon that Feather took back to the Cavendish in 1930 (see chapter 4).

32. *New York Times*. "$100,000 Radium Treatment Test to Save Bremner's Life."

33. *Trenton Evening Times*. "New Jersey Representative Succumbs in Baltimore Sanatorium."

34. Quoted in Aronowitz J. N., and R. F. Robison. "Howard Kelly establishes gynecologic brachytherapy."

35. The term linac for linear accelerator did not come into widespread use until 1950.

36. Tuddenham W. J., and A. Soiland. "Pioneer radiologist."

37. Jacobs C. D. *Henry Kaplan*, 111–112.

38. Cobalt-60 is an artificial radioisotope that is produced by bombarding the stable isotope cobalt-59 with neutrons. Uptake of an additional neutron into the cobalt-59 nucleus produces cobalt-60, which has an unstable nucleus due to an excess of neutrons (33) relative to protons (27). (See "Radioactive Decay" in chapter 3 for an explanation of how neutron excess results in radioactivity.) The nuclear instability of cobalt-60 is relieved through radioactive decay; this occurs when the parent cobalt-60 transmutes into the progeny nickel-60, with the concurrent emission of two high-energy gamma rays. Nickel-60 is a stable isotope.

39. Lymph nodes are small olive-shaped organs. Hundreds are distributed widely throughout the body, serving as traps and filters for circulating foreign bodies. They have a major role in immunity and harbor specific types of white blood cells (lymphocytes and macrophages). The nodes typically swell when fighting an infection, but shrink back to normal again when the infection abates. Persistent swelling of lymph nodes in the absence of infection can be sign of lymph-node cancer (lymphoma).

40. As opposed to Hodgkin's disease (Hodgkin's lymphoma), with its well-defined progression pattern, non-Hodgkin's lymphoma is actually a heterogeneous group of over 30 types of lymphoid cancers that differ in their microscopic appearance, biological behavior, and prognosis.

41. Cure rates cannot be exclusively attributed to radiation therapy advances. For advanced-stage Hodgkin's disease, radiation therapy is typically combined with chemotherapy (i.e., cancer therapy that uses drugs).

42. There were other important players in the success of radiation therapy for cancer that cannot be adequately recognized here. Advances came from all over the world. The following article reports on just the contributions of Americans to the field: Brady L. W., et al. "Radiation oncology."

43. For an explanation of cancer chemotherapy and its history, see Mukherjee S. *The Emperor of All Maladies*.

44. Treating disease with the goal of relieving pain and suffering, but without the intention of cure, is called *palliative therapy*.

CHAPTER 7: LOCATION, LOCATION, LOCATION

1. We will use the term "atomic bomb" to specifically mean a nuclear *fission* bomb, and not any other type of nuclear device.

2. Tibbets P. W. *Return of the Enola Gay*, 161.

3. The flight risks were considered the most uncertain part of the bombing mission. In contrast, scientists had estimated the risk of the bomb failing to detonate as less than 1 in 10,000. The scientists' confidence was supported by a successful test detonation, code named Trinity, in the Nevada desert a month earlier.

Even though the design of the test bomb was different (plutonium) than the Hiroshima bomb (uranium), the Hiroshima detonation mechanism was considered more simplistic in design and less likely to fail (Tibbets P. W. *Return of the Enola Gay*, 185).

4. This chapter primarily reports measurements in British units since they were in widespread use at the time, even among scientists, and were used in all original historical sources. The metric unit equivalent is given in parentheses when a value is stated for the first time.

5. The hypocenter is the exact location where a bomb detonates. All forms of damage initially radiate out from this spot. The hypocenter of the Hiroshima bombs was at a height of 1,890 feet (576 meters) above ground level. (The point on the ground directly under the hypocenter is called ground zero.) The eight-mile estimate was provided by William Sterling Parsons (1901–1954), associate director of the Los Alamos Laboratory in charge of atomic bomb ballistics (Tibbets, P. W. *Return of the Enola Gay*, 162–163).

6. For a complete description of the mathematics Tibbets used to calculate the angle of departure, see: http://user.xmission.com/~tmathews/b29/155degree/155degreemath.html.

7. Walker R. I., et al. *Medical Consequences of Nuclear Warfare,* 6–10.

8. Sasaki is one of the characters from John Hersey's nonfictional story *Hiroshima*. Hersey's account of the human casualties in the aftermath of the first atomic bombing was initially published in *The New Yorker* magazine on August 31, 1946, one year after the Hiroshima and Nagasaki bombings. The magazine story was later produced as a book. It remains the most poignant account of the human tragedy surrounding the Hiroshima bombing.

9. Rotter A. J. *Hiroshima*, 194.

10. Ikeda's watch was later donated, by his wife, to the Atomic Bomb Artifacts Collection of the Hiroshima Peace Memorial Museum, 1-2 Nakajima-cho, Naka-ku, Hiroshima.

11. Hersey J. *Hiroshima*, 81–82.

12. McRaney W., and J. McGahan. *Radiation Dose Reconstruction U.S. Occupation Forces in Hiroshima and Nagasaki, Japan, 1945–1946.*

13. Hersey J. *Hiroshima*, 4.

14. Hersey J. *Hiroshima*, 24.

15. Hersey J. *Hiroshima,* 46.

16. It would be a week or more before the general public in Hiroshima learned that the bomb was atomic. (Hersey J. *Hiroshima*, 62.)

17. Some estimates put the brightness of the bomb at 10 times the sun's. (Tibbets P. W. *Return of the Enola Gay*, 228.)

18. Crew members were also issued cyanide capsules in case their plane was shot down and they were captured.

19. The *Enola Gay* crew didn't lose sight of the mushroom cloud over Hiroshima until the plane was 415 miles away, the approximate distance between Washington, DC and Boston (Rotter A. J. *Hiroshima*, 192).

20. Rotter A. J. *Hiroshima*, 204.

21. In early September (four weeks after the bombing), the death rate among

Hiroshima survivors was approximately 100 deaths per day (Weisgall J. M. *Operation Crossroads*, 7).

22. Hersey J. *Hiroshima*, 72.

23. A syndrome is a collection of symptoms that characterizes a specific disease state.

24. The dose ranges given for each syndrome are approximations, not absolute boundaries. Also, there is some overlap of the syndromes at the doses near the cusps. So the minimum and maximum doses given for each syndrome should not be considered to be definitive thresholds, but rather just the approximate dose boundaries within which the symptoms of the syndrome predominate. Nevertheless, knowledge of the whole-body dose that a patient received is the single most important factor contributing to the patient's prognosis, including expected time to death or likelihood of survival.

25. Walker R. I., and T. J. Cerveny. *Medical Consequences of Nuclear Warfare*, 22.

26. Hersey J. *Hiroshima*, 78.

27. White blood cells can undergo a form of *programmed cell death* (also known by the Greek word *apoptosis*), which is more rapid than the necrotic death that most other cell types undergo.

28. After the Chernobyl nuclear power plant accident that took place in the Ukraine in 1986, some victims were treated with a medical procedure called bone marrow transplantation (BMT), which replaces the victim's destroyed bone marrow with healthy marrow from a donor. BMT is a highly sophisticated medical procedure that rescues some patients who would otherwise have died from the hematopoietic syndrome. But it has several drawbacks. A successful BMT requires a tissue-matched donor and highly trained personnel with sophisticated medical facilities, and entails the risk that a suboptimal donor match will itself cause death by *graft versus host disease*. This is a condition where the donor's immune system, harbored within the donated bone marrow (the graft), will actually attack the patient's body tissues (the host), mistaking them for foreign invaders. In practice, BMT is only a viable option for patients within a very narrow range of whole-body doses (approximately 8,000 to 10,000 mSv). For all of the reasons above, BMT is not the panacea for radiation sickness that it was once thought to be.

29. If the dose is spread out over an extended time, the threshold for radiation sickness is even higher (>3,000 mSv) because cells can repair damage that accumulates at a slower rate. Doses that are received instantaneously, however, as from the atomic bomb, are more damaging because they overwhelm the cells' repair systems.

30. The drop-off of radiation dose with distance follows the inverse-square law, meaning that the dose decreases in inverse proportion to the square of the distance from the source. For example, the dose at two meters from a radiation source (i.e., a point source) will be one quarter of the dose at one meter (dose = $1/d^2 = 1/2^2 = 1/4$). Doubling the distance will quarter the dose, and quadrupling the distance (i.e., 4 times the distance) will result in 1/16 of the dose. Thus, radiation sickness cases will be clustered between restricted perimeter bands that encompass a narrow range of doses from approximately 1,000 to 10,000 mSv. For

the atomic bomb of Hiroshima, that band was approximately 800 to 1,800 yards from ground zero (Nagamoto T., et al. "Thermoluminescent dosimetry of gamma rays").

31. Rotter A. J. *Hiroshima*, 198.

32. Hersey J. *Hiroshima*, 81; Weisgall J. M. *Operation Crossroads*, 210; Rotter A. J. *Hiroshima*, 222.

33. Rotter A. J. *Hiroshima*, 194.

34. Rotter A. J. *Hiroshima*, 201.

35. According to Tibbets, the US military was unaware that 23 downed US fliers were being held captive in Hiroshima Castle at the time of the bombing (Tibbets, P. W. *Return of the Enola Gay*, 202).

36. Los Alamos physicists Hans Bethe (1906–2005) and Klaus Fuchs (1911–1988) were largely responsible for designing the implosion mechanism for the plutonium bomb. Fuchs turned out to be a Soviet spy, and diagrams of the implosion design ultimately made their way to Soviet bomb scientists by way of Soviet agents, Julius and Ethel Rosenberg. The Rosenbergs were executed for espionage in 1953 in Sing Sing Prison's electric chair.

37. Cullings H. M. et al. "Dose estimates"; see figure 5 of this article.

38. The statement can be traced back to Aeschylus (525–456 BC), the Greek dramatist, but is most widely attributed to Senator Hiram Johnson (1866–1945), who in 1918, at the height of both World War I and the Spanish flu epidemic, said, "The first casualty, when war comes, is the truth." Coincidentally, Johnson died on August 6, 1945, the day of the Hiroshima bombing.

39. Rotter A. J. *Hiroshima*, 206.

40. Rotter A. J. *Hiroshima*, 223.

41. Hersey J. *Hiroshima*, 89.

42. Walker R. I., et al. *Medical Consequences of Nuclear Warfare,* 5.

43. Walker R. I., et al. *Medical Consequences of Nuclear Warfare,* 8.

CHAPTER 8: SNOW WARNING

1. Hoffman M. "Forgotten Atrocity of the Atomic Age."

2. Although the fishermen initially thought they had witnessed a meteor hitting Earth, they presently deduced what had actually occurred and kept radio silence during their return, because they were afraid that they would be arrested at sea by the American military due to the classified nature of what they had seen (Ōishi M. *The Day the Sun Rose in the West*, 5).

3. Retrieval of the radioactive catch was largely successful, and the fish from the *Lucky Dragon No. 5* were buried in a large pit dug on a vacant lot bordering the entrance to Tokyo Central Wholesale Market. Today there is a Tokyo Metro Subway stop near the entrance to the market, with a "Tuna Memorial" plaque marking the approximate gravesite of the contaminated fish (Ōishi M. *The Day the Sun Rose in the West*, 126–133).

4. Mahaffey J. *Atomic Accidents*, 82.

5. Ivy Mike, the first hydrogen bomb tested, was detonated on Enewetak Atoll in the Marshall Islands (near Bikini Atoll) on November 1, 1952.

6. We will consistently use the term "hydrogen bomb" when referring to bomb designs based on nuclear *fusion*. Be aware, however, that this type of bomb is also sometimes called an *H bomb* or a *thermonuclear weapon*.

7. The distance between the Capitol Building and the White House is about two miles (3.2 kilometers). The distance between Washington, DC's mall and Baltimore's Inner Harbor is about 40 miles (64.4 kilometers).

8. When the Paris fashion designer Louis Réard introduced his new women's swimsuit design for the summer of 1946, he called it the bikini, and suggested that its revealing design was just as shocking as the Bikini Atoll atomic bomb tests that were capturing the public's attention that summer.

9. This atoll was spelled Eniwetok until 1974, when the United States government changed its official spelling to Enewetak so that the word's phonetics more closely matched the pronunciation of the Marshall Islanders.

10. Nuclear bombs can be of either the fission or fusion type. In this book, we will use the term *atomic bomb* to mean a fission bomb, and the term *hydrogen bomb* to mean a fusion bomb.

11. Japanese artist Isao Hashimoto has created a time-lapse video map of the 2,053 nuclear explosions that took place between 1945 and 1998. The March 1, 1954, hydrogen bomb test (called Castle Bravo) on Bikini occurs 3 minutes and 4 seconds into the 14:24 minute video: https://www.youtube.com/watch?v=LLCF7vPanrY.

12. Ōishi M. *The Day the Sun Rose in the West*, 35.

13. Kuboyama's death galvanized the antinuclear movement in Japan. Although many Japanese had felt that they needed to share some of the culpability for the bombings of Hiroshima and Nagasaki because their own government had been an aggressor with an avowed strategy of "total war," they saw Kuboyama as an innocent martyr of the United States' aggressive and irresponsible participation in a nuclear arms race conducted during peacetime. More than 400,000 people attended Kuboyama's funeral. Further fanning the attendees' outrage, nuclear physicist Edward Teller, the developer of the hydrogen bomb, was alleged to have insensitively retorted, "It's unreasonable to make such a big deal over the death of a fisherman" (Ōishi M. *The Day the Sun Rose in the West*, 126).

14. Although some have proposed that the purpose of the high altitude detonation was to minimize fallout, this is not true. Aboveground detonation was chosen because it evenly distributed and maximized the shock wave effects that, along with fire, were among the most damaging aspects of the atomic bombs (see chapter 7).

15. The isotopes of hydrogen are hydrogen-1 (the major natural isotope), hydrogen-2 (deuterium), and hydrogen-3 (tritium). Of the three, only hydrogen-3 is radioactive.

16. The lithium-6 of lithium deuteride undergoes fission to produce tritium (hydrogen-3), which then fuses with the deuterium (hydrogen-2) to release energy.

17. Ironically, lithium-7, with its three protons and four neutrons, was the very atom that the Cavendish scientists first split with protons, demonstrating the feasibility of nuclear fission (see chapter 5). But lithium-7 was not thought to be susceptible to fission by neutrons. What the bomb physicists did not know was

that lithium-7 could efficiently be converted to lithium-6, by a one neutron in, two neutrons out, reaction (n, 2n). Thus, lithium-7's conversion into lithium-6 further fueled the fusion reaction beyond all expectations.

18. Enrico Fermi, the physicist who had performed the original Chicago Pile experiment demonstrating the feasibility of attaining criticality, had warned Slotin that his team's screwdriver approach to measuring criticality was dangerous. At one point Fermi had cautioned Slotin, "Keep doing that experiment that way and you'll be dead within a year!" Fermi's prediction was correct (Welsome E. *The Plutonium Files*, 184).

19. The Soviet Union would eventually test detonate five bombs larger than Shrimp.

20. An atoll is a circular ring of coral islands formed by coral growth along the rim of an extinct volcano that is gradually sinking into the ocean. Atoll islands, common in the Pacific, were made famous by the voyages of Charles Darwin, who visited many and speculated on their geological origin.

21. On both Bikini Atoll and nearby Enewetak Atoll, the United States performed a total of 67 test detonations of nuclear devices, totaling 100 megatons of TNT. This is the equivalent of detonating one Hiroshima bomb every day for 19 years. It was more explosive power than has ever been detonated during all wars combined, since the beginning of human history.

22. Weisgall J. M. *Operation Crossroads*, 107.

23. Figueroa R., and S. Harding. *Science and Other Cultures*, 106–125.

24. Weisgall J. M. *Operation Crossroads*, 36.

25. Ōishi M. *The Day the Sun Rose in the West*, 30.

26. After the accident, the *Lucky Dragon No. 5* was decontaminated and used for many years as a training vessel for the Tokyo University of Fisheries. It was ultimately sold for scrap, but before it could be demolished, a public campaign was waged to save it as a memorial. It is now on permanent display at the Tokyo Metropolitan Museum's Daigo Fukuryū Maru Exhibition Hall.

27. According to a commission set up in 1994 by President Clinton to study this incident and other human-subject radiation projects conducted by the US government, Project 4.1 was created for the following clinical and research purposes: "(1) evaluate the severity of radiation injury to the human beings exposed; (2) provide for all necessary medical care; and (3) conduct a scientific study of radiation injuries to human beings." See *Final Report of the Advisory Committee on Human Radiation Experiments*, chapter 12, part 3, "The Marshallese." Available online at http://biotech.law.lsu.edu/research/reports/ACHRE. Scholars have noted that the affected islanders were enrolled automatically into the research study in violation of clinical and bioethical principals, even as judged against the practices and requirements of medical ethics at that time (Johnston B. R., and H. M. Barker. *Consequential Damages of Nuclear War*). Given the stated link between clinical care and research in Project 4.1's stated goals, treatment was likely conditioned on participation in the study, giving affected islanders no reasonable avenue to opt out of the research.

28. Ōishi M. *The Day the Sun Rose in the West*, 38.

29. It is suspected that the Soviets learned about deuterium-tritium fusion through their spy, physicist Klaus Fuchs, who had worked on the Manhattan

Project. But it is not certain exactly what fusion secrets Fuchs actually passed on. In any event, the Soviets had been working on fusion bombs even before they had developed the fission bombs needed to trigger them.

30. Ōishi M. *The Day the Sun Rose in the West*, 31.

31. Rowland R. E. *Radium in Humans.*

32. Quoted in Mullner R. *Deadly Glow*, 134.

33. Mullner R. *Deadly Glow*, 135.

34. Iodine is usually added to table salt as a dietary supplement to prevent iodine insufficiency, the major cause of goiter—a disease of the thyroid gland.

35. Some marine organisms can concentrate fission products as much as 100,000 times. During the Baker fission bomb test of Operation Crossroads, in 1946, the hulls of observer ships remained highly radioactive even after moving to nonradioactive waters. It was soon found that algae and barnacles on the exterior of the hulls had concentrated the fission products to such an extent that the sailors' bunks had to be moved away from the interior hull walls so they wouldn't be overexposed to radiation while they slept (Weisgall J. M. *Operation Crossroads*, 232–233).

36. Baumann E. "Ueber das normale Vorkommen von Jod."

37. Advisory Committee on Human Radiation Experiments. *The Human Experiments*, 371.

38. Figueroa R., and S. Harding. *Science and Other Cultures*, 116.

39. This is usually the case. However, cesium also sticks to clay, and if soil has a fairly high clay content, the cesium will remain much longer.

40. Ōishi M. *The Day the Sun Rose in the West*, 43.

41. Ōishi M. *The Day the Sun Rose in the West*, 78.

42. The National Institute of Radiological Sciences in Chiba, Japan, administered the study, with funding from the United States.

43. Ōishi M. *The Day the Sun Rose in the West*, 64.

44. Johnston B. R., and H. M. Barker. *Consequential Damages of Nuclear War*, 115–121.

45. In 1985, the surviving islanders on Rongelap were still concerned about their safety and requested to be relocated; they were moved to Kwajalein.

46. Weisgall J. M. *Operation Crossroads*, 315.

47. Watkins A., at al. *Keeping the Promise.*

48. *Marshall Islands Nuclear Claims Tribunal: Financial Statement and Independent Auditor's Report, FY 2008 & 2009*, Deloitte & Touche LLP (March 25, 2010). The tribunal was authorized to make prorated payments of awards based on the annual funds available and, if the trust fund were depleted by the end of its statutory 15-year term, awards were not required to be paid in full. See Compact of Free Association, Title II of Pub. L. No. 99–239, 99 Stat. 1770, January 14, 1986; and the Marshall Islands Nuclear Claims Tribunal Act of 1987 (42 MIRC Ch. 1), Marshall Islands Revised Code 2004, Title 42—Nuclear Claims.

49. Barker H. M. *Bravo for the Marshallese*, 165.

50. Weisgall J. M. *Operation Crossroads*, 266.

51. There never was a Fourth Punic War, but it is unlikely that the salting of Carthage had anything to do with it.

52. Leó Szilárd's cobalt bomb idea is yet another example of his love-hate rela-

tionship with nuclear weapons. He was the first to fully recognize the potential for a fission bomb, and was actually the ghostwriter of Albert Einstein's famous 1939 letter to Roosevelt urging that the United States develop an atomic bomb. Nevertheless, after the bomb was developed, he actively lobbied against using it on the grounds that if the Soviets thought the bomb effort had been a failure, then an arms race might be averted. Lieutenant General Leslie Groves, military director of the Manhattan Project, found Szilárd's vacillations a "continually disruptive force," and Szilárd's behavior no doubt contributed to the widespread view of many military officials, as well as President Truman, that nuclear scientists were naive to think that they could produce a bomb that would never be used (Weisgall J. M. *Operation Crossroads*, 82–85; Tibbets P. W. *Return of the Enola Gay*, 181–182).

53. The distinction between a tactical and a strategic nuclear weapon is that the former has smaller destructive yields and, therefore, can be used within a battle to gain a military advantage, while strategic weapons are too large to have a role within a battle and can only be delivered by intercontinental ballistic missiles. Strategic nuclear weapons are instead used as part of a country's overall war deterrence strategy. Initially, atomic (fission) bombs, such as those dropped on Hiroshima and Nagasaki, were considered strategic weapons, but in the advent of hydrogen (fusion) bombs, such relatively low yield nuclear weapons are now considered to be of only tactical value.

CHAPTER 9: AFTER THE DUST SETTLES

1. Oe, K. *Hiroshima Notes*, 46.
2. *Spokesman-Review*. "Leukemia Claims A-Bomb Children."
3. Figueroa R., and S. Hardig. *Science and Other Cultures*, 117.
4. Cullings H. M., et al. "Dose estimates."
5. Statistical power can be defined as the probability that a study will produce a significant finding. The word *significant*, in statistics, means that the finding is not likely to be the result of mere chance. Significant should not be misconstrued as meaning "important." Significant findings can be unimportant, while nonsignificant findings may reveal something new and extremely important. Statistical significance can be calculated, but importance is in the eye of the beholder.
6. Lindee M. S. *Suffering Made Real*, 5.
7. About 40% of the atomic bomb survivors were still alive as of 2013 (Cullings H. M. "Impact on the Japanese atomic bomb survivors").
8. Douple E. B. et al. "Long-term radiation-related effects."
9. Bernstein P. L. *Against the Gods*, 82.
10. Johnson S. *The Ghost Map*.
11. Lindee M. S. *Suffering Made Real*, 43.
12. Dr. Matasubayashi's original report of his findings is currently on file in the Radiation Effects Research Foundation's library in Hiroshima.
13. Dando-Collins S. *Caesar's Legion*, 211–216.
14. If one cohort became severely depleted of soldiers, it might be merged with

another similarly depleted cohort to produce a single cohort with full fighting strength.

15. These alleged associations are fictitious and serve only as examples to illustrate the point.

16. For technical description of study designs, see: Rothman K. J. *Epidemiology*.

17. For discussion of biases, see: Szklo M., and F. J. Nieto. *Epidemiology*.

18. There is another type of study that the public often sees reported in the media, but that epidemiologists have very little faith in. It's called an *ecological study*, from the Greek-derived word *ökologie* (meaning "place of residence" or "community") because this type of study compares rates of exposures with rates of disease between different communities and looks for correlations between exposure and disease rates as evidence for an association. Although they have their place, ecological studies suffer from a conceptual fallacy that severely curtails their value for inferring individual risk (Morgenstern H. "Ecologic studies in epidemiology").

19. Lindee M. S. *Suffering Made Real*, 44.

20. Lindee M. S. *Suffering Made Real*.

21. The Radiation Effects Research Foundation, of Hiroshima and Nagasaki, is a private nonprofit foundation funded by the Japanese Ministry of Health, Labor and Welfare (MHLW) and the United States Department of Energy (DOE).

22. It is important to account for age and gender when comparing two populations for cancer risk because they are major risk determinants. Differences in age and gender distribution between two populations would bias the comparison and result in erroneous conclusions. For example, a population with fewer elderly women would obviously have a much lower breast cancer rate, but brain tumor rates would also be expected to differ because gender and age are risk factors for brain tumors as well.

23. Non-melanoma skin cancers, which are very common and are typically not fatal, are not included in this risk estimate. These cancers are more closely associated with exposure to sunlight than to ionizing radiation. Thus, it is difficult to determine non-melanoma cancer risk from ionizing radiation without also having a reliable assessment of sunlight exposure. Also, these cancers are of less concern because they rarely kill people. For these reasons, non-melanoma skin cancers are not included in the risk estimate per mSv. Henceforth, when we speak about cancer risk in this book, it will be understood to exclude non-melanoma skin cancers.

24. NAS/NRC. *Health Risks of Radon*.

25. E. B. Lewis was among the first scientists to propose that simple multiplication of the risk per unit dose by the dose received produced a reliable estimate of cancer risk (Lewis E. B. "Leukemia and ionizing radiation").

26. Whole-body spiral computed tomography (CT) scanning involves imaging all or most of the body with x-rays to produce many neighboring cross-sectional images that can be combined by computers to produce a three-dimensional image of the body's internal organ structures. Such scans are sometimes suggested for healthy individuals who have no symptoms of illness as a screening tool to detect the early stages of disease (see chapter 13).

27. Mossman K. L. *Radiation Risks in Perspective*, 47–64.

28. Atomic bomb survivors also received their doses at an extremely high dose

rate, which is known to be more damaging than lower dose rate exposures by at least twofold.

29. Kleinerman R. A., et al. "Self-reported electrical appliance use."

30. Boice J. D. "A study of one million U.S. radiation workers and veterans."

31. A close second would be cigarette smoking, for which a tremendous amount of data exists. Cigarette smoke, however, contains a variety of different chemical compounds; none of these, individually, has been as thoroughly characterized a carcinogen as radiation has been.

CHAPTER 10: BREEDING SEASON

1. Inheritable and heritable are synonyms. Both adjectives mean the ability to be inherited. Since most people are familiar with the verb "to inherit," we will use the adjective "inheritable" in our discussion here. (In scientific writing, the adjective of choice is "heritable.")

2. Henig R. M. *The Monk in the Garden.*

3. In genetics, the term *hybrid* can mean somewhat different things, depending upon the context. For the purposes of this chapter, *hybrid cross* means mating individuals with different traits to produce offspring with a mixture of the genes that govern those traits. A *trait* is a distinguishing physical characteristic of an organism.

4. Unfortunately, Mendel never saw his findings validated. He had already been dead six years when his ratios were rediscovered.

5. Genes normally exist in different forms, just as cars come in different models. Each of these different gene forms is called a *variant*, and each variant may be associated with a slightly different version of the same biological trait (e.g., different eye color, straight or curly hair, short or tall stature, etc.). Without variant genes, all people would look the same.

6. Shine I., and S. Wrobel. *Thomas Hunt Morgan.*

7. Sturtevant A. H. *A History of Genetics*, 45–50.

8. Shine I., and S. Wrobel. *Thomas Hunt Morgan*, 16–30.

9. Morgan T. H., et al. *The Mechanism of Mendelian Heredity.*

10. True to his personality, Morgan remained as unmoved by his attainment of Nobel laureate status as he was about his family's high social standing. He skipped the Nobel award ceremony in Stockholm, saying that he wasn't much on giving speeches and couldn't afford the time away from his laboratory.

11. Kohler R. E. *Lords of the Fly.*

12. Carlson E. A. *Genes, Radiation, and Society*, 40–50.

13. Muller was very conscious of the many parallels between Rutherford's discovery of artificial transmutation (Rutherford's terminology) of atomic nuclei by particulate irradiation (see chapter 4) and his own work mutating cell nuclei with x-irradiation. He even published an article establishing the precedence of his x-ray mutation findings entitled "Artificial transmutation of the gene," to craftily place his achievement in biology on par with Rutherford's in physics (Carlson E. A. *Genes, Radiation, and Society*, 147).

14. Between 1910, the first year that fruit flies were used for genetic research,

and 1926, the year that Muller decided he needed a model with higher mutation rates, there had only been 200 different fruit fly mutations discovered by genetics researchers worldwide.

15. In fruit flies, the Y chromosome carries only the genes necessary for sperm production. Mutant fruit flies that entirely lack their Y chromosome (i.e., X0 genotype, rather than XY) look like normal males, but are sterile due to lack of sperm. In contrast, mutant mammals missing their Y chromosome lack all male sex characteristics, and display the physical traits of a female.

16. To say that it increased exponentially with dose means that the mutation rate increased faster than the rate of the temperature change. Mathematically speaking, it increased by some power (i.e., exponent) of the temperature change.

17. Carlson E. A. *Genes, Radiation, and Society*, 142.

18. Carlson E. A. *Genes, Radiation, and Society*, 143.

19. Actually, in his first experiment, Muller did not have enough mutations to demonstrate linearity at low doses. He established the linear relationship later with the help of an associate, Frank Blair Hanson, who was working in Muller's laboratory while on sabbatical leave from Washington University in St. Louis. It was Hanson who upsized the experiment to achieve sufficient statistical power to measure mutation rates at lower doses. And it was Hanson who also showed that similar results could be obtained when radium, rather than x-rays, was used to irradiate the flies, suggesting that the same effect could be caused by different types of ionizing radiation, and was not specific to x-rays. (Carlson E. A. *Genes, Radiation, and Society*, 154.) Geneticist Nikolai Timofeef-Ressovsky (1900–1981) independently produced similar findings supporting the linearity of inheritable mutations in the 1930s.

20. Glad J. "Hermann J. Muller's 1936 letter to Stalin."

21. Muller H. M. *Out of the Night*.

22. Stanchevici D. *Stalinist Genetics*.

23. The idea that characteristics an individual acquired during its lifetime can be inherited by its offspring was described much earlier by Jean-Baptiste Lamarck (1744–1829). In 1936, the concept had long since been discredited by mainstream geneticists, but the Soviets, nevertheless, made *Lamarckism* the cornerstone of their agricultural genetics program, and instituted reforms based on it; these would ultimately precipitate famine.

24. Carlson E. A. *Genes, Radiation, and Society*, 320.

25. In the post-Stalin era, Vavilov's reputation was rehabilitated and he was honored on a Soviet postage stamp issued in 1977. A minor planet has been named after him, and his name has been given to a crater on the moon, an honor he shares with Roentgen (see chapter 2) and Becquerel (see chapter 3).

26. The aspect of Darwin's theory that some found most objectionable was his view that humans were just another type of animal under the same selective pressures as all others, as well as his conclusion that humans and the modern great apes must have descended from a common apelike ("monkey") ancestor. Hence, the trial of John Scopes, who allegedly violated school board policy and legal statutes by teaching Darwin's theory in school, was euphemistically called the "Monkey Trial." Scopes was found guilty and fined $100 in a ruling that was later set aside on appeal due to a legal technicality.

27. A *gene pool* is the stock of various genes, in all their variations, that are distributed among the individuals within an interbreeding population.

28. Graham L. R. "The eugenics movement in Germany and Russia in the 1920s."

29. The Nazi SS physician Horst Schumann experimented with radiation sterilization on male concentration camp prisoners, employing techniques similar to the sterilization of rams with radiation (see chapter 6). In April 1944, he sent a report to Heinrich Himmler entitled, *The Effect of X-Ray Radiation on the Human Reproductive Glands*. The report concluded that surgical castration was superior to radiation in that it was faster and had a more certain outcome.

30. Because of the well-known atrocities committed by the Nazis in the name of eugenics, some think eugenics was a Nazi idea, but that is far from the case. The eugenics movement had strong roots in America, and many prominent people had originally supported the movement, including luminaries such as Theodore Roosevelt, Woodrow Wilson, Oliver Wendell Holmes, and Alexander Graham Bell (Black E. *War Against the Weak*).

31. Graves J. L. *The Emperor's New Clothes*.

32. Carlson E. A. *Genes, Radiation, and Society*, 336.

33. Grobman A. B. *Our Atomic Heritage*, 7.

34. Muller's quotes in this paragraph come from: Carlson E. A. *Genes, Radiation, and Society*, 255 (emphasis is Muller's).

35. This project was kept secret during the war because it was believed, if people knew the military was heavily investing in such a radiation biology research project during wartime, they would suspect that an atomic bomb project was in the works. After the war ended, however, the project was declassified and its findings made public.

36. Most radiation workers in the Manhattan Project were men.

37. Charles D. R., et al. "Genetic effects of chronic x-irradiation."

38. It should be noted that the mSv is a dose equivalent unit intended exclusively for human radiation protection purposes. It really isn't proper to use it in the context of mouse irradiation experiments, where mGy would be the appropriate unit for dose to a mouse (or a fruit fly). However, under the irradiation circumstances of these experiments, the mGy doses delivered to the mice would correspond to the stated dose equivalents in mSv for humans. So, for the sake of simplicity and consistency of our narrative, we will state all mGy doses delivered to mice in terms of the comparable human mSv doses.

39. Lindee M. S. *Suffering Made Real*, 74.

40. Lindee M. S. *Suffering Made Real*, 74.

41. Hall E. J., and A. J. Giaccia. *Radiobiology for the Radiologist*, 162–164.

42. This might be one of the reasons that fruit flies had a much higher inheritable mutation rate than the mice. Muller found that lowering the dose rate did not lower the mutation rate in fruit flies, though it did in mice, for reasons that are still not clear (Carlson E. A. *Genes, Radiation, and Society*, 254).

43. Mossman K. L., and W. A. Mills. *The Biological Basis of Radiation Protection Practice*, 177–178.

44. Carlson E. A. *Genes, Radiation, and Society*, 128–129.

45. Carlson E. A. *Genes, Radiation, and Society*, 388.

46. Remarkably, Muller would be one of the last scientists in the world to learn of Watson's discovery. By 1953, Muller was in secluded retirement in Hawaii and was having some difficulty receiving his scientific journals. So it wasn't until fellow Nobel Laureate and friend Linus Pauling visited and told him, in late 1954, that Muller learned of the one-year-old discovery.

CHAPTER 11: CRYSTAL CLEAR

1. Dahm R. "Discovering DNA."
2. Jenkin J. *William and Lawrence Bragg*, 83 and 125.
3. Jenkin J. *William and Lawrence Bragg*, 164.
4. William Lawrence Bragg's collection of over 500 shell specimens currently resides at the Manchester Museum of the University of Manchester, UK.
5. Jenkin J. *William and Lawrence Bragg*, 246.
6. Jenkin J. *William and Lawrence Bragg*, 147–148.
7. Jenkin J. *William and Lawrence Bragg*, 146.
8. Jenkin J. *William and Lawrence Bragg*, 153.
9. Reeves R. *A Force of Nature*, 23.
10. Reeves R. *A Force of Nature*, 30.
11. Berger H. *The Mystery of a New Kind of Rays*, 34.
12. Jenkin J. *William and Lawrence Bragg*, 155.
13. A prior experiment by Avery, MacLeod, and McCarty in 1944, showing that purified DNA is sufficient to genetically transform bacteria, had suggested the same conclusion, but Hershey and Chase were the first to demonstrate, in 1952, that DNA was actually entering the host cells.
14. Jenkin J. *William and Lawrence Bragg*, 301.
15. Wilson's pulses are one of the first examples of scientists recognizing the dual nature of electromagnetic waves, in which they behave like both particles and waves. At this time, Einstein had been formalizing the particle characteristics of electromagnetic waves into a comprehensive quantum theory, where electromagnetic energy is described as traveling through space in particle-like packets, each containing small discrete units of energy (i.e., quanta).
16. Jenkin J. *William and Lawrence Bragg*, 330.
17. Jenkin J. *William and Lawrence Bragg*, 334.
18. Bragg W. H. *Concerning the Nature of Things*, 116–159.
19. To lessen our confusion, we will continue to refer to William Lawrence Bragg as "Willie," just as his friends and family did for the rest of his life.
20. The realization that DNA is the target of radiation did not come about as a single experimental discovery; rather, it developed due to the gradual accumulation of overwhelming supporting evidence contributed by many different scientists. That evidence is too extensive to be reviewed here. Interested readers may consult a radiation biology textbook, such as *Radiobiology for the Radiologist* by Hall and Giaccia, for accounts of that work.
21. A cellular phenomenon known as *bystander effect* has been demonstrated to occur in laboratory cell culture under some conditions. This is a phenomenon in which an unirradiated cell can be affected by radiation damage occurring to a

neighboring cell. The mechanism of this phenomenon is not understood, but likely involves a DNA damage signal in one cell being transmitted to its neighbors.

22. Jenkin J. *William and Lawrence Bragg*, 320.
23. Hall K. T. *The Man in the Monkeynut Coat.*
24. Van der Kloot W. "Lawrence Bragg's role in developing sound-ranging in World War I."
25. Jenkin J. *William and Lawrence Bragg*, 377.
26. Jenkin J. *William and Lawrence Bragg*, 374.
27. Jenkin J. *William and Lawrence Bragg*, 374–375.
28. http://www.findagrave.com/cgi-bin/fg.cgi?page=gr&GRid=34760878
29. Jenkin J. *William and Lawrence Bragg*, 340.
30. Jenkin J. *William and Lawrence Bragg*, 379.
31. Reeves R. *A Force of Nature*, 92.
32. Jenkin J. *William and Lawrence Bragg*, 402.
33. Crowther J. G. *The Cavendish Laboratory: 1874–1974.*
34. Rutherford mentored 11 scientists who later won the Nobel Prize.
35. Reeves R. *A Force of Nature*, 65.
36. German scientist Fritz Haber won the Nobel Prize for artificially synthesizing ammonia from nitrogen gas. Such fixation of nitrogen from atmospheric gas was previously only possible through the metabolic processes of anaerobic bacteria. Ammonia is an important ingredient in the production of fertilizers, so the discovery had a huge economic impact for world agriculture. Recruited by the German army to produce chemical weapons during World War I, Haber suffered postwar persecution from the Nazis because he was a Jew, and emigrated first to England (where Rutherford refused to shake his hand), and then to Palestine, where he died shortly before World War II broke out.
37. Kendrew J. C. *The Thread of Life*, 44.
38. Watson J. D. *The Double Helix*, ix, (emphasis added).
39. Watson J. D. *The Double Helix*, 33.
40. Watson J. D., and F.H.C. Crick. "A structure for deoxyribose nucleic acid."
41. Jenkin J. *William and Lawrence Bragg*, 435–436.
42. Hall K. T. *The Man in the Monkeynut Coat,* 11.
43. Sayre A. *Rosalind Franklin & DNA.*
44. Jenkin J. *William and Lawrence Bragg*, 436.
45. Watson J. D. *The Double Helix*, 34 and 69.
46. Watson J. D. *The Double Helix*, viii.
47. Watson J. D. *The Double Helix*, 140.
48. A more comprehensive and academically rigorous history of the discovery of the double helix, and all the scientific discoveries leading up to it, can be found in *The Path to the Double Helix,* by historian of science Robert Olby.
49. DeMartini D. G., et al. "Tunable iridescence in a dynamic biophotonic system."

CHAPTER 12: SILENT SPRING

1. Iqbal A. "Invisible Killer Invades Home."
2. Iqbal A. "Invisible Killer Invades Home."

3. It had been known since the 1970s that the air in houses built on radioactive tailings—the waste generated from the milling of uranium and radium—could contain high concentrations of radon, but it was not known that radon from natural ground sources could reach the levels found in the Watras home. In 1983, the US Environmental Protection Agency set an action level for radon in homes of 4 pCi/L (150 Bq/m^3), as it was mandated to do by the Federal Uranium Mill Tailings Radiation Control Act of 1978. An action level is the level above which some type of remediation is thought warranted.

4. Iqbal A. "Invisible Killer Invades Home."

5. Iqbal A. "Invisible Killer Invades Home."

6. NAS/NRC. *Health Risks of Radon.*

7. Lubin J., et al. "Design issues in epidemiological studies."

8. Haber's Rule has a notorious origin. The rule was developed by the same Fritz Haber who won the Nobel Prize in Chemistry for discovering a synthetic procedure to fix atmospheric nitrogen, and then went on to work on perfecting gas warfare weapons for the Germans during World War I. He formulated his famous rule in order to estimate the concentrations of mustard and chlorine gas needed to kill enemy troops under various battlefield exposure time scenarios. After he died, the Nazis used Haber's rule to determine doses of the gas Zyklon B needed to kill Jews in concentration camps. Sadly, some of Haber's relatives were among the Jews exterminated. The rule has since been found applicable to a variety of health effects from a number of different airborne agents, and is an important analytical tool routinely used by respiratory toxicologists for exposure assessments.

9. The WLM unit is technically defined as the amount of radon exposure that a miner would receive during a month of mining work (170 working hours) at a specific air radioactivity concentration called a Working Level (~200 pCi/L air).

10. Mossman K. L. *Radiation Risks in Perspective*, 53–64.

11. One of the rare times when LNT was put aside for determining low-dose carcinogen risk was the case of bladder cancer produced by the artificial sweetener saccharin. After being banned from the market for many years because high-dose studies in rats found that it could produce bladder cancer, subsequent studies of saccharin revealed that the cancer was the direct result of saccharin particles precipitating from the urine in the bladder. No cancer occurred without precipitates. Since saccharin precipitates cannot form at normal dietary concentrations no matter how much diet soda you drink, the high-dose situation in rats was deemed irrelevant to the very low doses humans are exposed to in artificially sweetened food and drinks. Therefore, the LNT model was dropped and saccharin was returned to the market (Rodricks J. V. *Calculated Risks*, 189–190).

12. Environmental Protection Agency. *EPA Assessment of Risks from Radon in Homes.*

13. Statisticians sometimes refer to this type of risk representation as an *effort-to-yield* measure.

14. Environmental Protection Agency. *EPA Assessment of Risks from Radon in Homes.*

15. The precise epidemiological definition of the number needed to harm is that it is the inverse (reciprocal) of the attributable risk, which, in turn, is defined as

the difference in rate of an adverse health effect between an exposed population and an unexposed population.

16. Gigerenzer G. *Risk Savvy*.

17. Odds ratios and relative risks are measures of association that epidemiologists commonly use to quantify risk levels determined through case-control studies and cohort studies, respectively.

18. Environmental Protection Agency. *EPA Assessment of Risks from Radon in Homes*.

19. American Cancer Society. *Cancer Facts and Figures 2014*.

20. This is the 2012 estimate for Americans over 18 years old, according to the Centers for Disease Control and Prevention (CDC).

21. Seasonal influenza deaths vary widely but average about 20,000 per year in the United States.

22. Slovik P. *The Perception of Risk*.

23. Kabat G. C. *Hyping Health Risks*, 111–145.

24. Kabat G. C. *Hyping Health Risks*, 143; the original source is Stat Bite. "Causes of lung cancer in nonsmokers."

25. Kabat G. C. *Hyping Health Risks*, 116.

26. Kabat G. C. *Hyping Health Risks*, 116.

27. Kabat G. C. *Hyping Health Risks*, 115.

28. Kabat G. C. *Hyping Health Risks*, 120–123.

29. Abelson P. H. "Radon today."

30. Abelson P. H. "Radon today."

31. The EPA continues to use the older unit of pCi/L, while most of the rest of the world uses the contemporary international standard unit of Bq/m^3. The use of the older unit for setting radon standards is unfortunate because it results in large rounded numbers, like 100 Bq/m^3, being converted into small, seemingly precise values, like 2.7 pCi/L. Using these smaller and apparently precise numbers suggests to the public that even trace quantities are dangerous and that the hazard level is known with a high degree of precision; however, neither is the case. For this reason, it is the author's opinion that the United States should abandon the old unit (pCi/L) and adopt the international unit (Bq/m^3) for purposes of home radon control.

32. World Health Organization. *WHO Handbook on Indoor Radon*.

33. Abelson P. H. "Radon today."

34. The Watras house had 16 working levels (WL) in the living room. Assuming 17 hours per day of occupancy and a 30-day month, this amounted to 575 WLM for the one year that the Watrases inhabited the home. This 575 WLM multiplied by EPA's lung cancer risk rate for smokers of 0.097% per WLM gives a risk level of 55.8%. The 16 WL radon concentration comes from measurements made by the Philadelphia Electric Company (Iqbal A. "Invisible Killer Invades Home").

35. Kabat G. C. *Hyping Health Risks*, 121.

CHAPTER 13: A TALE OF TWO CITIES

1. For our purposes we will use the convention that *radiography* refers to a radiological medical procedure, and *radiology* refers the radiological medical

profession, which includes both diagnostic and therapeutic radiologists. (To make things even more confusing, therapeutic radiologists are often called radiation oncologists because they treat cancer—*onco* means tumor in Greek.)

2. ICRP. 1990 Recommendations.
3. Gigerenzer G. *Risk Savvy*, 164.
4. Gigerenzer G. *Calculated Risk*, 111–114.
5. Radiation dose is dependent upon breast size; large-breasted women receive a higher dose than small-breasted women.
6. Differences in radiation sensitivity between tissues is another factor contributing to effective dose calculations, in addition to the tissue's fraction of total body weight (NCRP Report. *Limitations of Exposure,* 21–23; NAS/NRC. *Health Risks from Exposure*).
7. "Number needed to treat" gets its name from its origins in assessing the value of medical treatments. But the metric works for evaluating a variety of benefits, not just the benefits from medical treatments. For this reason, the author wishes that the metric had the more universal name of number needed to benefit (NNB) to clearly juxtapose it to its counter metric, the number needed to harm (NNH). Nevertheless, we will stick with the conventional terminology so as not to confuse things even more by adding yet another term. Readers should just keep in mind that NNH is always the harm metric, and NNT is the benefit metric. See http://www.thennt.com.
8. Beil L. "To screen or not to screen"; Harris R. P. "How best to determine the mortality benefit."
9. Kopans D. B., and A. J. Vickers. "Mammography screening and the evidence."
10. The 613 per 1,000 false-positive rate is specifically for women who begin annual screening at age 50, as in this example. Women who begin screening at age 60 have a lower false-positive rate.
11. Quoted in Beil L. "To screen or not to screen."
12. Welch H. G., et al. *Overdiagnosed*, chapter 5. (Source of original data is Sone S., et al. "Results of three-year mass screening programme.")
13. Some have estimated mean effective doses of 12 mSv, with 20 mSv being the upper end of the range of doses (Brenner D. J., and C. D. Elliston. "Estimating radiation risks").
14. AAPM Task Group 217. *Radiation Dose from Airport Scanners.*
15. Twombly R. "Full-body CT screening"; Douple E. B., et al. "Long-term radiation-related effects."
16. Gigerenzer G. *Calculated Risks*, 68–70.
17. Beil L. "To screen or not to screen."
18. NCRP Report. *Second Primary Cancers.*

CHAPTER 14: SORRY, WRONG NUMBER

1. Comments made by residents at the Scarborough Town Council meeting were obtained from a story by Michael Kelly in the *Scarborough Leader*, "Cell Tower Proposal Back to Square One."

2. Baan R., et al. "Carcinogenicity of radiofrequency electromagnetic fields."
3. Schüz J., et al. "Cellular telephone use and cancer risk."
4. INTERPHONE Study Group. "Brain tumor risk."
5. A decile is one of ten equal groups of individuals. Dividing individuals into deciles according to rank order is a method of splitting ranked data into subsections, such that individuals in each decile are similar for the variable being assessed. The top decile here means the group with the highest radio wave dose to their brains.
6. Lifetime risk rate is 0.6% for brain cancer or other nervous system cancer, according to SEER 2009–2011 data for the United States: http://seer.cancer.gov/statfacts/html/brain.html.
7. Source: Pew Research Center's Internet & American Life Project, April 17–May 19, 2013 Tracking Survey.
8. Deltour I., et al. "Time trends in brain tumor incidence rates"; Deorah S., et al. "Trends in brain cancer incidence"; Little M. P., et al. "Mobile phone use and glioma risk."
9. Hill A. B. "The environment and disease."
10. Kabat G. C. *Hyping Health Risks*, 144.
11. In particular, anticell phone activists often cite a specific report, by H. Lai and N. P. Singh ("Acute low-energy microwave exposure increased DNA single-strand breaks in rat brain cells") that claims radio waves with wavelengths similar to those used by cell phones produce DNA damage in the cells of rat brains. Yet, these same activists often fail to mention that a highly-competent team of well-respected radiation biologists at Washington University in St. Louis, subsequently could not reproduce the Lai and Singh findings under controlled experimental conditions. (Malyapa R. S., et al. "DNA damage in rat brain cells.") Additionally, the Washington University scientists found that the comet assay used by Lai and Singh to measure DNA damage in the isolated brain cells was greatly affected by the specific method by which the rats were euthanized, presumably due to differences in DNA degradation that occurs while the brains are being harvested from the animals. This suggests that the comet assay may be particularly prone to DNA degradation artifacts introduced during the cumbersome process of recovering the cells from the rat brains, thus raising further doubt about the validity of the findings in the Lai and Singh report.
12. Fagin D. *Toms River*, 150–151.
13. A small group of scientists advocates the concept that low doses of radiation are beneficial to health. The theory that high-dose toxins can actually be beneficial at low doses (as opposed to being merely benign) is called *hormesis*. The concept has failed to gain traction among mainstream scientists, particularly since its tenets share some resemblance to the now discredited field of homeopathy in medicine. Hormesis has been rejected by the National Academy of Sciences BEIR Committees, because adoption of its unproven principles by the radiation protection community could increase exposure limits to potentially dangerous levels.
14. Since the IARC ruling, two more case-control studies, from France and Sweden, have reported an association between cell phone use and brain cancer (Coureau G., et al. "Mobile phone use and brain tumours in the CERENAT case-

control study"; Hardell L., and M. Carlberg. "Mobile phone and cordless phone use and the risk for glioma—analysis of pooled case-control studies in Sweden, 1997–2003 and 2007–2009"). But both studies used patient interviews and questionnaires to assess cell phone exposures. As we've already seen, this approach to estimating brain doses is highly prone to recall bias (see chapter 9 and Kleinerman R. A, et al. "Self-reported electrical appliance use and risk of adult brain tumors"), and thus is likely to produce a false association. Also, in the French study, controls had a statistically significantly higher educational level than did the cases ($p < 0.001$), suggesting that cases represent a lower socioeconomic status than controls; this could, in turn, indicate other significantly different exposures between the two groups beyond just their cell phone usage. For these reasons, the INTERPHONE study, which the IARC largely based its 2011 decision on, remains the most reliable cell phone study to date, and these two weaker studies neither support nor refute the IARC's earlier ruling on the carcinogenic potential of cell phones.

15. Gaynor M. "The quietest place on earth."

16. Crockett C. "Searching for distant signals."

17. Röösli M. "Radiofrequency electromagnetic field exposure." See table 1 of this article for a listing of the double-blind studies.

18. In addition to x-rays and gamma rays (ionizing radiations), Hermann Muller also tested radio waves (nonionizing radiation). The military had asked him to conduct these tests because of rumors that radar workers were prone to sterility. These rumors were so rampant that, in England, some male soldiers specifically requested that they be assigned radar duties the week before a weekend leave. Muller found that radio waves produced neither sterility nor inheritable mutations in fruit flies, and condoms remained the most effective birth control option for soldiers (Carlson E. A. *Genes, Radiation, and Society*, 283).

CHAPTER 15: HOT TUNA

1. Brown E. "Radioactive Tuna from Fukushima? Scientists Eat it Up."

2. With the exception of C-14, all the naturally abundant isotopes of carbon and nitrogen are stable.

3. Narula S. K. "Sushiomics."

4. Madigan D. J., et al. "Pacific bluefin tuna transport Fukushima-derived radionuclides."

5. At least one nuclear bomb test has occurred since the Fukushima accident in 2011. On February 12, 2013, North Korea conducted an underground bomb test that was estimated to have an explosive power similar to that of the Hiroshima and Nagasaki bombs (approximately 10 KT each).

6. Many element symbols use letters that don't even appear in the English word. This is because most of the symbols are abbreviations from Latin words. In this case, Na for sodium comes from *natrium* (Latin), and K for potassium is an abbreviation of *kalium* (also Latin).

7. Keane A., and R. D. Evans. *Massachusetts Institute of Technology Report*.

8. Remember that cesium-134 and cesium-137 will have the same biological

half-life, but different radiological half-lives. This is because the biological half-life is determined by cesium's chemistry, which doesn't differ among cesium isotopes. Radiological half-lives, in contrast, are driven by nuclear forces that are specific to particular isotopes.

9. These kilogram units represent dry weights (i.e., after the sample has been desiccated to remove water), so the radioactivity has been concentrated in the dry tissues. This is done so that varying moisture content between samples does not affect the findings.

10. A more detailed description of the rationales, procedures, and calculations that the FDA uses to set limits for radioactivity contamination in food can be found in: *Accidental Radioactive Contamination of Human Food and Animal Feeds: Recommendations for State and Local Agencies*, by the US Department of Health and Human Services, Food and Drug Administration.

11. To demonstrate its commitment to food safety, in April 2012, Japan's Ministry of Health, Labour and Welfare reduced its limit for radioactive cesium in seafood from 500 to 100 Bq per kg (one-fifth of its previous limit, and one-twelfth of the United States' limit).

12. Humans have about 50 Bq per kg of K-40 in their bodies overall. (Strom D. J., et al. "Radiation doses to Hanford workers.") Assuming that the potassium is enriched in muscle and that muscle represents 55% of body mass, the maximum muscle concentration would be expected to be somewhere between 75 and 100 Bq per kg.

13. Medical pundits frequently dredge up the "banana dose equivalent" when trying to make dose comparisons to other radiation exposures. Usually the purpose is to downplay the risk of the exposure in question by comparing it to the seemingly nonthreatening banana. One television "expert" recently made a stir by saying that the radiation dose from a chest x-ray was comparable to the dose received from eating a banana. The truth, however, is that a typical chest x-ray has an effective dose of about 0.02 mSv, while eating a banana delivers an effective dose of zero mSv. It's true that 0.02 mSv is a very low dose (<1% of typical annual background dose), but it is not zero. Experts that use banana dose equivalents are either being disingenuous in their comparison or are ignorant about bananas.

14. Fisher N. S., et al. "Evaluation of radiation doses and associated risk."

15. Oddly enough, the biggest exposure to radioactivity from eating seafood comes not from the Cs-134, Cs-137, or K-40, but rather from polonium-210 (Po-210), one of the two elements that the Curies discovered in pitchblende. In seawater, Po-210 originates from uranium-238 decay on the sea floor. It is ubiquitous in all ocean waters, and is present in all sea life. In seafood, Po-210 delivers an effective dose to consumers that is hundreds of times greater than the dose from the cesium in the bluefin, but still well below safe levels as defined by federal regulations. Like K-40, Po-210 is just one other natural radioisotope that contributes to our annual background radiation dose. And the sea is not the only source of radioactivity in food. In fact, you would be hard pressed to find any food completely devoid of radioactivity. But there is no evidence that any of the natural radioactivity in food is dangerous, no matter how much you eat.

16. Foundation for Promotion of Cancer Research. *Cancer Statistics in Japan—2013.*

17. Rowland R. E. *Radium in Humans*, 7.

18. Chen J. "Review of radon doses."

19. NCRP Report. *Uncertainties in the Estimation of Radiation Risks and Probability of Disease Causation.*

20. The word *uncertainty* is thrown around by a lot of different people who take it to mean different things. Some scientists don't distinguish between uncertainty and *statistical variability*. Others consider statistical variability a subtype of uncertainty. For our purposes, statistical variability is something completely different than uncertainty. Here we define statistical variability as the natural spread or heterogeneity in data. Uncertainty, in contrast, has to do specifically with facts that are unknown. The difference is that variability is an inherent property of the data, described by statistical parameters such as *variance* and *standard deviation.* Uncertainty is best defined as a lack of precise knowledge. It is uncertainty that usually explains why measures of statistical variability alone can sometimes fail to accurately predict the true frequency of adverse events. Uncertainty causes relevant information to be missing from the statistical model, and it is the enemy of probabilistic risk assessment. Uncertainty needs to be reduced by adding more precise and reliable information about the determinants of risk to the model, so that statistical variability is the only imprecision that remains in risk predictions. (Note that this type of uncertainty has nothing to do with Werner Heisenberg's *uncertainty principle*, which is a quantum mechanical theory that addresses the constraints to simultaneously measuring both the momentum and position of an atomic particle.)

21. A *black swan* is a metaphor that comes from *black swan theory*. This theory relates to any event that comes as a complete surprise because it was previously beyond the realm of everyday experience. The name originates from the old English expression, "rare as a black swan," which was intended to connote something impossible because the English were not yet familiar with any species of black swan. Then, in 1697, a black swan species was discovered in Western Australia and the expression lost its original meaning. Now the term "black swan" is taken to mean the occurrence of any event that had erroneously been ruled out simply because it had not happened before. Black swan, as used in black swan theory, now connotes faulty logic about the probability of rare events (Taleb N. N. *The Black Swan*).

22. Greene K. "No Fukushima radiation found in coastal areas."

23. http://www.ourradioactiveocean.org/

24. Crowd funding is the practice of funding a research project by raising many small donations from a large number of contributors, typically via solicitation over the Internet.

25. This chapter did not address irradiated food because irradiated food does not contain any manmade radioactivity. Irradiated food is food that has been treated with x-rays or gamma rays for the purpose of sterilization, so that it is free of bacterial contamination. It's similar to how pasteurization kills bacteria in milk with heat, except that radiation is used to kill the bacteria instead. But that irradiation does not make the food radioactive. There has been some discussion

as to whether irradiation of food decreases its nutritional value, but we cannot speak to that issue here. Nevertheless, we can emphatically state that irradiated food poses absolutely no radiation risk to those who eat it.

CHAPTER 16: BLUE MOON

1. Carl Pillitteri's story of the earthquake and its aftermath was conflated from his firsthand accounts reported in a television interview ("Fukushima Survivor: I've Hardly Smiled This Whole Year" on the *PBS NewsHour*; March 9, 2012), and in a live monologue performance ("Fog of Disbelief" on *The Moth* podcast; March 11, 2014).

2. A 1-unit increase on the moment magnitude scale amounts to an increase of $10^{1.5}$ (32), rather than 10^1 (10).

3. It would later be determined that the quake damaged a total 383,429 buildings in Japan (Mahaffey J. *Atomic Accidents*, 390).

4. Hough S. *Predicting the Unpredictable.*

5. This description of earthquake risk is drawn largely from a highly cogent depiction of the current state of earthquake prediction science in *The Signal and the Noise* by Nate Silver (chapter 5: "Desperately Seeking Signal").

6. Birmingham L., and D. McNeill. *Strong in the Rain*, 7–8.

7. NBC News. "How the Quake Shifted Japan."

8. It's been estimated that the energy released by the earthquake exceeded one million kilotons (KT) of TNT. For comparison, the energy released by both the Hiroshima and Nagasaki bombs combined was no more than 30 KT (Birmingham L., and D. McNeill. *Strong in the Rain*, 6).

9. Council of the National Academies. *Lessons Learned from the Fukushima Nuclear Accident*, 112–119.

10. Council of the National Academies. *Lessons Learned from the Fukushima Nuclear Accident*, 118.

11. Mahaffey J. *Atomic Accidents*, 392.

12. Council of the National Academies. *Lessons Learned from the Fukushima Nuclear Accident,* 119–130.

13. Mahaffey J. *Atomic Accidents*, 393.

14. Recent archeological evidence has suggested that the Jōgan tsunami of 869 AD was even more destructive than the 1896 tsunami.

15. In 2008, TEPCO did trial calculations of possible tsunami heights for the Fukushima Daiichi plant using fault models for an earthquake postulated by the Headquarters for Earthquake Research Promotion (HERP) and estimated ocean waves 8.4–10.2 meters (27–33 feet) in height. Some further fortification of the plant against tsunamis occurred following these calculations. (Council of the National Academies. *Lessons Learned from the Fukushima Nuclear Accident*, 94–100.) Seismologists at Japan's National Institute of Advanced Industrial Science and Technology had proposed, in 2009, that the Jōgan earthquake be used as the standard for nuclear power plant safety. Nothing came of the proposal (Birmingham L., and D. McNeill. *Strong in the Rain*, 39).

16. Birmingham L., and D. McNeill. *Strong in the Rain*, 81.

17. This was the first time that the Japanese people had heard an address by their emperor broadcast nationwide since Hirohito publicly announced Japanese surrender in 1945, thereby officially ending World War II.

18. Mahaffey J. *Atomic Accidents*, 342–356.

19. US Nuclear Regulatory Commission. *Reactor Safety Study*.

20. U.S. Nuclear Regulatory Commission. *Risk Assessment Review*.

21. http://www.world-nuclear.org/info/current-and-future-generation/nuclear-power-in-the-world-today/

22. The 430 number is as of 2014 and comes from the World Nuclear Association.

23. Perrow C. *Normal Accidents*.

24. Perrow C. *Normal Accidents*.

25. Council of the National Academies. *Lessons Learned from the Fukushima Nuclear Accident*.

26. Mahaffey J. *Atomic Accidents*, 400.

27. Perrow C. *Normal Accidents*, 157–162.

28. Taleb N. N. *The Black Swan*.

29. US Nuclear Regulatory Commission, Advisory Committee on Reactor Safeguards, Subcommittee on Regulatory Policies and Practices: License Renewal, ACRS-T-1789, March 26, 1990, 153-154.

30. Smith J., and N. A. Beresford. *Chernobyl: Catastrophe and Consequences*.

31. Mahaffey M. *Atomic Accidents*, 357–375.

32. Cardis E., et al. "Estimates of the cancer burden in Europe from radioactive fallout."

33. Cardis E., et al. "Estimates of the cancer burden in Europe from radioactive fallout."

34. Reiners C., et al. "Twenty-five years after Chernobyl."

35. Tokonami S., et al. "Thyroid doses for evacuees from the Fukushima nuclear accident."

36. Mahaffey J. *Atomic Accidents*, 374.

37. Bardi J. "Chernobyl Cleanup Workers Had Significantly Increased Risk of Leukemia."

38. Mahaffey J. *Atomic Accidents*, 400.

39. The Japanese government startled some people when it upped the permissible dose limits for the radiation workers from 50 to 250 mSv during the Fukushima crisis, seemingly backpedaling on what was considered safe working conditions. Actually, it is standard policy for both Japan and the United States to raise the exposure limits for radiation workers in the case of a worker performing potentially lifesaving activities during a radiation emergency. In the United States, it would have been permissible to raise the worker limit as high as 500 mSv in such a situation. Even 500 mSv poses no risk of radiation sickness, and increases lifetime cancer risk by just 2.5% (i.e., 25% to 27.5%).

40. Birmingham L., and D. McNeill. *Strong in the Rain*, 87.

41. The hydrogen explosions that occurred in the reactors at Fukushima were chemical explosions from the combustion of hydrogen gas, not nuclear explosions caused by fusing hydrogen nuclei.

42. Birmingham L., and D. McNeill. *Strong in the Rain*, 144–155.
43. http://www.world-nuclear.org/info/Safety-and-Security/Safety-of-Plants/Fukushima-Accident/
44. There is probably not much sense in assuming that doses will remain constant for more than five years, because we would anticipate a substantial reduction of dose due to the combination of decay and dissipation. If it doesn't actually decrease (perhaps because of further radioactivity releases from the plant), the risk levels for later years will need to be rethought.
45. Fackler M. "Tsunami Warning, Written in Stone."
46. CBS News. "Ancient Stone Markers Warned of Tsunamis."

CHAPTER 17: THE THINGS THEY CARRIED

1. The last Mark 53 (B53) bomb in the United States nuclear stockpile was dismantled in 2011. It is currently on display at the National Museum of the Air Force at Wright-Patterson AFB, near Dayton, OH.
2. The Garrett County Historical Society maintains a collection of artifacts from the crash site, as well as memorabilia from the rescue efforts at their Grantsville Museum. There is also a map and self-directed driving tour to locations of the major events of the rescue, available at http://buzzonefour.org/pdf_files/BuzzOneFourOrg_Brochure1.pdf. (The crash site itself is on private property and is off limits to the public.)
3. Maggelet M. H., and J. C. Oskins. *Broken Arrow*, 195–202.
4. Levi M. *On Nuclear Terrorism*, 6–9.
5. Commission on the Prevention of Weapons of Mass Destruction Proliferation and Terrorism. *World at Risk*, 43–75.
6. Allison G. *Nuclear Terrorism*, 24–29.
7. Dobson J. *The Goldsboro Broken Arrow*.
8. Dobson J. *The Goldsboro Broken Arrow*, 89–90.
9. Today in Goldsboro, there is a roadside memorial plaque in the approximate vicinity of the plane crash. The crash site itself, where one of the two bombs still remains buried, is on land that is off limits to the public.
10. McGill E. J. *Jet Age Man*, 132–133.
11. McGill E. J. *Jet Age Man*, 159–163.
12. The Strategic Air Command's motto was "Peace Is Our Profession."
13. Moran B. *The Day We Lost the H Bomb*.
14. A fusion detonation of a plutonium core is only possible when all of the conventional explosives surrounding the core fire synchronously. Anything less than perfect synchrony will fail to accomplish this because of uneven compression of the core. Although this requirement for synchronous implosion detonation presents an engineering and munitions challenge, it also makes the bombs "one-point-safe," meaning detonation of a single conventional charge will not produce a supercritical mass. The fact that no nuclear explosions occurred in Palomares despite detonation of the conventional charges, suggests that the detonations were incomplete, or at least asynchronous.

15. Mahaffey J. *Atomic Accidents*, 314–324.
16. Mahaffey J. *Atomic Accidents*, 322.
17. Schlosser E. *Command and Control.*
18. *Stars and Stripes*. "Air Force Fires 2 Nuclear Missile Corps Commanders."
19. Actually, the original rule states that the longer that something continuously occurs (e.g., the sun rising in the morning), the more likely that it will occur tomorrow; the converse is posed here to facilitate its application to the probability of an accident.
20. Lowry I. S. "Postattack population of the United States."
21. The model assumed a mean lethal dose of radiation to be 9,000 mSv. We now know that a mean lethal dose for humans is approximately 5,000 mSv, so the model greatly underestimated the deaths caused by radiation.
22. The MAD nuclear strategy amounts to a crude form of *Nash equilibrium*, a type of noncooperative game theory that was first proposed by John Forbes Nash Jr., a Princeton University mathematician. MAD posits that there is nothing to be gained by initiating a nuclear strike when an equally damaging retaliatory strike is assured. Therefore, neither side is likely to initiate hostilities. The concept of Nash equilibrium has also had a profound influence on economic theory, which resulted in Nash's receiving the Nobel Prize in Economics in 1994. Nash is one of the most famous sufferers of schizophrenia. His personal story was told in *A Beautiful Mind*, a 1998 book by Sylvia Nasar that was made into a major motion picture in 2001.
23. Glasstone S., and P. J. Dolan. *The Effects of Nuclear Weapons.*
24. "Nuke Effects" is available through Apple's App Store.
25. NUKEMAP website: http://nuclearsecrecy.com/nukemap/.
26. Jones B. "This Scary Interactive Map Shows What Happens If A Nuke Explodes In Your Neighborhood."
27. Davis T. C. *Stages of Emergency*.
28. *Duck and Cover* was an official United States government civil defense film produced in 1951, geared toward children. It starred animated character Bert the Turtle portraying the act of ducking down and covering up his body as a demonstration of a useful method to protect oneself from an atomic bomb attack. On bomb shelters, see Rose K. D. *One Nation Underground*.
29. US Department of Defense. *The Defense Science Board Permanent Task Force on Nuclear Weapons Surety: Report on the Unauthorized Movement of Nuclear Weapons*, p. 7.
30. The USAF Counterproliferation Center is funded jointly by the Defense Threat Reduction Agency and the United States Air Force.
31. Spencer M., et al. "The Unauthorized Movement of Nuclear Weapons and Mistaken Shipment of Classified Missile Components: An Assessment."

EPILOGUE: N-RAYS

1. Klotz I. M. "The N-ray affair."
2. Blondlot actually used a "prism" made from aluminum rather than glass.

3. Sir William Thomson, 1st Baron Kelvin (1824–1907), more commonly known as Lord Kelvin, was a prominent mathematical physicist and engineer.

4. Wood R. W. "The N-rays."

5. We now know that waves have some characteristics in common with particles, but "corpuscles" were thought to be actual microscopic bits of matter that were flying through space.

BIBLIOGRAPHY

AAPM Task Group 217. *Radiation Dose from Airport Scanners*. College Park, MD: American Association of Physicists in Medicine, June 2013.

Abelson, P. H. "Radon Today: The Role of Flimflam in Public Policy." *Regulation: The Cato Review of Business and Government*. Fall 1991:95–100, 1991.

Advisory Committee on Human Radiation Experiments. *The Human Experiments: Final Report of the President's Advisory Committee on Human Radiation Experiments*. New York: Oxford University Press, 1996.

Allison, G. *Nuclear Terrorism: The Ultimate Preventable Catastrophe*. New York: Holt, 2004.

American Cancer Society. *Cancer Facts and Figures 2014*. Atlanta, GA: American Cancer Society, 2014.

Aronowitz, J. N., and R. F. Robison. "Howard Kelly Establishes Gynecologic Brachytherapy in the United States." *Brachytherapy*, 9:178, 2010.

Baan, R., Y. Grosse, B. Lauby-Secretan, F. El Ghissassi, V. Bouvard, L. Benbrahim-Tallaa, N. Guha et al. "Carcinogenicity of Radiofrequency Electromagnetic Fields." *Lancet*, 12:624–625, 2011.

Badash, L. *Radioactivity in America: Growth and Decay of a Science*. Baltimore, MD: Johns Hopkins University Press, 1979.

Bardi, J. "Chernobyl Cleanup Workers Had Significantly Increased Risk of Leukemia." *Univ. of California at San Francisco News*, November 8, 2012, http://www.ucsf.edu/news/2012/11/13087/chernobyl-cleanup-workers-had-significantly-increased-risk-leukemia.

Barker, H. M. *Bravo for the Marshallese: Regaining Control in a Post-Nuclear, Post-Colonial World*, 2nd ed. Case Studies in Contemporary Social Issues. Belmont, CA: Wadsworth Cengage Learning, 2013.

Baumann, E. "Ueber das normale Vorkommen von Jod im Thierkörper." *Hoppe-Seylers Zeitschrift für physiologische Chemie*, 21:319–330, 1895.

Beil, L. "To Screen or Not To Screen: After 40 Years, Mammography's Limits Becoming Clearer." *Science News*, June 28, 2014, 23–26.

Bernstein, P. L. *Against the Gods: The Remarkable Story of Risk*. New York: John Wiley & Sons, 1996.

Berger, H. *The Mystery of a New Kind of Rays: The Story of Wilhelm Conrad Roentgen and His Discovery of X-rays*. North Charleston, SC: CreateSpace, 2012.

Birmingham, L., and D. McNeill. *Strong in the Rain: Surviving Japan's Earthquake, Tsunami, and Fukushima Nuclear Disaster.* New York: Palgrave, 2012.
Black, E. *War Against the Weak: Eugenics and America's Campaign to Create a Master Race.* Expanded ed. Washington, DC: Dialog Press, 2012.
Boice, J. D. "A Study of One Million U.S. Radiation Workers and Veterans: A New NCRP Initiative. Boice Report No. 6." *Health Physics News*, November 2012.
Brady, L. W., S. Kramer, S. H. Levitt, R. G. Parker, and W. E. Powers. "Radiation Oncology: Contributions of the United States in the Last Years of the 20th Century." *Radiology*, 219:1–5, 2001.
Bragg, W. H. *Concerning the Nature of Things.* London: G. Bell & Sons, 1929.
Brenner, D. J., and C. D. Elliston. "Estimating Radiation Risks Potentially Associated with Full-Body CT Screening." *Radiology*, 232:735–737, 2004.
Brooks, M. *13 Things That Don't Make Sense: The Most Baffling Scientific Mysteries of Our Life.* New York: Vintage Books, 2008.
Brown, D. E. *Inventing Modern America: From Microwave to the Mouse.* Cambridge, MA: MIT Press, 2002.
Brown, E. "Radioactive Tuna from Fukushima? Scientists Eat It Up." *Los Angeles Times*, May 8, 2013.
Brown, P. *American Martyrs to Science through the Roentgen Rays.* Baltimore, MD: Thomas Books, 1936.
Buchholz, M. A., and M. Cervera. *Radium Historical Items Catalog.* Oak Ridge, TN: US Nuclear Regulatory Commission, 2008.
Campbell, J. *Rutherford: Scientist Supreme.* Christchurch, New Zealand: AAS Publications, 1999.
Cardis, E., D. Krewski, M. Boniol, V. Drozdovitch, S. C. Darby, E. S. Gilbert, S. Akiba, et al. "Estimates of the Cancer Burden in Europe from Radioactive Fallout from the Chernobyl Accident." *International Journal of Cancer*, 199:1224–1235, 2006.
Carlson, E. A. *Genes, Radiation, and Society: The Life and Work of H.J. Muller.* Ithaca, NY: Cornell University Press, 1981.
Carlson, W. B. *Tesla: Inventor of the Electrical Age.* Princeton, NJ: Princeton University Press, 2013.
Cathcart, B. *The Fly in the Cathedral: How a Group of Scientists Won the International Race to Split the Atom.* New York: Farrar, Straus & Giroux, 2006.
CBS News. "Ancient Stone Markers Warned of Tsunamis." April 6, 2011.
Charles, D. R., J. A. Tihen, E. M. Otis, and A. B. Grobman. "Genetic Effects of Chronic X-irradiation Exposure in Mice." *Genetics*, 46(1):5–8, 1961.
Chen, J. "A Review of Radon Doses." *Radiation Protection Management*, 22(4): 27–31, 2005.
Clark, C. *Radium Girls: Women and Industrial Health Reform, 1910–1935.* Chapel Hill: University of North Carolina Press, 1997.
Close, F. *Particle Physics: A Very Short Introduction.* Oxford, UK: Oxford University Press, 2004.
Commission on the Prevention of Weapons of Mass Destruction Proliferation and Terrorism. *World at Risk: The Report of the Commission on the Prevention of WMD Proliferation and Terrorism.* New York: Vintage Books, 2008.

Conant, J. *109 East Palace: Robert Oppenheimer and the Secret City of Los Alamos*. New York: Simon & Schuster, 2005.
Council of the National Academies. *Lessons Learned from the Fukushima Nuclear Accident for Improving Safety of U.S. Power Plants*. Washington, DC: National Academy Press, 2014.
Coureau, G., G. Bouvier, P. Lebailly, P. Fabbro-Peray, A. Gruber, K. Leffondre, J. S. Guillamo, et al. "Mobile Phone Use and Brain Tumours in the CERENAT Case-Control Study." *Occupational and Environmental Medicine*, 71:514–522, 2014.
Crockett, C. "Searching for Distant Signals." *Science News*, 186(3):22–27, 2014.
Crowther, J. G. *The Cavendish Laboratory 1874–1974*. New York: Science History Publications, 1974.
Cullings, H. M. "Impact on the Japanese Atomic Bomb Survivors of Radiation Received from the Bombs." *Health Physics*, 106(2):281–293, 2014.
Cullings, H. M., S. Fujita, S. Funamoto, E. J. Grant, G. D. Kerr, and D. L. Preston. "Dose Estimates of Atomic Bomb Survivor Studies: Its Evolution and Present Status." *Radiation Research*, 166:219–254, 2006.
Dahm, R. "Discovering DNA: Friedrich Miescher and the Early Years of Nucleic Acid Research." *Human Genetics*, 122:565–581, 2008.
Dando-Collins, S. *Caesar's Legion: The Epic Saga of Julius Caesar's Elite Tenth Legion and the Armies of Rome*. New York: Wiley & Sons, 2002.
Davis, E. A., and I. J. Falconer. *J. J. Thomson and the Discovery of the Electron*. London: Taylor & Francis, 1997.
Davis, T. C. *Stages of Emergency: Cold War Nuclear Civil Defense*. Durham, NC: Duke University Press, 2007.
Deltour, I., C. Johansen, A. Auvinen, M. Feychting, L. Klaeboe, and J. Schüz. "Time Trends in Brain Tumor Incidence Rates in Denmark, Finland, Norway, and Sweden, 1974–2003." *Journal of the National Cancer Institute*, 101:1721–1724, 2009.
DeMartini, D. G., D. V. Krogstad, and D. E. Morse. "Membrane Invaginations Facilitate Reversible Water Flux Driving Tunable Iridescence in a Dynamic Biophotonic System." *Proceedings of the National Academy of Science*, 110(7): 2552–2556, 2013.
Deorah, S., C. F. Lynch, Z. A. Sibenaller, and T. C. Ryken. "Trends in Brain Cancer Incidence and Survival in the United States: Surveillance, Epidemiology, and End Results Program, 1973 to 2001." *Neurosurgical Focus*, 20(4):E1, 2006.
Dobson, J. *The Goldsboro Broken Arrow*. Raleigh, NC: Lulu Press, 2011.
Douple, E. B., K. Mabushi, H. M. Cullings, D. L. Preston, K. Kodama, Y. Shimizu, S. Fujiwara, and R. E. Shore. "Long-Term Radiation-Related Effects in a Unique Human Population: Lessons Learned from the Atomic Bomb Survivors of Hiroshima and Nagasaki." *Disaster Medicine and Public Health Preparedness*, 5:S122–S133, 2011.
Einstein, A. *Relativity: The Special and General Theory*. New York: Crown Publishers, 1961.
Engelmann, W. "Radium Emanation Therapy." *Lancet*, 1 (May 3, 1913):1225–1258, 1913.
Environmental Protection Agency. *EPA Assessment of Risks from Radon in*

Homes. Washington, DC: US Environmental Protection Agency, Office of Radiation and Indoor Air, 2003.
Essig, M. *Edison & the Electric Chair: The Story of Light and Death*. New York: Walker, 2003.
Fackler, M. "Tsunami Warning, Written in Stone." *New York Times,* April 20, 2011.
Fagin, D. *Toms River: A Story of Science and Salvation*. New York: Bantam Books, 2013.
Figueroa, R., and S. Harding, eds. *Science and Other Cultures: Issues in Philosophies of Science and Technology*. New York: Routledge, 2003.
Fisher, N. S., K. Beaugelin-Seiller, T. G. Hinton, Z. Baumann, D. J. Madigan, and J. Garnier-Laplace. "Evaluation of Radiation Doses and Associated Risk from the Fukushima Nuclear Accident to Marine Biota and Human Consumers of Seafood." *Proceedings of the National Academy of Sciences*, 110(26):10670–10675, 2013.
Foundation for Promotion of Cancer Research. *Cancer Statistics in Japan—2013*. Tokyo, Japan, 2013
Frame, P. W. "Tales from the Atomic Age." *Health Physics Society Newsletter*, July 1996.
Gaynor, M. "The Quietest Place on Earth." *Washingtonian Magazine*, 50(4):60–67, 2015.
Gigerenzer, G. *Calculated Risks: How to Know When the Numbers Deceive You*. New York: Simon & Schuster, 2002.
Gigerenzer, G. *Risk Savvy: How to Make Good Decisions*. New York: Viking, 2014.
Glad, J. "Hermann J. Muller's 1936 letter to Stalin." *The Mankind Quarterly*, 43(3):305–319, 2003.
Glasstone, S., and P. J. Dolan. *The Effects of Nuclear Weapons*. Washington, DC: US Department of Defense, and the Energy Research and Development Administration, 1977.
Gleick, J. *Isaac Newton*. New York: Vintage, 2004.
Gottschall, J. *The Storytelling Animal: How Stories Make Us Human*. New York: Mariner, 2012.
Graham, L. R. "The Eugenics Movement in Germany and Russia in the 1920s." *American Historical Review*, 82(5):1133–1164, 1977.
Graves, J. L. *The Emperor's New Clothes: Biological Theories of Race at the Millennium*. New Brunswick, NJ: Rutgers University Press, 2001.
Greene, K. "No Fukushima Radiation Found in Coastal Areas." *Berkeley Lab News Center*, September 2, 2014.
Greenwood, V. "My Great-Great-Aunt Discovered Francium. And It Killed Her." *New York Times* (Magazine), December 3, 2014.
Grobman, A. B. *Our Atomic Heritage*. Gainesville: University of Florida Press, 1951.
Grubbe, E. *X-Ray Treatment: Its Origins, Birth, and Early History*. St. Paul, MN: Bruce Publishing, 1949.
Hacker, B. C. *The Dragon's Tail: Radiation Safety in the Manhattan Project, 1942–1946*. Berkeley: University of California Press, 1987.

Hall, E. J., and A. J. Giaccia. *Radiobiology for the Radiologist*, 7th ed. New York: Lippincott Williams & Wilkins, 2011.

Hall, K. T. *The Man in the Monkeynut Coat: William Astbury and the Forgotten Road to the Double Helix*. Oxford, UK: Oxford University Press, 2014.

Hardell, L., and M. Carlberg. "Mobile Phone and Cordless Phone Use and the Risk for Glioma—Analysis of Pooled Case-Control Studies in Sweden, 1997–2003 and 2007–2009." *Pathophysiology*, 22:1–13, 2015.

Harris, R. P. "How Best to Determine the Mortality Benefit from Screening Mammography: Dueling Results and Methodologies from Canada." *Journal of the National Cancer Institute*, 106(11): dju317, November 12, 2014. doi:10.1093/jnci/dju317.

Henig, R. M. *The Monk in the Garden: The Lost and Found Genius of Gregor Mendel, the Father of Genetics*. New York: Mariner, 2000.

Hersey, J. *Hiroshima*. New York: Vintage, 1989.

Hill, A. B. "The Environment and Disease: Association or Causation?" *Proceedings of the Royal Society of Medicine*, 58:295–300, 1965.

Hodges, P. C. *The Life and Times of Emil H. Grubbe: The Biography of a Pioneering Chicago Radiologist*. Chicago, IL: University of Chicago Press, 1964.

Hoffman, M. "Forgotten Atrocity of the Atomic Age," *Japan Times*, August 28, 2011, p. 11.

Hough, S. *Predicting the Unpredictable: The Tumultuous Science of Earthquake Prediction*. Princeton, NJ: Princeton University Press, 2010.

ICRP (International Commission on Radiation Protection). *The 1990 Recommendations of the International Commission on Radiation Protection*. Publication 60. *Annals of the ICRP*, 21:1–3, 1991.

ICRP (International Commission on Radiation Protection). *The 2007 Recommendations of the International Commission on Radiological Protection*. Publication 103. *Annals of the ICRP*, 37:2–4, 2007.

INTERPHONE Study Group. "Brain Tumour Risk in Relation to Mobile Telephone Use: Results of the INTERPHONE International Case-Control Study. *International Journal of Epidemiology*, 39:675–694, 2010.

Iqbal, A. "Invisible Killer Invades Home: Seeping Radon Gas Increases Family's Lung-Cancer Risk." *Chicago Tribune*, October 8, 1987.

Isaacson, W. *Einstein: His Life and His Universe*. New York: Simon & Schuster Paperbacks, 2007.

Israel, P. *Edison: A Life of Invention*. New York: John Wiley & Sons, 1998.

Jacobs, C. D. *Henry Kaplan and the Story of Hodgkins's Disease*. Stanford, CA: Stanford General Books, 2010.

Jenkin, J. *William and Lawrence Bragg, Father and Son: The Most Extraordinary Collaboration in Science*. Oxford, UK: Oxford University Press, 2011.

Johnson, S. *The Ghost Map: The Story of London's Most Terrifying Epidemic—and How It Changed Science, Cities, and the Modern World*. New York: Riverhead, 2007.

Johnston, B. R., and H. M. Barker. *Consequential Damages of Nuclear War: The Rongelap Report*. Walnut Creek, CA: Left Coast Press, 2008.

Jones, B. "This Scary Interactive Map Shows What Happens If a Nuke Explodes in Your Neighborhood." *Business Insider*, July 16, 2013.

Jonnes, J. *Empires of Light: Edison, Tesla, Westinghouse, and the Race to Electrify the World.* New York: Random House Trade Paperback, 2004.

Kabat, G. C. *Hyping Health Risks: Environmental Hazards of Daily Life and the Science of Epidemiology.* New York: Columbia University Press, 2008.

Kean, S. *The Disappearing Spoon: And Other True Tales of Madness, Love, and the History of the World from the Periodic Table of Elements.* New York: Back Bay Books, 2011.

Keane, A., and R. D. Evans. *Massachusetts Institute of Technology Report*, 952–5 (Part II), *S*–1 to *S*–10. Cambridge, MA, 1968.

Kelly, C. C. *The Manhattan Project: The Birth of the Atomic Bomb in the Words of Its Creators, Eyewitnesses, and Historians.* New York: Black Dog & Leventhal, 2007.

Kelly, H. A., and C. F. Burnam. "A Resume of Results of Three Hundred and Forty-Seven Cases of Cancer of the Uterus and Vagina." *American Journal of Obstetrics and Diseases of Women and Children*, 74:436, 1916.

Kelly, M. "Cell Tower Proposal Back to Square One," *Scarborough Leader,* July 25, 2014.

Kendrew, J. C. *The Thread of Life: An Introduction to Molecular Biology.* Cambridge, MA: Harvard University Press, 1967.

Kevles, B. H. *Naked to the Bone: Medical Imaging in the Twentieth Century.* New Brunswick, NJ: Rutgers University Press, 1997.

Kleinerman, R. A., M. S. Linet, E. E. Hatch, R. E. Tarone, P. M. Black, R. G. Selker, W. R. Shapiro et al. "Self-reported Electrical Appliance Use and Risk of Adult Brain Tumors." *American Journal of Epidemiology*, 161(2):136–146, 2005.

Klotz, I. M. "The N-ray Affair." *Scientific American*, 242(5):170–175, 1980.

Kohler, R. E. *Lords of the Fly: Drosophila Genetics and the Experimental Life.* Chicago, IL: University of Chicago Press, 1994.

Kopans, D. B., and A. J. Vickers. "Mammography Screening and the Evidence: From One "Hominem" to the Other." *Medscape*, January 3, 2011.

Lai, H., and N. P. Singh. "Acute Low-Energy Microwave Exposure Increases DNA Single-Strand Breaks in Rat Brain Cells." *Bioelectromagnetics*, 16(3):207–210, 1995.

Lapp, L. E., and H. L. Andrews. *Nuclear Radiation Physics*, 4th ed. Englewood Cliffs, NJ: Prentice-Hall, 1972.

Larson, E. *Thunderstruck.* New York: Three Rivers, 2006.

Lazarus-Barlow W. S. "On the Disappearance of Insoluble Radium Salts from the Bodies of Mice after Subcutaneous Injection." *Archives of the Middlesex Hospital*, 30:92–94, 1913.

Levi, M. *On Nuclear Terrorism.* Cambridge, MA: Harvard University Press, 2007.

Lewis, E. B. "Leukemia and Ionizing Radiation." *Science*, 125:965, 1957.

Lindee, M. S. *Suffering Made Real: American Science and the Survivors of Hiroshima.* Chicago, IL: University of Chicago Press, 1994.

Lindley, D. *Uncertainty: Einstein, Heisenberg, Bohr, and the Struggle for the Soul of Science.* New York: Anchor, 2008.

Linton, O. "Francis H. Williams and William H. Rollins." *Journal of the American College Radiology*, 3(6):478–479, 2006.

Little, M. P., P. Rajaraman, R. E. Curtis, S. S. Devesa, P. D. Inskip, D. P. Check, and M. S. Linet. "Mobile Phone Use and Glioma Risk: Comparison of Epidemiological Study Results with Incidence Trends in the United States." *British Medical Journal*, 344:e1147, 2012. doi: 10.1136/bmj.e1147.

Lorenz, E. "Radioactivity and Lung Cancer: A Critical Review of Lung Cancer in Miners in Schneeberg and St. Joachimsthal." *Journal of the National Cancer Institute*, 5:1–4, 1944.

Lowry, I. S. "The Postattack Population of the United States." Memorandum RM-5115-TAB, The RAND Corporation, December 1966.

Lubin, J., J. Samet, and C. Weinberg. "Design Issues in Epidemiological Studies of Indoor Exposure to Rn and Risk of Lung Cancer." *Health Physics*, 59:807–817, 1990.

Madigan, D. J., Z. Baumann, and N. S. Fisher. "Pacific Bluefin Tuna Transport Fukushima-Derived Radionuclides from Japan to California." *Proceedings of the National Academy of Sciences*, 109(24):9483–9486, 2012.

Maggelet, M. H., and J. C. Oskins. *Broken Arrow: The Declassified History of U.S. Nuclear Weapons Accidents*. Raleigh, NC: Lulu Press, 2008.

Mahaffey, J. *Atomic Accidents: A History of Nuclear Meltdowns and Disasters from the Ozark Mountains to Fukushima*. New York: Pegasus Books, 2014.

Malyapa, R. S., E. W. Ahern, W. L. Straube, M. LaRegina, W. F. Pickard, and J. L. Roti Roti. "DNA Damage in Rat Brain Cells after In Vivo Exposure to 2450 MHz Electromagnetic Radiation and Various Methods of Euthanasia." *Radiation Research*, 149(6):637–645, 1998.

Mangrum, H. "1418 Eutaw Place and Her Mystery," *Baltimore Afro-American*, July 20, 1957, p. 11.

Marshall, J. L., and V. R. Marshall. "Ernest Rutherford, the 'True Discoverer' of Radon." *Bulletin of the History of Chemistry*, 28(2):76–83, 2003.

McGill, E. J. *Jet Age Man: SAC B-47 and B-52 Operations in the Early Cold War*, reprint. Solihull, UK: Helion, 2014.

McNichol, T. *AC/DC: The Savage Tale of the First Standards War*. San Francisco: Jossey-Bass, 2006.

McRaney, W., and J. McGahan. *Radiation Dose Reconstruction U.S. Occupation Forces in Hiroshima and Nagasaki, Japan, 1945-1946*. Report No. DNA5512F. McLean, VA: Science Applications, August 6, 1980.

Moran, B. *The Day We Lost the H-Bomb: Cold War, Hot Nukes, and the Worst Nuclear Weapons Disaster in History*. New York: Ballantine, 2009.

Morgan, T. H., A. H. Sturtevant, H. J. Muller, and C. B. Bridges. *The Mechanism of Mendelian Heredity*. New York: Henry Holt, 1915.

Morgenstern, H. "Ecologic Studies in Epidemiology: Concepts, Principles, and Methods." *Annual Review of Public Health*, 16:61–81, 1995.

Mossman, K. L. *Radiation Risks in Perspective*. New York: Taylor & Francis, 2007.

Mossman, K. L., and Mills W. A., eds. *The Biological Basis of Radiation Protection Practice*. Baltimore, MD: Williams & Wilkins, 1992.

Mukherjee, S. *The Emperor of All Maladies: A Biography of Cancer*. New York: Scribner, 2011.

Muller, H. M. *Out of the Night: A Biologist's View of the Future.* New York: Vanguard Press, 1935.

Mullner, R. *Deadly Glow: The Radium Dial Worker Tragedy.* Washington, DC: American Public Health Association, 1999.

Nagamoto, T., Y. Ichikawa, and M. Hoshi. "Thermoluminescent Dosimetry of Gamma Rays Using Ceramic Samples from Hiroshima and Nagasaki: A Comparison with DS86 Estimates." *Journal of Radiation Research*, Supplement, 48–57, 1991.

Narula, S. K. "Sushiomics: How Bluefin Tuna Became a Million-Dollar Fish." *The Atlantic*, January 5, 2014. http://www.theatlantic.com/international/archive/2014/01/sushinomics-how-bluefin-tuna-became-a-million-dollar-fish/282826/

NAS/NRC (National Academy of Sciences/National Research Council). *Health Risks of Radon and Other Internally-Deposited Alpha-Emitters, BEIR IV.* Washington, DC: National Academy Press, 1988.

NAS/NRC (National Academy Press/National Research Council). *Health Effects of Exposure to Radon, BEIR VI.* Washington, DC: National Academy Press, 1999.

NAS/NRC (National Academy of Sciences/National Research Council). *Health Risks from Exposure to Low Levels of Ionizing Radiation, BEIR VII Phase 2.* Washington, DC: National Academy Press, 2006.

Nature. "Curie Laid to Rest with France's Heroes." 374(6525):751, 1995.

Nature. "X-rays, not Radium, May Have Killed Curie." 377(6545):96, 1995.

NBC News. "How the Quake Shifted Japan," March 12, 2011. http://photoblog.nbcnews.com/_news/2011/03/12/6256280-how-the-quake-shifted-japan?lite.

NCRP (National Council on Radiation Protection). *Some Aspects of Strontium Radiobiology.* Report No. 110. Bethesda, MD: National Council on Radiation Protection and Measurements, 1991.

NCRP (National Council on Radiation Protection). *Limitations of Exposure to Ionizing Radiation.* Report No. 116. Bethesda, MD: National Council on Radiation Protection and Measurements, 1993.

NCRP (National Council on Radiation Protection). *Ionizing Radiation Exposure of the Population of the United States.* Report No. 160. Bethesda, MD: National Council on Radiation Protection and Measurements, 2009.

NCRP (National Council on Radiation Protection). *Second Primary Cancers and Cardiovascular Disease after Radiation Therapy.* Report No. 170. Bethesda, MD: National Council on Radiation Protection and Measurements, 2011.

NCRP (National Council on Radiation Protection). *Uncertainties in the Estimation of Radiation Risks and Probability of Disease Causation.* Report No. 171, Bethesda, MD: National Council on Radiation Protection and Measurements, 2012.

NBC News. "How the Quake Shifted Japan," March 12, 2011.

New York Times. "Coney Elephant Killed," January 5, 1903.

New York Times. "$100,000 Radium Treatment Test to Save Bremner's Life: Eleven Tubes Buried in Shoulder of New Jersey Congressman Stricken with Cancer," December 27, 1913.

Niedenthal, J. *For the Good of Mankind: A History of the People of Bikini and Their Islands*, 2nd ed. Majuro, Marshall Islands: Bravo, 2001.

Oe, K. *Hiroshima Notes*. New York: Grove Press, 1996.

Ōishi, M. *The Day the Sun Rose in the West: Bikini, the Lucky Dragon, and I*. Honolulu: University of Hawai'i Press, 2011.

Olby, R. C. *The Path to the Double Helix: The Discovery of DNA*. Mineola, NY: Dover, 1994.

Oransky, I. "Lauriston Taylor." *Lancet*, 365(9455):210, 2005.

Paulos, J. A. *Innumeracy: Mathematical Illiteracy and Its Consequences*. New York: Hill and Wang, 2001.

Perrow, C. *Normal Accidents: Living with High-Risk Technologies*. Princeton, NJ: Princeton University Press, 1999.

Poole, M., J. Dainton, and S. Chattopadhyay. "Cockcroft's Subatomic Legacy: Splitting the Atom." *CERN Courier*, December 2007.

Quinn, S. *Marie Curie: A Life*. New York: Addison-Wesley, 1996

Reeves, R. *A Force of Nature: The Frontier Genius of Ernest Rutherford*. New York: Atlas, 2008.

Regaud, C., and R. Ferroux. "Discordance des effets de rayons X, d'une part dans le testicule, par le fractionnement de la dose." *Comptes Rendus Societe Biologique*, 97:431, 1927.

Reiners, C., J. Biko, H. Haenscheid, H. Hebestreit, S. Kirinjuk, O. Baranowski, R. J. Marlowe, et al. "Twenty-Five Years after Chernobyl: Outcome of Radioiodine Treatment in Children and Adolescents with Very-high-Risk Radiation-Induced Differentiated Thyroid Carcinoma." *Journal of Clinical Endocrinology and Metabolism*, 98(7):3039–3048, 2013.

Robison, R. F. "Howard Atwood Kelly (1858–1943): Founding Professor of Gynecology at Johns Hopkins Hospital & Pioneer American Radium Therapist." *NOWOTWORY Journal of Oncology*, 60(1)1:21e–35e, 2010.

Rodricks, J. V. *Calculated Risks: The Toxicity and Human Health Risks of Chemicals in Our Environment*, 2nd ed. Cambridge, UK: Cambridge University Press, 2007.

Roentgen, W. C. "On a New Kind of Rays," A. Stanton, trans. *Science*, 3:227–231, February 14, 1896.

Röösli, M. "Radiofrequency Electromagnetic Field Exposure and Non-Specific Symptoms of Ill Health: A Systematic Review." *Environmental Research*, 107:277–287, 2008.

Rose, K. D. *One Nation Underground: The Fallout Shelter in American Culture*. New York University Press, 2001.

Rothman, K. J. *Epidemiology: An Introduction*. Oxford, UK: Oxford University Press, 2012.

Rotter, A. J. *Hiroshima: The World's Bomb*. Oxford, UK: Oxford University Press, 2009.

Rowland, R. E. *Radium in Humans: A Review of U.S. Studies* (ANL/ER-3). Argonne, IL: Argonne National Laboratory, 1994.

Rutherford, E. R. *Radio-Activity*, 2nd ed. Lexington, KY: Juniper Grove, 2007. (First published in 1904 by Cambridge University Press.)

Sayre, A. *Rosalind Franklin & DNA*. New York: Norton, 1975.

Schlosser, E. *Command and Control: Nuclear Weapons, the Damascus Accident, and the Illusion of Safety*. New York: Penguin, 2013.

Schuttmann, W. "Schneeberg Lung Disease and Uranium Mining in the Saxon Ore Mountains (Erzebirge)." *American Journal of Industrial Medicine*, 23 (2):355–368, 1993.

Schüz, J., R. Jacobsen, J. H. Olsen, J. D. Boice, J. K. McLaughlin, and C. Johansen. "Cellular Telephone Use and Cancer Risk: Update of a Nationwide Danish Cohort." *Journal of the National Cancer Institute*, 98:1707–1713, 2006.

Shine, I., and S. Wrobel. *Thomas Hunt Morgan: Pioneer of Genetics*. Lexington: University Press of Kentucky, 1976.

Silver, N. *The Signal and the Noise: Why So Many Predictions Fail—but Some Don't*. New York: Penguin, 2012.

Slovik, P. *The Perception of Risk*. London: Earthscan, 2000.

Smith J., and N. A. Beresford. *Chernobyl: Catastrophe and Consequences*. Chichester, UK: Springer Praxis, 2005.

Sone, S., F. Li, Z. Yang, T. Honda, Y. Maruyama, and S. Takashima. "Results of Three-Year Mass Screening Programme for Lung Cancer Using Mobile Low-Dose Spiral Computed Tomography Scanner." *British Journal of Cancer*, 84(1):25–32, 2001.

Spencer, M., A. Ludin, and H. Nelson. *The Unauthorized Movement of Nuclear Weapons and Mistaken Shipment of Classified Missile Components: An Assessment*. The Counterproliferation Papers Future Warfare Series No. 56, January 2012. Maxwell Air Force Base, Alabama: USAF Counterproliferation Center, 2012.

Spokesman-Review. "Leukemia Claims A-Bomb Children." August 12, 1965.

Stanchevici, D. *Stalinist Genetics: The Constitutional Rhetoric of T.D. Lysenko*. Amityville, NY: Boxwood, 2010.

Stars and Stripes. "Air Force Fires 2 Nuclear Missile Corps Commanders," November 4, 2014.

Stat Bite. "Causes of Lung Cancer in Nonsmokers." *Journal of the National Cancer Institute*, 98 (May 17, 2006):664, 2006.

Strom, D. J., T. P. Lynch, and D. R. Weier. *Radiation Doses to Hanford Workers from Natural Potassium-40*. Report no. PNNL-18240. Richland WA: Pacific Northwest Laboratories, US Department of Energy, February 2009.

Stross, R. *The Wizard of Menlo Park: How Thomas Alva Edison Invented the Modern World*. New York: Three Rivers, 2007.

Sturtevant, A. H. *A History of Genetics*. Cold Spring Harbor, NY: Cold Spring Harbor Laboratory Press, 2000.

Szklo, M., and F. J. Nieto. *Epidemiology: Beyond the Basics*. Gaithersburg, MD: Aspen, 2000.

Taleb, N. N. *The Black Swan: The Impact of the Highly Improbable*. 2nd ed. New York: Random House Trade Paperbacks, 2010.

Taylor, L. S. *Organization for Radiation Protection: The Operations of the ICRP and NCRP 1928–1974*. Washington, DC: Assistant Secretary for the Environment, Office of Health Research, and the Office of Technical Information, US Department of Energy, 1979.

Tibbets, P. W. *Return of the Enola Gay*. Columbus, OH: Mid Coast Marketing, 1998.

Tokonami, S., M. Hosoda, S. Akiba, A. Sorimachi, I. Kashiwakura, and M. Bolonov. "Thyroid Doses for Evacuees from the Fukushima Nuclear Accident." *Scientific Reports* 2012; 2:507. DOI: 10.1038/SREP00507. Epub 2012 Jul 12.

Trenton Evening Times. "New Jersey Representative Succumbs in Baltimore Sanatorium after Long Fight Against Disease: Radium Treatment Fails to Save Him," February 5, 1914.

Tuddenham, W. J., and A. Soiland. "Pioneer Radiologist." *Radiology*, 131(3):578, 1979.

Tuniz, C. *Radioactivity: A Very Short Introduction*. Oxford, UK: Oxford University Press, 2012.

Twombly, R. "Full-Body CT Screening: Preventing or Producing Cancer?" *Journal of the National Cancer Institute*, 96:1650–1651, 2004.

Upton, A. C. "The First Hundred Years of Radiation Research: What Have They Taught Us?" *Environmental Research*, 59:36-48, 1992.

U.S. Department of Defense. *The Defense Science Board Permanent Task Force on Nuclear Weapons Surety: Report on the Unauthorized Movement of Nuclear Weapons*. Washington, DC: Office of the Under Secretary of Defense for Acquisition, Technology, and Logistics, Department of Defense, February 2008 (revised April 2008).

U.S. Department of Health and Human Services. *Accidental Radioactive Contamination of Human Food and Animal Feeds: Recommendations for State and Local Agencies*. Rockville, MD: Food and Drug Administration, Radiation Programs Branch, August 13, 1998.

U.S. Nuclear Regulatory Commission. *Reactor Safety Study: An Assessment of Accident Risks in U.S. Commercial Nuclear Power Plants*. Report No. NUREG 75/014 (WASH-1400). Washington, DC, 1975.

U.S. Nuclear Regulatory Commission. *Risk Assessment Review Group Report to the U.S. Nuclear Regulatory Commission*. Report no. NUREG/CR-0400. Washington, DC, 1978.

Van der Kloot, W. "Lawrence Bragg's Role in the Development of Sound-Ranging in World War I." *Notes and Records of the Royal Society*, 2005, 59(3):273–284, 2005. doi: 10.1098/rsnr.2005.00952005.

Walker, R. I., and T. J. Cerveny, eds. *Medical Consequences of Nuclear Warfare*. Textbook of Military Medicine, Part I, Vol. 2. Falls Church, VA: TMM Publications, 1989.

Waterbury Observer. "After Glow: 90 Years Ago Workers at the Waterbury Clock Company Began Dying after Painting Radium on Clock Dials." October 30, 2011.

Watkins, A., A. Kamji, K. Smaczniak, and L. Zhang. *Keeping the Promise: An Evaluation of Continuing U.S. Obligations Arising out of the U.S. Nuclear Testing Program in the Marshall Islands*." Cambridge, MA: Harvard Law School, 2006.

Watson, J. D. *The Double Helix*. New York: Mentor, 1969.

Watson, J. D., and F.H.C. Crick. "A Structure for Deoxyribose Nucleic Acid." *Nature*, 171:737, 1953.

Weisgall, J. M. *Operation Crossroads: The Atomic Tests at Bikini Atoll*. Annapolis, MD: Naval Institute Press, 1994.

Welch, H. G., L. M. Schwartz, and S. Woloshin. *Overdiagnosed: Making People Sick in the Pursuit of Health*. Boston, MA: Beacon Press, 2011.

Welsome, E. *The Plutonium Files: America's Secret Medical Experiments in the Cold War*. New York: Dell, 2000.

Williams, F. H. *The Roentgen Rays in Medicine and Surgery: As an Aid in Diagnosis and as a Therapeutic Agent*. New York: Macmillan, 1901.

Wood, R. W. "The N-rays." *Nature*, 70(1822):530–531, 1904.

World Health Organization. *WHO Handbook on Indoor Radon: A Public Health Perspective*. Geneva: WHO Press, 2009.

Yoshinaga, S., K. Mabuchi, A. J. Sigurdson, M. M. Doody, and E. Ron. "Cancer Risks among Radiologists and Radiologic Technologists: Review of Epidemiologic Studies." *Radiology*, 233(2):313–321, 2004.

Zoellner, T. *Uranium: War, Energy, and the Rock That Shaped the World*. New York: Penguin, 2010.

Zueblin, E. "The Present Status of Radioactive Therapy in Medicine." *Maryland Medical Journal*, 57:108–151, 1914.

INDEX

Page numbers in italics refer to figures.